CANADIAN CRIMINAL JUSTICE

A PRIMER

THIRD EDITION

CANADIAN CRIMINAL JUSTICE

A PRIMER

THIRD EDITION

CURT T. GRIFFITHS
SIMON FRASER UNIVERSITY

NELSON / EDUCATION

NELSON / EDUCATION

Canadian Criminal Justice: A Primer
Third Edition
by Curt T. Griffiths

Associate Vice-President,
Editorial Director:
Evelyn Veitch

Senior Marketing Manager:
Lenore Taylor

Senior Developmental Editor:
Rebecca Rea

Photo Researcher:
Lisa Brant

Permissions Coordinator:
Lisa Brant

Production Editors:
Karri Yano/Anne Macdonald

Copy Editor:
Matthew Kudelka

Proofreader:
Gail Marsden

Indexer:
Andrew Little

Production Coordinator:
Susan Ure

Design Director:
Ken Phipps

Interior Design:
Andrew Adams

Cover Design:
Andrew Adams

Cover Image:
Gordon Caruso

Compositor:
Interactive Composition Corporation

Printer:
Transcontinental

Library and Archives Canada
Cataloguing in Publication Data

Griffiths, Curt T. (Curt Taylor), 1948–
  Canadian criminal justice : a primer /
Curt T. Griffiths. —3rd ed.

First ed. written by Alison Hatch
Cunningham, Curt T. Griffiths.
2nd ed. written by Curt T. Griffiths,
Alison Hatch Cunningham.

Includes bibliographical references and
index.
ISBN-10: 0-17-640715-4

1. Criminal justice, Administration of—
Canada—Textbooks. I. Title.

HV9960.C2G75 2006   364.971
C2005-907862-6

To the kids – Collin, Lacey, Daniel, and Jordan
For making life a joy and an adventure!

and

To Yvon, Vivienne, Ron, and Lyle
For their friendship

Contents

Chapter 8: Release and Re-entry

Acknowledgments

I would like to acknowledge the many people throughout the criminal justice system who have contributed ideas and information that have been incorporated into this book. A special thanks to Doug LePard, Deputy Chief Constable, and to Ryan Prox, Criminal Intelligence Analyst, both of the Vancouver Police Department, Morgan Andreassen of the Correctional Service of Canada, and to my former colleagues on the British Columbia Board of Parole. Also, my thanks to Dr. Robert Gordon, Director, School of Criminology, Simon Fraser University for his ongoing support of my work.

I would also like to thank the reviewers for their invaluable comments, criticisms, and suggestions—Peter Maher, Georgian College; John Martin, University College of the Fraser Valley; David Douglass, Canadore College; Oliver Stoetzer, Fanshawe College; and Mary-Anne Nixon, Barrister and Solicitor, who completed a legal review of the material. And, a special thanks to Matthew Kudelka for his outstanding editing work on the manuscript.

As always, it has been a pleasure to work with the professionals at Thomson Nelson: Joanna Cotton, Rebecca Rea, Gail Marsden, Lisa Brant, Karri Yano, and Anne Macdonald. Their energy and enthusiasm helped make it happen.

Preface

The Canadian criminal justice system is a complex, dynamic, and ever-changing enterprise. How the various components of the system operate and the extent to which they succeed in preventing and responding to crime and criminal offenders affects not only the general public but criminal justice personnel and offenders as well. This edition of *Canadian Criminal Justice: A Primer* is designed with the same basic objectives as the first two: to present in a clear and concise fashion materials on the criminal justice system in Canada and to highlight the key issues surrounding this country's responses to crime and offenders. This book is not an exhaustive examination of all facets of the criminal justice process. Rather, its intent is to present, with broad brush strokes, information on the structure and operations of the criminal justice system while at the same time identifying some of the more significant challenges and controversies that arise at each stage of the justice process. Where relevant, it also compares the criminal justice process in Canada with that of other Western countries.

In the years since the second edition of this book was published, much has changed not only in Canadian society generally, but in the arena of crime and criminal justice as well. The criminal justice system in the early twenty-first century operates in an ever-changing world. The advent of a number of technologies and the emergence of the global community are presenting opportunities to develop more effective responses to crime and social disorder, but they are also presenting the justice system with significant challenges. There is little doubt that the terrorist attacks on the United States on September 11, 2001, have ushered in a new era in which security and safety have assumed even more importance. These attacks, and the responses of Canada and other Western governments to what has become a pervasive threat of terrorism, have also brought to the fore the inherent tension between the rights of citizens and the requirements of due process on the one hand, and the need to protect and safeguard communities and their residents on the other. In the face of increasing global uncertainty, it is likely that these types of issues will continue to present Canadians, their politicians, and the criminal justice system with difficult challenges and choices. It is also likely that the costs of providing security will continue to escalate; in turn this may affect the resources available for the more traditional elements of the justice system.

Increasing concerns about public safety and security have often overshadowed other developments in the justice system. Recent years have witnessed a continued expansion of restorative and community justice programs, a continuing evolution of community policing, increasing protections and rights for the victims of crime, and innovative correctional programs for female offenders. On the less positive side of the ledger, there continues to be considerable disparity in the sentences meted out in the criminal courts; public confidence in the criminal justice system (with the exception of the police) remains rather low; Aboriginal people continue to be greatly overrepresented at all stages of the justice system; and inmate populations continue to face high rates of infectious diseases, including HIV/AIDS and hepatitis C. Also, the provincial/territorial and federal governments are continuing to confront fiscal crises that have resulted in significant reductions in monies available for criminal justice programs and services. This has accelerated the exploration of public–private sector partnerships for service delivery as well as the trend toward contracting out many programs.

A final word: Any attempt to capture the dynamics of the criminal justice system can be no more than a snapshot taken against a shifting landscape. In this third edition I have tried to provide you with information on who's who and what's what in the justice system, but also to capture the human dynamics of what is one of the more unique facets of Canadian society.

As always, I encourage your feedback on the book generally and on any specific materials in it. Feel free to contact me at griffith@sfu.ca with any comments, questions, or suggestions for future editions of the text. And please take a moment to complete the enclosed card. The information will be used to improve subsequent editions of the text.

Thanks.

Curt Taylor Griffiths
Vancouver, British Columbia

December 2005

CANADIAN CRIMINAL JUSTICE

A PRIMER

THIRD EDITION

chapter

1

The Criminal Justice System: An Overview

learning objectives

After reading this chapter, you should be able to

- Identify the components of the criminal justice system.

- Discuss the roles and responsibilities of governments in the criminal justice system.

- Identify the agencies and organizations that collectively make up the criminal justice system.

- Define the elements of criminal law.

- Discuss the major challenges confronting the criminal justice system.

- Compare and contrast the due process and crime control models of law and justice.

- Identify the major trends in criminal justice.

key terms

Canadian Charter of Rights
    and Freedoms   8
civil law   25
conflict model   27
Constitution Act, 1867   6
crime   24
crime control   30
Criminal Code   8

criminal law   24
discretion   30
due process   30
interoperability   23
restorative justice   41
rule of law   24
task environment   31
value consensus model   27

You are about to embark on the study of one of the most controversial, and fascinating, topics in Canadian society. Criminal justice—the way we respond to those who are alleged to have committed and/or who have been convicted of criminal offences—is the subject of emotionally charged debates and arguments in the home, in classrooms, in legislatures, and in the media. As a potential participant in these discussions and/or a future employee in the justice system, you should have a basic understanding of the structure and operations of the Canadian criminal justice system as well as the key issues, controversies, and trends. This text will provide you with that understanding.

The basic components of the criminal justice system are the police, the courts, and the systems of corrections. Another integral part of any study of the criminal justice system is the community, which not only influences politicians, legislatures, and justice system personnel but also involves itself directly in preventing and responding to crime.

As you work your way through this book, keep in mind that the criminal justice system is first and foremost a human enterprise. (See photos.) The decisions of police officers, judges, probation/parole officers, and parole boards are based not on scientific formulas but on professional judgment, experience, and intuition. There is much that remains to be understood about the causes of criminal behaviour and the most effective responses to it. As well, offenders often struggle to understand and correct their behaviour. The discussions that follow will reveal that although preventing and responding to crime is a challenging and often frustrating exercise, there are many exciting initiatives that hold considerable promise.

This chapter takes you through the criminal justice system from the point at which a victim reports a crime to the end of an offender's sentence (in some cases, beyond the end of the sentence). Once they learn of a crime, the police have several response options, which may or may not include formal action through investigation and charge (Chapters 4 and 5). Crown attorneys are involved in determining whether a charge should be laid and are also faced with many decisions about each case (Chapter 6). Most cases are resolved with a guilty plea; a few go to trial.

In the criminal justice process, judges and justices of the peace make key decisions—for example, they issue warrants that allow the police to conduct investigations (Chapter 5). Judges are involved in adjudicating cases in the criminal courts (Chapter 5) and in sentencing those who have been convicted (Chapter 6). The most common sanction is probation, which allows convicted offenders to remain in the community. A relatively small percentage of offenders receive a period of confinement in a

Defence lawyer in front of courthouse, Yorkton, Saskatchewan

Jeff McIntosh/CP Photo Archive

Chuck Stoody/CP Photo Archive

British Columbia provincial court judge

Andrew Vaughan/CP Photo Archive

Nova Scotia legislature sits for the fall session in Halifax.

An offender is escorted from the courthouse after being sentenced to a life term for murder in Belleville, Ontario.

A Toronto police officer checks the communication system.

correctional institution (Chapter 7). Most offenders who are sentenced to a period of incarceration serve time in provincial/territorial correctional facilities, and most provincial/territorial and federal offenders serve at least a portion of their sentence under supervision in the community (Chapter 8).

An overview of all this—of "who's who" and "what's what" in the criminal justice system—would be useful before examining this process in depth. The materials in this chapter describe the structure of the criminal justice system, identify the contexts in which criminal justice is administered, highlight several challenges facing the system, and review important trends in Canadian criminal justice. This overview begins with an examination of the role governments play in criminal justice, and then goes on to consider the system's main components: the police, the courts, and corrections.

THE ROLE AND RESPONSIBILITIES OF GOVERNMENTS IN CRIMINAL JUSTICE

The criminal justice system is operated and controlled almost entirely by governments. Elected officials—most notably members of Parliament and their provincial counterparts—play a small but significant role in the criminal justice process. It is in Parliament and in provincial legislatures that elected officials enact and amend laws, establish annual budgets for criminal justice agencies, determine fiscal allocations, and—where necessary—conduct investigations and inquiries into various activities of the justice system.

Most of the tens of thousands of people who work in the criminal justice system are government employees. In the federal and provincial governments, senior civil servants—all of whom are appointed rather than elected—wield considerable power in carrying out the mandates of their respective governments. It is at the civil service level that professionals—most of whom are unknown to the general public—make many of the day-to-day decisions with respect to criminal justice policy and practice.

But not everyone who works in the justice system is a government employee. Across the country, thousands of citizen volunteers are aiding the victims of crime, staffing community police offices, working with at-risk youth, sponsoring programs in correctional institutions, and providing support for offenders who have been released from correctional facilities. In addition, many NGOs (nongovernmental organizations) and other not-for-profit groups deliver a variety of services on a contract basis—for example, they supervise offenders on bail, assist victims, provide community-based and institutional programming, and supervise parolees. These groups include, among others, the John Howard Society, the Elizabeth Fry Society, the St. Leonard's Society, and various Aboriginal organizations. Provincial governments often contract organizations like these to deliver such services. Also, more and more private security firms are being contracted to protect property and maintain order in business districts. This practice raises a number of important questions, especially this one: What are the proper limits to private sector involvement in criminal justice?

Each level of government in Canada—federal, provincial, and municipal— plays a role in the justice system. The division of responsibilities between the federal and provincial governments was spelled out in the **Constitution Act, 1867**. The

Constitution Act, 1867
Constitutional authority for the division of responsibilities between the federal and provincial governments

FIGURE 1.1   Division of Responsibilities for Criminal Justice

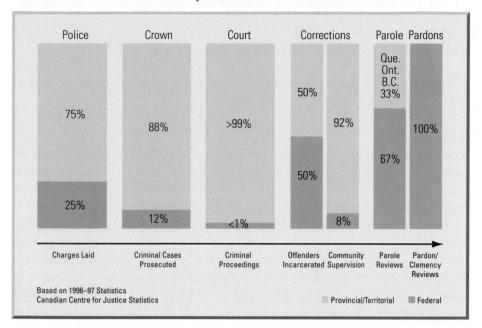

Source: Steering Committee on Integrated Justice Information, *Integrated Justice Information Action Plan* (Ottawa: Strategic Policy and Integrated Justice, Solicitor General Canada, 1999), 16.

basic division is that the federal government decides which behaviours constitute criminal offences, while the provincial/territorial governments are responsible for law enforcement and for administering the justice system (see Figure 1.1). The federal government plays a major role in policing through agreements with the RCMP, which is involved in federal policing as well as provincial and municipal policing under contract and Native band councils. The municipal governments of cities and towns play a lesser role, primarily related to policing and bylaw enforcement.

The following discussion sketches out the three levels of government and the justice services each operates.

Federal Government

Because of the Constitution, Parliament in Ottawa has the absolute power to create, amend, and repeal criminal law for the entire country. It also sets the procedures for prosecuting persons charged with criminal offences and establishes the punishments that will be imposed for all federal offences. In addition, the federal government:

- operates a national police force (the RCMP);
- operates the Security Intelligence Review Committee (SIRC) and the Canadian Security Intelligence Service (CSIS);
- prosecutes some federal offences, including narcotics offences;
- appoints some judges;
- manages some courts;

- operates correctional institutions for those offenders who receive sentences totalling two years or more;
- operates a parole board for federal inmates and for provincial/territorial inmates in jurisdictions other than Quebec, Ontario, and British Columbia (which have their own provincial parole boards); *and* enters into international and bilateral agreements with other countries designed to improve public security and crime response.

Federal Offences

From "abduction" to "witchcraft," the behaviours that are crimes in Canada are listed in the **Criminal Code**, as are many of the procedures for responding to crimes (http://laws.justice.gc.ca/en/C-46). The Criminal Code also sets out the procedures for arrest and prosecution and the penalties that may be imposed by a judge following conviction. The original version of the Criminal Code dates from 1869. It was initially consolidated in 1892 and since that time has been amended and revised. The most recent version, composed of 849 sections in 28 parts, is three times longer than the original version. Readers are urged to acquire a current copy of the Criminal Code to use in conjunction with the materials presented in this text.

> **Criminal Code**
> federal legislation that sets out criminal laws, procedures for prosecuting federal offences, and sentences and procedures for the administration of justice

Federal offences are set out in various acts or statutes. For example, under the Controlled Drugs and Substances Act it is a federal offence to possess, traffick, make, import, or grow certain drugs or narcotics. Controlled drugs include marijuana, "ecstasy" (MDMA), methamphetamines, cocaine, and heroin. In 2005, amidst growing concern with the expanding use of "crystal meth," methamphetamines were moved to Schedule I of the Controlled Drugs and Substances Act, which allows for the imposition of the stiffest maximum penalties under the act. When sentencing offenders, judges now view crystal meth as harshly as cocaine and heroin. This act spells out the penalties for the various offences it contains; these sentences vary with the type of drug and the circumstances of the offence. For example, selling drugs to children can result in a more severe penalty than selling drugs to adults.

Other Federal Statutes

The Criminal Code is the most important statute governing the prosecution of criminal offences. However, the following federal laws also affect the criminal justice process:

> **Canadian Charter of Rights and Freedoms**
> a component of the Constitution Act that guarantees basic rights and freedoms

- *Canadian Charter of Rights and Freedoms.* Part of the Constitution Act, 1982, the Charter (http://laws.justice.gc.ca/en/charter) is the primary law of the land. With very limited restrictions, it guarantees fundamental freedoms, legal rights, and equality rights for any person in Canada (citizens and non-citizens), including those persons accused of crimes. All of the basic and functional components of the criminal justice system must operate in such a way as not to violate the rights guaranteed to Canadians in the Charter. Since the decisions of the Supreme Court of Canada are binding on all Canadian courts, one cannot overstate the impact of the Charter on the Canadian justice system.
- *Canada Evidence Act* (1985, c. C-5) (http://laws.justice.gc.ca/en/C-5). This act covers most aspects of the testimony of witnesses. For example, it

addresses oaths and solemn affirmations, as well as oath taking as it pertains to witnesses whose mental capacity is challenged.

- *Criminal Records Act* (1985, c. C-47) (http://laws.justice.gc.ca/en/C-47). This act defines the procedures by which offenders secure pardons. A person who has been conviction-free and "of good conduct" for a set period can apply to have the conviction "vacated."
- *Corrections and Conditional Release Act* (1992, c. 20) (http://laws.justice. gc.ca/en/C-44.6). This act governs many aspects of the operations of federal correctional institutions and parole boards.
- *Youth Criminal Justice Act* (2002, c. 1) (http://laws.justice.gc.ca/en/Y-1.5). This act, passed in 2002 as a replacement for the Young Offenders Act, sets out the philosophy, process, and sanctions for responding to young offenders.

Other legislation is designed to help the justice system respond more effectively to crime:

- *DNA Identification Act* (1998, c. 37) (http://laws.justice.gc.ca/en/D-3.8). This statute and a legislative amendment established the national DNA data bank and define the procedures for the collection, storage, use, and destruction of DNA samples taken from criminal offenders. The same statute identifies those offenders who are required to submit a DNA sample.
- *Anti-Terrorism Act* (2001, c. 41) (http://laws.justice.gc.ca/en/A-11.7). This legislation gives the justice system broad powers to identify, prosecute, convict, and punish terrorist groups and individuals.
- *Sex Offender Information Registration Act* (2004, c. 10) (http://www.canlii.org/ ca/as/2004/c10). This legislation established a national sex offender database containing information on convicted sex offenders.

These and other federal statutes—collectively referred to as "the Statutes"—and documents called "regulations" can be accessed through the Department of Justice Canada website, http://laws.justice.gc.ca/en.

The Portfolio of Public Safety and Emergency Preparedness

Created in 2003, this portfolio includes the key federal agencies responsible for public safety and security. It is directed by a federal minister and includes the following:

- Department of Public Safety and Emergency Preparedness (http://www. psepc.gc.ca), which provides strategic policy advice and a range of programs and services relating to public security and safety;
- Royal Canadian Mounted Police;
- Canadian Security Intelligence Service (CSIS) and the Security Intelligence Review Committee (SIRC);
- Correctional Service of Canada;
- National Parole Board;
- Canada Firearms Centre;
- Canadian Border Services Agency; *and*

FIGURE 1.2   Portfolio of Public Safety and Emergency Preparedness

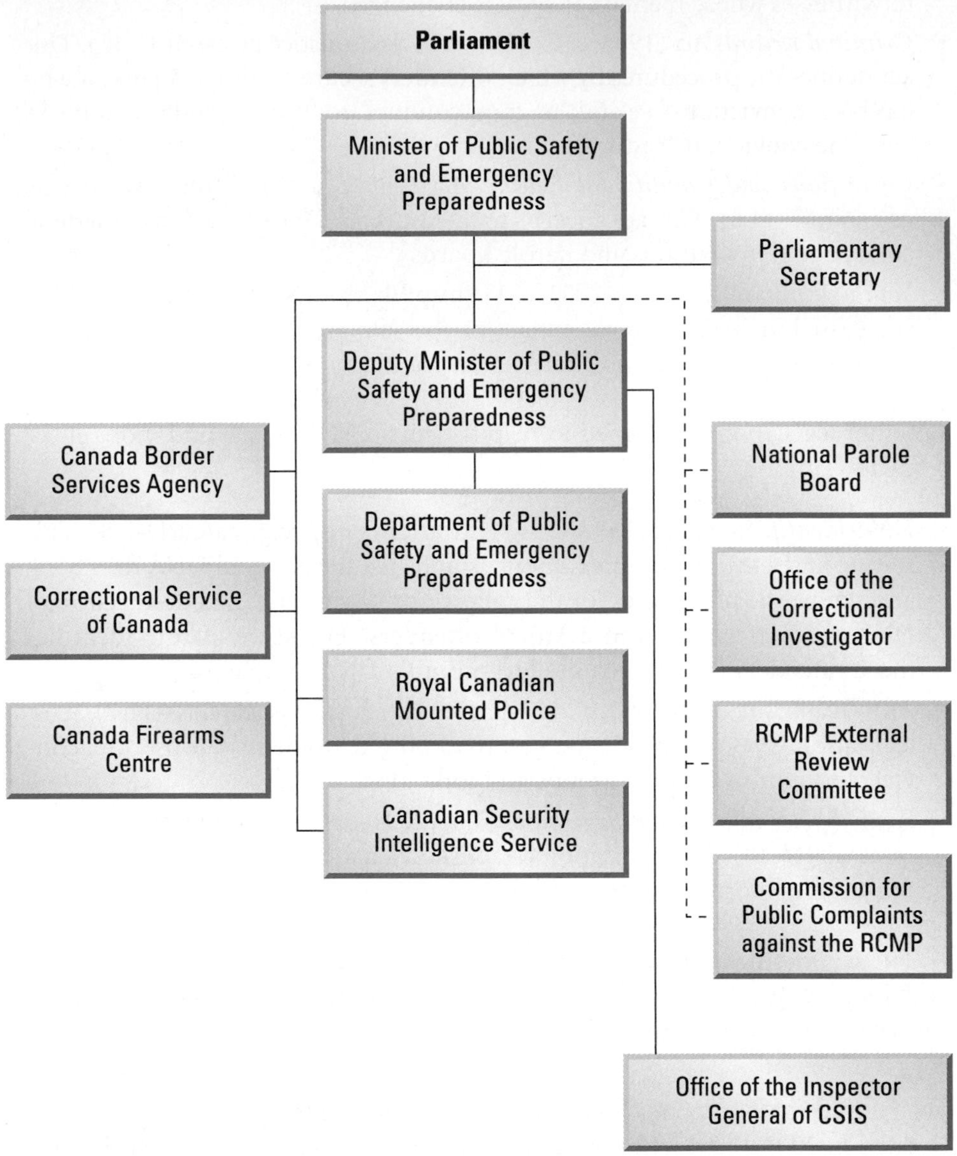

Source: Portfolio of Public Safety and Emergency Preparedness
http://psepc-sppcc.gc.ca/abt/wwwa/index-en.asp. Reproduced with the permission of the
Minister of Public Works and Government Services Canada.

- three review agencies: the RCMP External Review Committee, the Commission for Public Complaints Against the RCMP, and the Office of the Correctional Investigator (see Figure 1.2).

Federal Police

The RCMP operates as a federal police force across Canada. It also operates the Canadian Police College, the Canadian Police Information Centre (CPIC), Forensic Laboratory Services, and a variety of other specialized services (http://www.rcmp.gc.ca). Note that another federal police force is Military Police of the Canadian Armed Forces.

Federal Prosecutors, Courts, and Judges

The Federal Prosecution Service (FPS), located in the Department of Justice (http://canada.justice.gc.ca) is staffed by full-time prosecutors and lawyer agents recruited from the private bar. It conducts prosecutions across Canada on behalf of the Attorney General of Canada. These prosecutions are carried out under a wide range of federal statutes, including the Controlled Drugs and Substances Act, the Customs Act, the Excise Act, and the Income Tax Act. As the sole prosecution authority in Yukon, the Northwest Territories, and Nunavut, the FPS prosecutes violations of the Criminal Code in those jurisdictions. The FPS also plays an important advisory role on criminal law matters and acts on behalf of the Attorney General of Canada on significant criminal law issues in the appellate courts.

Through the Office of the Commissioner for Federal Judicial Affairs, the federal government appoints judges to the various federal courts, such as the Supreme Court of Canada and the Federal Court of Canada. The federal Minister of Justice appoints all of the judges in the superior trial courts in the provinces/territories.

Federal Corrections

The Correctional Service of Canada (CSC) (http://www.csc-scc.gc.ca) oversees the operation of federal correctional institutions, which house inmates serving sentences of two years or more. This responsibility extends to the post-release supervision of federal offenders. The release of federal inmates from correctional institutions is, in most cases, decided by the National Parole Board (NPB) (http://www.npb-cnlc.gc.ca).

Canadian Border Services Agency

This agency manages Canada's border at airports, seaports, and highway entrances. Its personnel are involved in a wide range of activities, including inspection, enforcement, and surveillance; they work closely with their counterparts in other countries as well as with Canadian police services (http://www.cbsa-asfc.gc.ca).

Department of Justice Canada

The primary mandate of the Department of Justice is to assist the federal government in making and reforming the criminal law and in developing criminal justice policy. This Department is also the government's law firm; that is, it provides legal advice to the federal government as well as legal services (for example, through the FPS, discussed earlier). The Minister of Justice and Attorney General of Canada is an elected MP and is appointed to this cabinet post by the prime minister. Although Parliament has the final say in creating and amending criminal laws, new laws are first drafted in the Department of Justice, often with public input. These drafts, called bills, are debated in committees in the House of Commons before they are brought before Parliament. Some justice bills that have recently drawn public attention include those on the mandatory registration of firearms, on higher sentences for crimes motivated by hate, and on the creation of a national DNA data bank. The Anti-Terrorism Act also generated strong public debate.

The justice department provides funding for various organizations such as the Canadian Society of Forensic Science and the Canadian Association of Provincial Court Judges. Through cost-sharing arrangements with the provincial governments, the department partially funds legal aid, criminal injury compensation, and the Native Court Worker Program. Fully one-third of the department's budget is passed on to the provinces to fund programs for young offenders.

The federal government also operates the Canadian Centre for Justice Statistics, a division of Statistics Canada (http://www.statcan.ca) that produces statistical reports on the operations and costs of some components of the justice system, as well as on such topics as the victims of crime, youth crime, criminal justice expenditures, and crimes reported to the police.

Provincial and Territorial Governments

Provisions in the Constitution Act make the ten provinces and three territories responsible for the administration of justice, although (as noted earlier) provincial legislatures are increasingly active in passing quasi-criminal legislation. Provincial/territorial governments pass laws, oversee police services, prosecute offences, manage courthouses, employ some judges, run programs for offenders, and operate correctional institutions. Each jurisdiction is unique, so what follows provides a general overview.

Provincial/Territorial Offences

Provincial/territorial legislators can pass laws in the areas that come under their constitutional jurisdiction. Common provincial offences include underage drinking, speeding, selling cigarettes to minors, scalping tickets, attending unlicensed raves and, in British Columbia, "aggressive panhandling."

Provincial governments are playing an increasingly proactive role in the enactment of legislation to address specific types of crime. This has been due, in part, to the perception in some provinces that the federal government has moved too slowly in taking legislative action. Several recent examples of enacted (or pending) crime-related legislation are provided in Box 1.1.

Provincial Police

There are three provincial police services: the Ontario Provincial Police, the Sûreté du Québec, and the Royal Newfoundland Constabulary. Provincial governments also oversee independent municipal police departments wherever they exist. This includes most major cities, such as Montreal, Toronto, Edmonton, and Vancouver. Provincial police statutes set out policies that address training standards, the public complaints process, and other aspects of police practice.

Provincial Prosecutors, Courts, and Judges

Every province and territory has an attorney general, the chief law enforcement officer, who is responsible (figuratively) for laying charges and prosecuting cases. In some provinces, such as Manitoba, the justice minister also serves as attorney general. All Criminal Code and provincial offences are prosecuted by provincial Crown attorneys.

focus

BOX 1.1

Selected Provincial Crime Legislation

Sex offender registry In 2003 the Ontario government took the initiative and established a provincial Sex Offender Registry to strengthen the ability of the police to monitor sex offenders in the community. A year later the federal government created a national sex offender registry.

Pit bulls In 2005, Ontario enacted legislation (the Public Safety Related to Dogs Statute Law Amendment Act) that bans pit bull–type dogs from the province, places restrictions on the current owners of pit bulls, and prescribes penalties for offences committed under the act, including a fine of up to $10,000, and/or six months imprisonment, and/or restitution to the victim.

Street racing and motor vehicle offences British Columbia and Manitoba have amended their provincial motor vehicle acts to allow police officers to seize and impound cars that are involved in street racing. Similar legislation had been tabled in Ontario as of mid-2005. In Nova Scotia, legislation was tabled in 2005 that would enable courts to automatically treat sixteen- and seventeen-year-olds as adults in cases involving serious provincial motor vehicle offences.

Marijuana "grow ops" Under a proposed (2005) law in Alberta (which may subsequently be determined to be constitutionally invalid), proceeds from the sale of real estate, vehicles, and other goods seized from indoor pot farms would be spent on law enforcement and crime prevention and on compensating victims.

Proceeds of crime Legislation in Ontario allows authorities to seize cash and property related to crime without the burden of having to prove a criminal conviction. The alleged offender must show in civil proceedings that any assets were legitimately obtained—otherwise these assets are subject to seizure. The first priority for any money that is recovered is compensation for victims. As of 2005, several other provinces had tabled similar legislation.

Check the website for your province/territory to see what other crime-related legislation has been proposed or enacted in recent years.

Each jurisdiction has provincial or territorial courts, as the case may be, as well as courts of appeal. Provincial and territorial courts hear the majority of cases. Judges are usually appointed after being recommended by an independent tribunal.

Provincial/Territorial Corrections

Provincial/territorial governments operate institutional and community corrections. Non-custodial programs include probation, fine option programs, and victim–offender reconciliation programs. There are provincial institutions for offenders whose sentences total less than two years. Quebec, Ontario, and British Columbia have parole boards for provincial offenders. In the remaining

focus

BOX 1.2

Ontario Ministries Responsible for Criminal Justice

Ministry of the Attorney General
http://www.attorneygeneral.jus.gov.on.ca

The attorney general, an elected member of the provincial legislature, is Ontario's chief law enforcement officer and oversees (officially if not actually) the laying of criminal charges and their prosecution. There are a number of divisions in the Ministry, including Court Services, Criminal Law, Family Justice, Legal Services, the Ontario Native Affairs Secretariat, and Victims and Witnesses. This last division houses the Victim/Witness Assistance Program (VWAP), which provides assistance to Crown witnesses when they testify in court. Several independent boards and commissions are associated with this ministry, including the Special Investigations Unit and the Criminal Injuries Compensation Board.

Ministry of Community Safety
and Correctional Services
http://www.mpss.jus.gov.on.ca

This ministry is responsible for a broad range of areas related to criminal justice, including policing services, correctional services, and public safety. In the policing area, this includes the Ontario Provincial Police (OPP), the Ontario Police College, the Criminal Intelligence Service, and the Ontario Civilian Commission on Police Services. In corrections, the ministry is also responsible for adult institutional and community correctional services as well as the Ontario Parole and Earned Release Board. The ministry is also home to Emergency Management, which coordinates the provincial government's responses to serious incidents and natural disasters.

Ontario Women's Directorate
http://www.citizenship.gov.on.ca

The Women's Directorate, located in the Ministry of Citizenship and Immigration, focuses on the social, economic, and justice-related issues that confront women in the province. This includes violence against women and children; the directorate helps support programs that address and respond to this issue. It also coordinates programs and services for women victims of violence that are funded by the Ministry of Community and Social Services.

Ministry of Community and Social Services
http://www.cfcs.gov.on.ca

This ministry provides a broad range of programs and services to adults and youth throughout the province. These include:

- the Violence Against Women program, which supports transition houses, rape crisis centres, domestic violence courts, and victim witness assistance; *and*

- the Aboriginal Healing and Wellness Strategy, which has a mandate to reduce family violence and improve access to services in Aboriginal communities. Such services include shelters for abused women and their children, treatment centres, and traditional healing lodges.

Ministry of Children and Youth Services
http://www.children.gov.on.ca

This ministry provides justice services for youth from twelve to seventeen within the framework of the federal Youth Criminal Justice Act. These services include open and security custody facilities, treatment services, educational programs, and community supervision.

Centre for Addiction and Mental Health (CAMH)
http://www.camh.net

Located in Toronto, the CAMH is an addiction and mental health teaching hospital. The physicians and staff are involved in clinical practice, health promotion, education, and research. The Assessment and Triage Unit provides forensic assessments of individuals who are facing charges under the Criminal Code—that is, it judges their "fitness" to stand trial. It also conducts pre-sentence evaluations.

Ministry of Health and Long-Term Care
http://www.health.gov.on.ca

Ontario's justice system is supported by many other agencies besides the ones described above. Mental health organizations are key among them; these provide addiction assessment and treatment services, sexual assault treatment centres, and mental health assessments. The health ministry also administers the Ontario Criminal Code Review Board, a tribunal that oversees the transfer and release of patients who were detained in psychiatric facilities because they were declared not criminally responsible on account of mental disorder.

Ombudsman Ontario
http://www.ombudsman.on.ca

The primary mission of the ombudsman is to investigate and resolve complaints from citizens about services of the provincial government and to ensure that services are delivered in an efficient, economical, and effective way. As part of this mandate, the office reviews the complaints of inmates confined in provincial correctional institutions.

provinces/territories, decisions to release incarcerated offenders are made by the National Parole Board. The provinces/territories also operate probation services, custody facilities, and related programs for young offenders.

In every province/territory the criminal justice system is comprised of a number of ministries, agencies, boards, and commissions. Box 1.2 sets out the ministries responsible for criminal justice in Ontario. Other provinces/territories have similar agencies. Throughout this text, websites are provided so that you can learn more about the arrangements for justice services in other provinces and in the three territories.

The provinces also involve themselves in international justice. In an effort to capitalize on the increasing globalization of criminal justice, the Ontario Ministry of the Attorney General operates Ontario Justice International, a consulting firm that provides services to international client governments in the areas of law and policy, court administration, dispute resolution, access to justice (for example, legal aid), courthouse architecture, and criminal justice training.

Municipal Governments

Municipal governments may enact local bylaws, which are valid only within the city limits. An example is bans on smoking in bars, restaurants, and commercial buildings. Minor penalties—principally fines—are levied for violating these bylaws. Note that no municipally enacted bylaw can encroach on the jurisdiction of another level of government. In cities with independent municipal police services, police budgets are controlled by police service boards and/or the city council. These same bodies also hire and dismiss the police chief.

THE COSTS OF CRIME AND CRIMINAL JUSTICE

It is possible to assign a dollar figure to the costs of policing, the courts, and corrections; it is difficult to put a price on the damage crime does to victims, to their families, to communities, and to society.

The Costs of Crime

It is estimated that Canada's total annual crime bill—which covers loss of property, medical costs, the pain and suffering of victims, and other impacts—is around $70 billion. Many observers consider this a conservative figure (Leung, 2004). Figure 1.3 illustrates the broad range of crime's "hidden" costs—and there are certainly others as well. Note that the impacts of crime extend into all sectors of the community.

The costs of crime have been viewed by some as contributing to economic prosperity, in that higher crime rates tend to result in increased spending on policing services, on case processing in the courts, and on security measures; all this extra spending in turn increases the gross national product. The contrasting view is that high crime rates are a drag on society, in that they siphon off resources that could be invested more productively in other sectors of society.

Figure 1.4 is a recent (2003) estimate of the financial costs of crime by sector. It indicates that a large portion ($47 billion, or 67 percent) of the costs of crime are borne by crime victims; the justice system ($13 billion, or 19 percent) and

FIGURE 1.3   The Hidden Costs of Crime

Source: C.G. Nicholl, *Community Policing, Community Justice, and Restorative Justice: Exploring the Links for the Delivery of a Balanced Approach to Public Safety* (Washington, DC: Office of Community Oriented Policing Programs, U.S. Department of Justice, 1999), 44.

FIGURE 1.4   Costs of Crime by Sector (in billions of dollars)

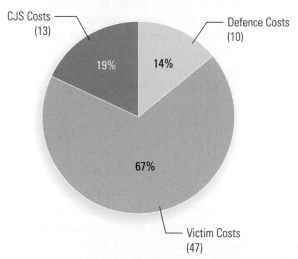

Costs by Sector
(Billions of Dollars $)

CJS Costs
(13)

Defence Costs
(10)

19%

14%

67%

Victim Costs
(47)

Source: K. Li, Costs of Crime in Canada: An update. *JustResearch*, 2004, http://www.canada.
justice.gc.ca/en/ps/rs/rep/justresearch/jr112_007.html. Department of Justice. Reproduced with
the permission of the Minister of Public Works and Government Services Canada.

FIGURE 1.5   Costs of Crime by Crime Category (in billions of dollars)

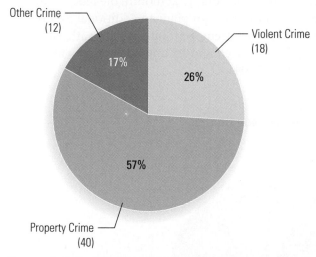

Costs by Crime Category
(Billions of Dollars $)

Other Crime
(12)

Violent Crime
(18)

17%

26%

57%

Property Crime
(40)

Source: K. Li, Costs of Crime in Canada: An update. *JustResearch*, 2004, http://www.canada.
justice.gc.ca/en/ps/rs/rep/justresearch/jr112_007.html. Department of Justice. Reproduced with
the permission of the Minister of Public Works and Government Services Canada.

defensive crime prevention/protection measures taken by residents and businesses ($10 billion, or 14 percent) account for far less of the costs of crime.

Figure 1.5 breaks down the financial costs of crime directly attributable to a
crime category. These figures indicate that property crimes—the most difficult
type of crime for the police to respond to—inflict the greatest costs ($40 billion)

on Canadians annually. Violent crimes against the person are estimated to cost $18 billion a year, other crimes $12 billion.

A recent report (GPI Atlantic, 2005) calculated that crime in the province of Nova Scotia costs $550 million per year in economic losses to victims, public expenditures on police, courts, and prisons, and private-sector spending on security devices and theft insurance. In 1997 this amounted to $1,650 per household in the province. When additional "hidden costs" of crime (see Figure 1.3) were added in, the costs increased substantially, to $3,500 per household, or $1.2 billion annually. And these figures do not include the costs associated with the deaths, injuries, and property damage that result from crime, or the costs of lawyers.

The Costs of Criminal Justice

The total expenditures for operating and maintaining the criminal justice system are conservatively estimated as at least $11 billion (Statistics Canada, 2003). Figure 1.6 shows the percentage of expenditures in 2000–01 allocated for policing, courts, legal aid plans, criminal prosecutions, and adult corrections. No current figures are available for youth corrections. As you can see, the biggest slice of the budget pie (61 percent) goes to policing. In total, justice spending represents about 3 percent of overall government spending; compare this with social services (31 percent), debt charges (15 percent), education (14 percent), and health (14 percent) (Besserer and Tufts, 1999: 2).

The costs of criminal justice have been inflated in recent years by the increased focus on safety and security following the terrorist attacks in the United States on September 11, 2001. The following month, the federal government introduced a $280 million anti-terrorism plan; an additional $7.7 billion was later committed to this plan to strengthen intelligence gathering and policing, to improve security in air travel, and to upgrade security along the Canada–U.S. border (Taylor-Butts, 2004).

FIGURE 1.6   Criminal Justice Expenditures

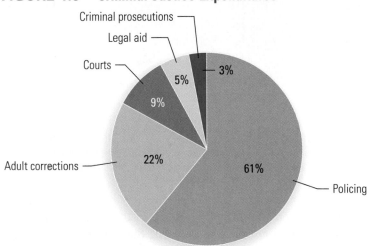

The provincial/territorial focus on deficit reduction in recent years has meant that fewer resources have been available for developing and sustaining innovative justice policies and programs. At the same time, justice agencies are being held to higher standards of fiscal accountability and job performance. These challenges have led many justice agencies to adopt private-sector practices (for example, "best practices," corporate plans, and performance measures).

THE FLOW OF CASES THROUGH THE JUSTICE SYSTEM

A flow chart illustrating how cases proceed through the criminal justice system is presented in Figure 1.7. You will want to refer to this figure often as you read the following chapters. The figure is useful in telling us who is where in the system at any given point in the process; however, it provides little insight into the actual *dynamics* of criminal justice—that is, how decisions are made by justice personnel, the challenges they face in preventing and responding to crime, the role of crime victims, and the new initiatives that are being undertaken in an effort to make the justice system more effective. These and other issues will be addressed in later chapters.

Case Attrition in the Justice System

The criminal justice system responds to lawbreaking with investigation, prosecution, and (when appropriate) punishment. It does not, however, respond to every breach of the law. Only a portion of the criminal acts committed come to the attention of the police, and a much smaller percentage of these are heard in the courts or lead to a sentence of incarceration. So dramatic is the attrition of cases in Canadian criminal justice that it is often represented graphically by a crime funnel (see Figure 1.8). Figure 1.9 illustrates the funnel effect using Canadian crime data.

Is the Criminal Justice System a "System"?

An entity functions as a system when its individual components work together in an organized and methodical manner using agreed-upon principles and when they (perhaps) share a common goal. In fact, the criminal justice "system" does not always function like a true system. Although its various components often work together, on many occasions they tend to work somewhat independently or even at cross-purposes. This divergence ensures that there are checks and balances but also limits the ability of the justice system to respond to offenders in a consistent way. The need for cooperation and coordination among criminal justice agencies has assumed even greater importance with the increasing concerns over public safety and security in the post-9/11 world. Yet a report of the Federal Auditor General (2003) found that there were serious problems in information sharing and in integrating security systems across the country.

The factors preventing criminal justice agencies and personnel from operating as a system include the following: multiple mandates, obstacles to information sharing, and the diversity and complexity of the system.

FIGURE 1.7   The Criminal Justice Process

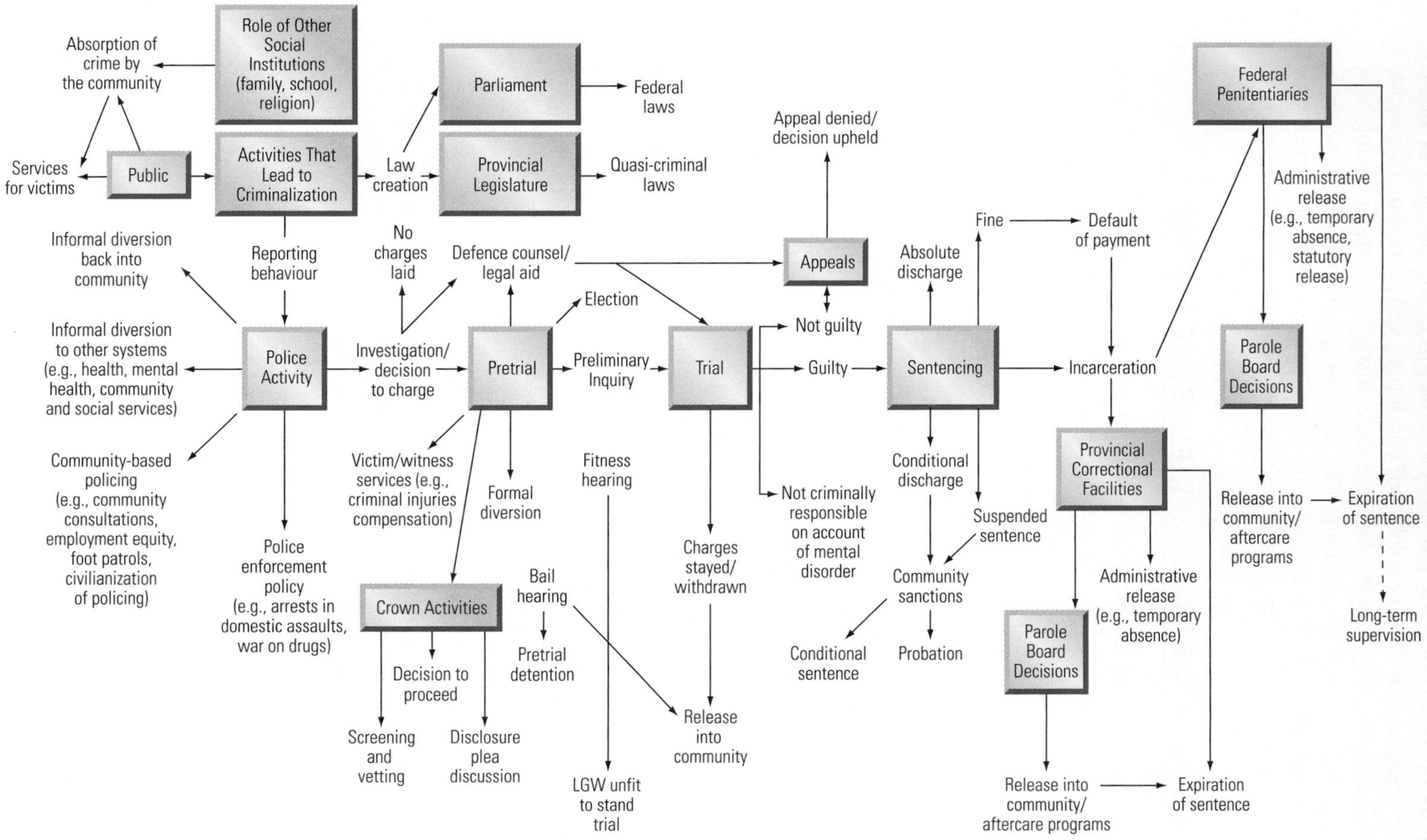

Source: C.T. Griffiths, *Canadian Corrections*, 2nd ed. (Scarborough, ON: Thomson Nelson, 2004), 71. Report on the Commission on Systemic Racism in the Ontario Criminal Justice System: A Community Summary, Gittens, M. and D. Cole. © Queen's Printer for Ontario, 1995. Reproduced with permission.

FIGURE 1.8   The Crime Funnel

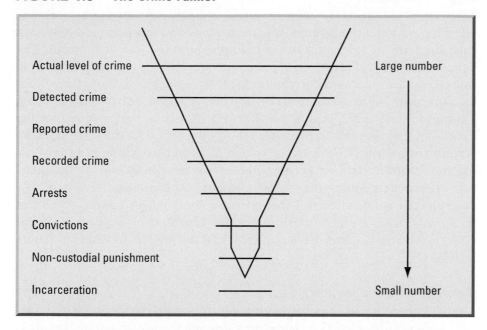

Source: I.M. Gomme, *The Shadow Line: Deviance and Crime in Canada* (Toronto: Harcourt Brace Jovanovich, 1993), 158. © 1993. Reprinted with permission of Nelson, a division of Thomson Learning: www.thomsonrights.com, FAX 800 730-2215.

FIGURE 1.9   The Criminal Justice Funnel

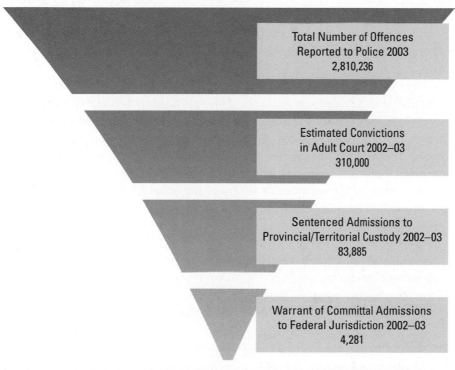

Source: Corrections Statistics Committee, *Corrections and Conditional Release Statistical Overview* (Ottawa: Public Safety and Emergency Preparedness, 2004), 17. Reproduced with the permission of the Minister of Public Works and Government Services Canada.

Multiple Mandates

A persistent feature of the criminal justice system is that its various components have different mandates. A "system" should have a common mandate. Each department at Microsoft, for example, has the same goal: profit. The same cannot be said of the criminal justice system. The police, the criminal courts, and the systems of corrections are interdependent and operate within the same criminal law framework. While justice system personnel may share a common goal of responding to offenders in such a way as to reduce the likelihood that the offensive behaviour will be repeated, at the same time the various components of the justice system differ in their views as to how this common objective is best accomplished. This is reflected in the "sniping" that often occurs among the various components of the justice system. For example, the police often feel that criminal court judges are too lenient in their sentencing practices. Through the Canadian Professional Police Officers Association, they have also criticized the release decisions of parole boards.

Obstacles to Information Sharing

Although there is a set structure within which persons charged with and convicted of criminal offences are processed through the justice system, information about them does not always follow along. The police, the courts, and the corrections system each keep their own files. Will a parole board know what the police learned in their investigation of the crime? Will it know whether the victim testified in court? Will it know what was in the judge's mind in selecting a particular sentence? The answer may be "no." This is related in part to resource constraints that may hinder the duplication and distribution of materials and in part to privacy laws that may hinder information sharing. Also, there are some constitutional protections that may limit the sharing of information, particularly in cases involving youth.

Barriers to information sharing can affect the completeness of information available to key decision makers. This was a subject of critical comment in several recent coroner's inquests and inquiries that examined the circumstances in which parole boards released offenders from prison without knowing all the facts about the offender. Deadly outcomes have occurred when justice officials were not aware of the offender's criminal history—or, for that matter, of the number of offences the offender had committed that had not resulted in charges although the authorities knew about them. Sadly, it is often only after a heinous crime has been committed that review boards are able to piece together an accurate picture of the offender's background, and to examine all of the decisions made by mental-health and criminal justice practitioners.

Other consequences of inadequate information sharing among the components of the justice system include the following:

- flawed police investigations of serious crimes;
- incomplete information on offenders (regarding, for example, backgrounds and risk assessments);

- the staying (that is, suspension or discontinuance) of nearly 100,000 criminal cases a year in Canada owing in part to failures to get the cases to court within a reasonable period;
- delays in sending to correctional authorities important information that could be used in planning treatments; *and*
- potential revictimization resulting from a failure to notify crime victims that offenders are being released into the community.

Many of these difficulties are due at least in part to the fundamental principle that a person is innocent until proven otherwise. The importance of erring on the side of caution is illustrated by cases in which innocent persons have been found guilty and sent to prison for crimes it turned out they did not commit. The issue of wrongfully convicted persons is discussed in Chapter 6.

In recent years there have been a number of initiatives aimed at increasing offender-related information sharing among the various components of the justice system. The current buzzword in criminal justice circles to address these issues is **interoperability**, defined as the ability of hardware and software from multiple databases from multiple agencies to communicate with one another. Operationally, this means that the police, for example, will be able to exchange criminal justice information with their counterparts in corrections and parole using common data standards. Similarly, the courts will have access to information generated by correctional agencies and parole boards.

> **interoperability**
> the ability of hardware and software from multiple databases from multiple agencies to communicate with one another

The interoperability initiative is being coordinated by the federal government through the Department of Public Safety and Emergency Preparedness and includes a number of specific initiatives designed to integrate justice information across the country. These include Real Time Identification, designed to enhance the response to fingerprint identification requests, and the Child Exploitation Tracking System (CETS), a database developed jointly by law enforcement agencies and Microsoft to facilitate the gathering and sharing of information to fight Internet child pornography. As well, the Correctional Service of Canada has developed a massive database on criminal offenders (IntelPol) that is available to police officers across the country.

Diversity and Complexity of the System

A third obstacle to cooperation and mutual understanding among criminal justice agencies is the sheer complexity of the system and the enormous amount of information one must assimilate in order to understand the system as a whole. New research findings are published monthly. Laws change. The decisions of some criminal court judges become judicial precedent. It is difficult enough for justice personnel—police officers, Crown attorneys, and correctional officers, to name a few—to keep abreast of the current developments that pertain to their own positions and responsibilities, let alone those of other parts of the system.

CRIMES AND THE CRIMINAL LAW

rule of law
the foundation of the Canadian legal system

All legal systems in Canada are governed by the principle of the **rule of law**, which can be traced back to the English Magna Carta of 1215. The rule of law has several components:

- All citizens are supposed to obey the law, to expect punishment if they break the law, and to look only to the legal system to respond to transgressions committed against them.
- The legal system, in turn, must be fair and impartial in its responses.
- Only elected officials can decide what is against the law.
- A law must be written clearly so that everyone—or at least lawyers—can understand which action or activity is being referred to.
- A law must apply equally to everyone—for example, to men and women, and to rich and poor.
- To give the law "teeth," a penalty must be defined for when the law is broken.

Criminal Law

criminal law
that body of law which deals with conduct considered so harmful to society as a whole that it is prohibited by statute and prosecuted and punished by the government

There are two general categories of law: public law and private law. Public law deals with those matters which affect society as a whole in its interactions with governments. There are four types of public law: criminal law, constitutional law, administrative law, and taxation law. **Criminal law** can be defined as "that body of law that deals with conduct considered so harmful to society as a whole that it is prohibited by statute, prosecuted and punished by the government" (http://www.duhaime.org). The criminal law defines which acts (or omissions) are against the law and sets out the available penalties. It also sets out the rules that police and judges must follow in criminal matters, including procedures for making arrests, gathering evidence, and presenting evidence in court.

Private law, by contrast, regulates relationships between individuals other than the state and is used to resolve disputes between private citizens.

In Canadian society, criminal law:

- acts as a mechanism of social control;
- maintains order;
- defines the parameters of acceptable behaviour;
- reduces the risk of personal retaliation (that is, vigilantism, or people taking the law into their own hands);
- assists in general and specific deterrence;
- serves as a means of punishment;
- criminalizes behaviour;
- protects group interests; *and*
- prevents crime and serves as a deterrent to criminal behaviour.

What Is a Crime?

crime
an act or omission that is prohibited by criminal law

A **crime** is generally defined as an act or omission that is prohibited by criminal law. Every jurisdiction sets out a limited series of acts (crimes) that are prohibited and punishes the commission of these acts by a fine or imprisonment or some other type of sanction. In exceptional cases, an omission to act can constitute a crime. Examples are failing to give assistance to a person in peril or failing to report a case of child abuse.

Two critical ingredients of a crime are the commission of an act (*actus reus*) and the mental intent to commit the act (*mens rea*). A crime occurs when a person:

- commits an act or fails to commit an act when under a legal responsibility to do so;
- has the intent, or *mens rea*, to commit the act;
- does not have a legal defence or justification for committing the act; *and*
- violates a provision in criminal law.

Criminal Law versus Civil Law

As one among several legal systems that exist in Canada, the criminal justice system concerns itself only with offenders who are criminally liable for wrongdoing. The police investigate the crime, lay an information or charges, and compel the accused person to appear in court. At that point the matter is prosecuted in the name of the Crown, ostensibly on behalf of the community. The correctional system metes out the punishment. Except for those accused who can afford to hire a lawyer, the entire process is paid for by the public's tax dollars. The victim's principal role is to testify for the prosecution at trial. The accused is considered innocent until proven guilty, and the Crown must prove the case beyond a reasonable doubt.

One of the key distinctions in law is between **criminal law** and **civil law**. Although certain acts may constitute a violation of both civil and criminal law, and although both systems of law involve the imposition of sanctions, there are important differences between the two. When a crime is committed:

civil law
a general category of laws relating to contracts, torts, inheritances, divorce, custody of children, ownership of property, and so on

- the government assumes the responsibility for prosecuting the alleged offender;
- the criminal courts, on behalf of the victim and the community, undertake the task of determining the guilt or innocence of the offender;
- the criminal courts impose a sanction, which may involve supervision of the offender in the community or a period of confinement in a correctional facility; *and*
- there may be a financial penalty attached to the disposition handed down by the judge in the criminal court.

In contrast, civil law is essentially concerned with disputes between individuals. Civil law includes all statute law other than criminal law, such as divorce and human rights law. In civil suits:

- the person who feels wronged by the alleged behaviour brings suit in civil court; *and*
- the wronged party seeks damages, which usually involve the payment of monetary compensation by the wrongdoer if there is a finding of liability.

The most common types of civil suits involve divorce, disputes over inheritance, and breaches of contract. However, increasingly, the civil law is being used to combat crime. In 2002, Ontario became the first province to allow civil courts to seize, and freeze, any assets that were obtained as proceeds of crime.

One significant difference between criminal law and civil law relates to the standard of proof required to convict a person of wrongdoing. In a criminal trial, the prosecutor must prove that the defendant is guilty "beyond a reasonable doubt." In a civil trial, liability is determined using the standard of "the balance of probabilities." The standard is one of "reasonable probability" or "reasonable belief" rather than "proof beyond a reasonable doubt." This reasonable probability is a much lower standard of proof.

Where Do Criminal Laws Come From?

Have you ever thought about why marijuana use in most instances is illegal, but drinking alcohol is not? To say the least, there is not always agreement about what should be against the law. Murder? Yes. Impaired driving? Yes. Bank robbery? Sure. Homosexuality? Three-card monte? Possessing crime comics? Charging as much as 28 percent interest for credit card loans? Laws can reflect social values and priorities, but they can also be affected by morality. Thus there is no single answer to the question of where laws come from.

Most of Canada's criminal law was inherited from Victorian England, and many of those offences can be traced even further back, some to biblical times or earlier. Some of our laws seem archaic, and a perusal of the Criminal Code reveals laws that are no longer enforced: alarming the Queen (s. 49), inciting to mutiny (s. 53), duelling (s. 71), obstructing a clergyman from performing any function in connection with his calling (s. 176), setting man traps (s. 247)—wherein a homeowner sets a trap in an attempt to prevent a break and enter—and feigning marriage (s. 292), among others.

When a law is created, there is some reason behind it, even when the logic is not readily apparent or when the motive seems sinister by today's standards. Why did lawmakers at one time create the criminal offence of discharging a stink bomb in a public place (s. 178)? Perhaps the answer to that question will remain forever lost in history. But researchers have been conducting some interesting historical studies of law reform. In so doing, they have been examining the social and political climate of times past. For example, laws against opium use first passed in the early 1900s have been linked to anti-Asian prejudice among Euro-Canadians of the day.

There was a time when lotteries operated entirely through underground channels and participants were liable to criminal prosecution. Now that governments are operating lotteries, it is a different story. But private lottery schemes continue to be illegal. The same is true when it comes to betting on horseraces, operating casinos, and organizing prizefights. As long as the government is regulating these activities, they are legal. But outside government regulation, all of these activities remain criminal acts.

As a society changes, behaviours can be criminalized. In Canada there have been ongoing efforts to ensure that the criminal law reflects changes in technology.

As a result of recent additions to the Criminal Code, the following are now offences:

- destroying or altering computer data (s. 430[1.1]);
- using the Internet to distribute child pornography (s. 163.1) and to communicate with a child for the purposes of facilitating the commission of certain sexual offences (s. 172.1); *and*
- deliberately misappropriating cable television service (s. 326).

Similarly, to combat the rise in organized crime, various amendments to the Criminal Code have addressed money laundering and outlaw motorcycle groups. Also, it is now an offence to be involved with a criminal organization. However, it is likely that the constitutionality of this law will be challenged. In 2005, the B.C. Supreme Court (*R. v. Accused No. 1* [2005] B.C.J. No. 2702) upheld the application of an accused who argued that s. 467.13 of the Criminal Code was constitutionally invalid because it was too broad and too vague.

Conversely, some activities have been decriminalized over the years; that is, the laws against them have been repealed or struck down. Threatening cattle is no longer an offence. The same is true of attempted suicide. Laws that were applied against homosexuals and Chinese immigrants no longer exist. The Supreme Court of Canada has used the Charter of Rights and Freedoms to strike down laws that are inconsistent with the Charter's provisions and protections.

The Origins and Application of Criminal Law

There are two main explanations of the origins and application of criminal law. The first, the **value consensus model**, views crime and punishment as reflecting society's commonly held opinions as well as its limits of tolerance. This view assumes that there is a consensus on what should be against the law. Indeed, there probably *is* consensus that murder should be against the law. Incest is another act that is widely condemned. Such offences, called *mala in se* (wrong in themselves), are perceived as so inherently evil as to constitute a violation of "natural law."

In contrast, some actions are "criminal" only because they happen to be against the law, not because there is universal abhorrence to them. At the turn of the century, narcotics were widely used in patent medicines and soft drinks, and both were freely available at the corner drugstore. Now the situation is reversed. Instead of being *mala in se*, such offences are *mala prohibita* (acts considered criminal only because they violate a criminal statute).

The **conflict model**, the second theory of the origins and application of criminal law, draws our attention to the fact that some groups are better able than others to influence which behaviours are criminalized. In particular, conflict theorists see the rich and privileged as having an advantage in influencing law reform. Scholars who conduct research using a conflict perspective might ask the following questions:

- Why does a person who steals less than $100 from a convenience store often receive a much more severe sentence than someone who fraudulently poses as a doctor, a lawyer, or a university professor?

value consensus model
the view that crime and punishment reflect commonly held opinions and limits of tolerance

conflict model
the view that crime and punishment reflect the power some groups have to influence the formulation and application of criminal law

- Why are crimes committed by corporations (such as overpricing goods, failing to create and maintain healthy and safe working environments, and illegally disposing of hazardous wastes) most often dealt with through civil suits and the imposition of fines rather than under criminal law?

- Why are Canadian correctional institutions populated by large numbers of Aboriginal people and by those with low education and skill levels, high rates of alcohol and drug addiction, and dysfunctional family backgrounds? Are these groups actually more criminal than other groups in society?

- What role do interest groups play in influencing the enactment of criminal legislation or in decriminalizing certain behaviour?

Conflict theorists highlight some of the inequities and paradoxes in the system. If someone takes money from a bank at gunpoint, it is called robbery. A business decision that causes a bank collapse, thus depriving thousands of account holders of their money, is called a bad day on the stock market. Conflict theorists believe that our attention is wrongly focused on street crime when the greater risk to most people lies in the actions of elites, including corporations that dump toxic waste, fix prices, condone unsafe workplaces, and evade taxes.

The Role of Interest Groups in Law Reform

Canadians live in a pluralistic society in which people hold diverse views and opinions. You need to think only of the issue of abortion to realize that there are many topics about which there is no societal consensus. Assisted suicide for the terminally ill, which is a crime, is another issue about which there is widespread disagreement.

During the 1920s and early 1930s, the prohibition against the consumption of alcohol in the United States was an unpopular law that was broken by many otherwise law-abiding citizens. Prohibition was championed by *moral entrepreneurs* who believed that criminal law can and should be used to enforce moral values and standards. Laws regarding gambling, prostitution, soft-drug use, homosexuality, and abortion have also been traced to the influence of moral entrepreneurs.

Many groups across the country are involved in efforts to influence law reform. Some of the high-profile ones are listed below:

- The National Association of Women and the Law (http://www.nawl.ca) and the National Action Committee on the Status of Women (http://www.nac-cca.ca) lobby for the rights of women.

- The Canadian Resource Centre for the Victims of Crime (http://www.crcvc.ca) works to improve the treatment crime victims receive from the criminal justice system.

- The Canadian Professional Police Association (http://www.cppa-acpp.ca) submits briefs in support of legislation to increase penalties for criminal behaviour and to extend the powers of the police to respond to crime.

- Mothers Against Drunk Driving (http://www.madd.ca) lobbies governments for more severe impaired-driving laws and for lengthier sentences for persons convicted of impaired driving offences.

- The National Organization for Reform of Marijuana Laws (NORML; http://www.norml.ca) advocates the decriminalization and, ultimately, the legalization of marijuana.
- The National Firearms Association (http://www.nfa.ca) supports the property rights of Canadians, especially as these rights relate to gun ownership.

THEMES IN CRIMINAL JUSTICE

There are a number of recurring themes in any examination of the criminal justice system.

Individual Rights and Public Protection: A Delicate Balance

There is a fundamental and ever-shifting balance to be struck between (1) giving criminal justice agencies such as the police and prosecutors the unfettered power to apprehend and prosecute offenders (crime control); and (2) protecting citizens from the potential abuses of that power (due process). Below are listed some of the more recent initiatives that highlight the issue of individual rights versus public safety.

National DNA Data Bank

The challenges of maintaining an appropriate balance are illustrated by the national DNA bank, which is managed by the RCMP and holds the genetic profiles of more than 3,000 offenders. The DNA samples are taken from people who have been convicted of certain serious crimes, including murder and sexual assault. These samples are matched with samples of blood, hair, bone, or semen taken from crime scenes or from the bodies of crime victims. With the approval of a judge, DNA samples can be taken from convicted offenders without their consent. New advances in DNA technology will allow the identification of a suspect's ancestry, and this has raised concerns about genetic profiling.

Sex Offender Registries

Another area where the interests of the general public have been determined to outweigh the rights of the individual is in policies relating to sex offenders. The federal government operates a national sex offender registry, and several provinces have provincial registries as well. These registries generally require designated sex offenders who are released from correctional facilities to register within fifteen days of their release (or upon conviction if they have received a non-custodial sentence). The provincial registry in Ontario, for example, includes the offender's name, date of birth, current address, and photograph. This information is in addition to the power given to Ontario police to notify the public that an offender considered to be a significant risk is being released into the community (more on this in the discussion of the release and reentry of incarcerated offenders in Chapter 8). The federal government's registry for federal offenders requires convicted sex offenders to tell police where they live and when they move.

Organized Crime and Terrorist Legislation

In recent years, the powers of the police have been extended in an attempt to better prevent and respond to organized crime and terrorist activity. Under the Anti-Terrorism Act, a peace officer may arrest a person without a warrant and have that person detained in custody if the officer suspects, on reasonable grounds, that the person's detention is necessary in order to prevent a specific terrorist activity.

To enhance the capacity of the justice system to respond to organized crime, federal legislation has been enacted that gives the police who are working under-cover the authority to break most criminal laws (exceptions: those relating to murder, assault causing bodily harm, sexual offences, and obstruction of justice) and that reduces the burden of proof for prosecutors attempting to convict members of criminal organizations. Legislative enactments have also criminalized mere involvement with a criminal organization, whether or not the individual actually breaks the law.

Hate Crimes

There may also be tensions between the right of individuals to free speech and the need to protect the equality and rights of minorities. This was evident in a recent, high-profile case in which an Aboriginal leader was charged with a hate crime under the Criminal Code (see Box 1.3).

Due Process versus Crime Control

The conflicting opinions expressed by the majority and dissenting Supreme Court justices are often discussed in terms of two conflicting orientations: **due process** and **crime control**. Neither view should dominate, and a delicate balance must be struck between the two. Too much emphasis on the crime control side would result in a police state in which laws are applied in an arbitrary fashion and average citizens have no protection from abuses of state power. Too much emphasis on due process would make the police powerless to protect us from lawbreakers.

The Criminal Justice System as a Human Enterprise

As noted at the beginning of the chapter, the criminal justice system is staffed by people who make decisions about other people. Criminal justice personnel respond to a wide variety of events in a wide variety of settings in conditions that often are not ideal. They carry out their tasks within the framework of written laws and policies; but they also exercise considerable **discretion** when making decisions. This can lead to inconsistencies in how laws are applied, how cases are processed in the courts, and what decisions are made about offenders by correctional officers, parole boards, and parole officers.

Many factors influence criminal justice personnel when they make discretionary decisions. At the centre of all these influences is the decision maker, a human being who brings to his or her work a unique combination of education, training, personal experiences, and perhaps religious beliefs. It would be naive to think that life experiences and community pressure do not sometimes influence the decisions made by criminal justice personnel. For example, the

What do you think?

What concerns, if any, do you have about the legislation that has been enacted to address violent offenders, sex offenders, organized crime, and terrorism?

due process

an orientation to criminal justice in which the legal rights of individual citizens, including those suspected of committing a crime against the State, are paramount

crime control

an orientation to criminal justice in which the protection of the community and the apprehension of offenders are paramount

discretion

the freedom to choose between different options when confronted with the need to make a decision

focus

BOX 1.3

Aboriginal Leader Convicted of a Hate Crime

The definition of "hate crime" in Canada evolved from sections 318 and 319 of the Criminal Code relating to "hate propaganda," which is defined as advocating genocide, publicly inciting hatred, or willfully promoting hatred against an identifiable group, including a group distinguished by colour, race, religion, sexual orientation, or ethnic origin (Silver, Mihorean, and Taylor-Butts, 2004: 4).

In December 2002, David Ahenakew, an Aboriginal leader in Saskatchewan and an appointee to the Order of Canada, stated in an interview with a newspaper reporter:

> How do you get rid of a disease like that, that's going to take over, that's going to dominate? The Jews damn near owned all of Germany prior to the war. That's how Hitler came in. He was going to make

damn sure that the Jews didn't take over Germany or Europe . . . That's why he fried six million of those guys, you know. Jews would have owned the God damned world. And look what they're doing. They're killing people in Arab countries (CTV. ca, 2002).

He was subsequently charged and, in 2005, convicted of a hate crime under the Criminal Code. In finding Mr. Ahenakew guilty of inciting hatred against an identifiable group under section 319 of the Criminal Code, the presiding judge stated: "To suggest that any human being or group of human beings is a disease is to invite extremists to take action against them." The judge fined Mr. Ahenakew $1,000; subsequently, he was removed from the Order of Canada (Harding, 2005: A11).

---

decision making of parole boards has come under intense scrutiny in recent years, due in large part to several high-profile cases in which offenders released by parole boards committed heinous crimes.

Even if discretion were not a factor, different justice system personnel would often make different decisions in a given situation. This is referred to as *disparity* in decision making. For example, a sentencing judge may order probation, even though another judge would have sent the same offender to prison. Similarly, a parole applicant's chances for release may depend on the composition of the parole board panel that particular day.

The Task Environments of Criminal Justice

A **task environment** in the context of the criminal justice system is the cultural, geographic, and community setting in which the criminal justice system operates and in which criminal justice personnel make decisions. These environments range from small Inuit villages in Nunavut to inner-city neighbourhoods in major urban centres such as Toronto.

The characteristics of a particular task environment influence the types of crime and social disorder that justice system personnel are confronted with, the decision-making options that are available, the effectiveness of justice policies and programs, and the potential for developing community-based programs

What do you think?

Was the government correct to charge Mr. Ahenakew with a hate crime? If not, why not? If the charge was appropriate, was the sentence as well? The maximum penalty under the Criminal Code is a $2,000 fine and up to two years in prison.

task environment
the cultural, geographic, and community setting in which the criminal justice system operates and justice personnel make decisions

and services. In addition, the same urban area may contain a variety of task environments, ranging from neighbourhoods with a high concentration of shelters for the homeless, to neighbourhoods with large populations of recently arrived immigrants, to exclusive, high-income neighbourhoods. Unique challenges are faced by criminal justice personnel in remote and northern areas of the country, where there are few resources and community-based programs for victims and offenders.

Crime manifests itself differently in remote Arctic villages than in Vancouver's skid row or in a wealthy suburban Montreal neighbourhood. City police may be faced with gang activity and traffic management; rural police may have to deal more with cigarette smuggling and marijuana cultivation. In short, the demographics of the area, the local economic conditions, and the ethnic mix combine to influence decisions. As these factors vary, so too will the types of crime or social disorder, community expectations regarding enforcement, the community's capacity to address local issues, and relations between the justice system and the citizens it serves.

Communities and the Criminal Justice System

Historically, the criminal justice system has been characterized by increasing centralization of responsibility for preventing and responding to crime. Police agencies, courts, and systems of corrections gradually replaced earlier bodies that had been the focal point for preventing and responding to crime and social disorder. Criminal justice professionals working in large agencies assumed responsibility for crime and criminal offenders. The community's exclusion from a substantive role in justice administration has contributed to the often negative perceptions that community residents have of the criminal justice system. Canadians' attitudes toward the criminal justice system are examined in Chapter 2.

The general failure of many justice policies and programs to make communities safer, to assist the victims of crime, and to address the needs of offenders has led many justice practitioners to explore ways for communities to participate in addressing crime and social disorder. Perhaps the most visible manifestation of this change in thinking is the emergence of community policing (see Chapter 4) as a model for delivering policing services. Community involvement is a means for the justice system to become more proactive in preventing crime and to develop more effective intervention strategies (perhaps by adopting a problem-solving approach to issues confronting the community).

Another facet of the community involvement has been the emergence of restorative justice, which allows crime victims and offenders, their families, interested community residents, and justice system personnel all to play a role in the response to crime. This approach is examined later in the chapter.

CHALLENGES IN CRIMINAL JUSTICE

The criminal justice system of the early twenty-first century faces a number of complex and diverse challenges. These include, but certainly are not limited to, the following.

Gender Issues

Over the past decade, considerable attention has been paid to gender as it relates to the system's operations. Police services are working harder to recruit women; regional correctional facilities for federally sentenced women are being established; and the special needs of women under supervision in the community are being given more consideration.

Despite these efforts, women continue to be vastly underrepresented in police services, in the courts as lawyers and judges, and in institutional and community corrections. Also, the RCMP's efforts to attract more female applicants by lowering recruiting standards have raised concerns (see Chapter 3). Another topic of debate (see Chapter 7) has been cross-gender staffing (that is, men working in women's institutions, and vice versa).

In recent years there has been growing awareness of the unique needs of both female victims of crime and female offenders. Programs and services for female crime victims and offenders are discussed in Chapters 2 and 8, respectively.

Criminal Justice in a Multicultural Society

A distinguishing feature of Canada is its multiculturalism. Visible minorities are now roughly 13 percent of the Canadian population; British Columbia has the highest percentage of visible minorities (21.6), followed by Ontario (19.1) and Alberta (11.2). The ethnocultural composition of these provinces places unique demands on their criminal justice systems.

In terms of sheer numbers, 54 percent of Canada's visible minority people live in Ontario. Toronto is the most diverse urban centre in the country; around 37 percent of Torontonians belong to a visible minority (Ontario Ministry of Finance, 2003). Projections are that by 2017:

- one in every five people in Canada will be a member of a visible minority;
- as is currently the case, the large majority (75 percent) of visible minority people will reside in Toronto, Vancouver, and Montreal;
- more than half of Canada's South Asians and about two-fifths of Canada's Chinese will be living in Toronto;
- Blacks will represent 27 percent of Montreal's visible minorities, and Arabs 19 percent; *and*
- most Vancouverites will be visible minorities, half of them Chinese (Statistics Canada, 2005).

The justice system has always been slow to address the challenges presented by ethnocultural diversity. There is no conclusive evidence of systemic racism in the criminal justice process; however, some visible minorities *perceive* that such racism exists, and there have been several high-profile incidents in which the courts determined that the police behaved in a discriminatory manner toward a visible minority person. The racial profiling of visible minorities by the police (see Chapter 4) has been the subject of considerable debate in recent years. In corrections, the numbers of visible minority offenders in the federal corrections system have been increasing over the past decade. Visible minorities as a group do not appear to be overrepresented among incarcerated federal offenders. However, Blacks are overrepresented in federal corrections generally; they account for 2 percent of the Canadian population but 6 percent of offenders in federal correctional institutions and 7 percent of those serving time in the community. Black offenders are most likely to be incarcerated in federal institutions in Ontario and Quebec (Trevethan and Rastin, 2004). These figures suggest the need for additional research on visible minorities in the justice system.

Most recently arrived immigrants and long-term residents who are visible minorities tend to settle in urban centres. Unfortunately, there are often few police officers, probation officers, court workers, judges, lawyers, or parole officers who speak a language other than English or French. The language barrier hinders the development of police–community partnerships, the supervision of youth and adults on probation, and the implementation of effective treatment programs. The problem is especially acute in rural areas, where there are few resources and programs (such as multilingual services).

Many new immigrants speak neither English nor French. Some have immigrated from countries where the criminal justice system is feared and may be reluctant—understandably so—to utilize justice services and resources. In the absence of any research findings, it is difficult to determine how reluctant immigrants are to report crimes to the police. In New York City, the large majority of residents in six ethnic communities told researchers they would not hesitate to report break-ins, assaults, family violence, and (to a lesser extent) drug trafficking (cf. Davis and Henderson, 2003). Residents of communities where there was a high degree of solidarity and where citizens worked together to solve problems—that is, where there was a high level of social cohesion— were more likely to say they would report crimes. This issue remains to be explored by researchers in Canada.

Observers have identified "gaps in justice"—that is, gaps between what the justice system provides and what the members of ethnocultural minorities expect. Strategies for bridging some of the more significant gaps include:

- addressing the need for public legal education and information about the law, legal resources, and sources of assistance to increase the accessibility of the justice system;
- requiring that cultural sensitivity be applied in the administration of justice and the delivery of justice services;
- increasing the numbers of ethnocultural persons working in the criminal justice system; *and*
- developing procedures for creating partnerships between ethnocultural communities and the justice system (Currie and Kiefl, 1994).

The issues that are likely to surround the development of legal structures specific to an ethnocultural group or community are highlighted by the controversy over the proposal in Ontario to use sharia law in family law cases in the Muslim community. Under this proposal these courts would have handled cases involving marriage, divorce, and other family-related issues and operate under the Islamic Institute of Civil Justice, although subject to Canadian law and to the Charter of Rights and Freedoms. The provincial government gave tentative approval to the creation of these courts in 2004; then, in 2005, Premier Dalton McGuinty stated that there would be no sharia law in Ontario and, furthermore, that all existing religious tribunals would be banned. In the words of the premier: "There will be one law for all Ontarians" (Leslie, 2005: 1). Opposition to the sharia courts had been expressed from a number of quarters, including the Canadian Council of Muslim Women, whose members feared that the courts would increase the vulnerability of Muslim women (http://www.ccmw.com). In 2004 the Quebec National Assembly struck down a proposal to establish sharia courts in that province.

Special Groups of Offenders and the Justice System

Many individuals who come into contact with the criminal justice system have unique needs and present special challenges to the police, the courts, systems of corrections, and community agencies and organizations. Among those groups requiring a coordinated and informed response by criminal justice and community-based personnel are sex offenders, violent offenders, elderly offenders, offenders with mental and developmental disorders, drug-addicted offenders, offenders with fetal alcohol spectrum disorder (FASD), and offenders with HIV/AIDS. The justice system is often slow to develop sufficient treatment resources for these high-needs offenders; for example, offenders who are addicted to crystal meth find it very difficult to access treatment programs.

Aboriginal People and Criminal Justice

Aboriginal people live in a wide variety of settings, from remote northern hamlets to reserves in rural areas to urban centres. Also, Aboriginal cultures are highly diverse; there are roughly 600 bands across the country and in the country's urban centres; for example, in the Vancouver/Richmond area alone there are Aboriginal people representing more than 200 bands (Joseph, 1999).

Two of the biggest challenges facing the criminal justice system are the high rates of crime and victimization in many Aboriginal communities and the overrepresentation of Aboriginal people at all stages of the justice process, from arrest to incarceration to parole. For example, although Aboriginal people are only 3.3 percent of the Canadian population, 18 percent of the men in federal prisons are Aboriginal, and 28 percent of the women. This is an especially disturbing figure given that federal offenders are incarcerated for more serious offences. The overrepresentation is even more acute in provincial correctional facilities (21 percent of all admissions), particularly in Saskatchewan (78 percent), Manitoba (68 percent), and Alberta (39 percent) (Johnson, 2004: 17).

Most explanations for this focus on socioeconomic factors such as chaotic family backgrounds characterized by abuse and poverty, the long-term impact of residential schools, low levels of educational attainment, high unemployment, and high rates of alcohol and substance abuse (Canadian Criminal Justice Association, 2000). It is estimated that 30 to 40 percent of all children in the care of children's aid societies in Canada are Aboriginal (Curry, 2005). Aboriginal women may be especially susceptible to hardships, including abusive relationships; it does not help that they tend to live in remote and rural areas where services are typically scarce or nonexistent.

Throughout the text, the discussion will highlight the initiatives being taken by Aboriginal bands and communities and by the justice system—often in partnership—to prevent and respond to crime and to address the needs of Aboriginal offenders. Carol LaPrairie (2002) has found that cities in the Prairie provinces contribute disproportionately to the overrepresentation of Aboriginal people in the justice system. The findings from her study and others should inform initiatives that are undertaken. It is of concern that despite the development of policies and programs over the past three decades designed to reduce high levels of Aboriginal involvement in the justice system, the rates are increasing.

The High Rates of Crime and Violence in Remote and Northern Communities

Largely "out of sight and out of mind" to most Canadians, communities in Yukon, the Northwest Territories, and Nunavut and in the northern regions of the provinces face the highest rates of crime and violent crime in the country. Many of these communities have higher rates of violent crime than major Canadian and American cities, and this has a severe impact on all facets of community life. Women are particularly vulnerable to violence, and too often, few programs and resources are available to assist them. These women typically must leave their own community to seek safety in a women's shelter, often hundreds of kilometres from home. The remote location and small population of northern communities means that the police are most often the only nearby representatives of the justice system. Court services are delivered via "fly-in" or "drive-in" circuit courts composed of a judge, a Crown attorney, a defence lawyer, and court officials. Corrections services are often nonexistent, and correctional institutions are usually hundreds or thousands of kilometres away.

Throughout the text, the efforts that have been made to improve the delivery of justice services to remote and northern communities will be highlighted.

Responding to Organized Crime in the Global Community

As conflict theorists point out, the criminal justice system has been designed to respond to "traditional" types of crimes such as assault, break and enter, robbery, and homicide. The overall rates of these common varieties of crime are declining; meanwhile the justice system is being increasingly challenged by more sophisticated types of criminal activity, including international drug trafficking, illegal immigration and human trafficking, sexual slavery, money laundering, identity theft, high-tech and computer crime, cyberterrorism, and industrial espionage. Another emerging area of concern is the vulnerability of children to Internet predators; it is estimated that 19 percent of children between fourteen and seventeen have been the targets of online sexual solicitation (Kirkey, 2001).

Hells Angels motorcycle jacket

Much of this "non-traditional" criminal activity is carried out by organized crime syndicates, including outlaw motorcycle gangs such as the Hells Angels, whose network of criminal activities extends to the international level. Investigating and prosecuting these types of crime is expensive, time consuming, and resource intensive. For example, it cost an estimated $5.5 million to secure the convictions of four associates of the Rock Machine, a motorcycle club in Quebec: $2.5 million for the police investigation—it was a drug-dealing case—and $3 million for the prosecution and four-month trial (Cherry, 2001).

At the provincial level, joint enforcement units are being established to combat organized crime:

- Integrated Market Enforcement Teams (IMETs) operate in urban financial centres and focus on fraud and other crimes in the capital markets. These teams are composed of RCMP investigators, lawyers, forensic accountants, and other investigative experts.

- The Manitoba Integrated Organized Crime Task Force is a collaborative effort of Manitoba Justice and the RCMP that focuses on organized crime and street gangs.

- Combined Forces Special Enforcement Units (CFSEUs) are now operating in Toronto, Montreal, Quebec City, and British Columbia. In Ontario the unit has enlisted the following organizations: Toronto Police Service, Ontario Provincial Police, York Regional Police, Peel Regional Police Service, Royal Canadian Mounted Police, Citizenship and Immigration Canada, and the Criminal Intelligence Service Ontario.

focus

BOX 1.4

Texas Rangers in British Columbia? The Increasing Presence of American Police in Canada

In 2004 a motorist (who happened to be an off-duty Vancouver police officer) was stopped east of the city by a Texas State Trooper, who was working with the RCMP to detect motorists driving under the influence of marijuana. Although the subsequent civil suit was settled out of court, it raised the larger issue of foreign police presence in Canada and the extent of their policing activities. As part of various collaborative policing arrangements, FBI and DEA (Drug Enforcement Agency) agents are involved in intelligence-gathering activities, although officially these agents are not permitted to involve themselves in case investigations. In defending this arrangement, a senior RCMP officer stated: "There are no borders when it comes to crime." This has sparked debate as to the role of foreign police agents in Canada.

Source: K. Bolan (2005, August 20), "The Long Arm of Uncle Sam," *Vancouver Sun*, A1, A4. Reprinted by permission of the *Vancouver Sun*.

The effectiveness of these strategies in combating organized crime is discussed in Chapter 4.

At the international level, Canada is participating in the International Criminal Court, which is located in The Hague and operates under the auspices of the United Nations. Closer to home, there are Canada–U.S. Integrated Border Enforcement Teams (IBETs), a collaborative effort of American and Canadian police services designed to combat cross-border criminal activity and terrorist threats. There has also been an increasing American enforcement presence in Canada, as part of efforts to combat terrorism and organized crime and for training purposes (see Box 1.4).

What do you think?

Do you support or oppose the presence of American law enforcement agents in Canada as part of a collaborative effort to fight organized crime and terrorist activity? What limits, if any, do you think should be placed on these agents?

TRENDS IN CRIMINAL JUSTICE

A number of trends are evident today that may have significant effects on the administration of justice.

Increasing Workloads

Over the past several decades, the demands made on the criminal justice system have increased exponentially, testing the limits of the system and the personnel working in it. The increased workloads are reflected in delays in police response to calls for service, a backlog of cases in many provincial courts, probation caseloads as high as 150 per officer, double-bunking in correctional facilities, and overextended resources for tracking offenders released into the community. Placing further stress on the system is the increasing number of specialized client groups, including sex offenders, mentally disordered offenders, visible minority offenders, Aboriginal offenders, and offenders who are members of criminal networks. All of these offenders have attributes that require specialized responses from the justice system. Although crime rates have remained relatively level in recent years—and in some categories have declined—the types of issues presented by offenders have become more acute. At the same time, there have been reductions in programs and services in many jurisdictions, and this has exacerbated the workload issues.

One indicator of the increasing complexity of cases dealt with by the criminal courts is that in 2003–04, for the first time in a decade, multiple-charge cases accounted for the majority of cases heard in adult criminal courts (Thomas, 2004). The increasing demands have prompted criminal justice agencies to devise strategies to "work smarter." This has involved utilizing computer technologies to gather and share information. This is part of efforts to move away from a paper-based justice system.

Community Involvement in Crime Prevention and Responses to Crime

The history of the criminal justice system has been one of increasing centralization of the authority and responsibility for preventing and responding to crime. The transfer of responsibility for crime prevention and responses to criminal justice professionals has been accompanied by a diminishing role for community residents. Until recently the only remaining significant involvement of community residents in the criminal justice process was serving on juries in criminal trials. Figure 1.10 identifies some of the negative consequences when communities depend on the justice system to respond and solve a wide range of problems. The vicious circle that results from unmet expectations is depicted in Figure 1.11.

Of particular concern is the limited ability of the criminal justice system to solve problems rather than simply react to them. Generally, the criminal justice system defines problems according to the law rather than on the basis of how they are experienced by people. This tendency is reflected in the comments of a jury member following a trial in which the accused had been charged with unlawful possession of a firearm and discharging a firearm in a public place:

> At the end of the case, although we found him guilty, we felt we were no nearer to understanding why the man did what he did. He might have been mentally deranged, a drug dealer, an upset father, or high on drugs. We will never know because the case gave us less than 10% of the information that we needed to make a sensible judgement. (Nicholl, 1999: 48)

From the perspective of traditional criminal justice, the extent to which "justice" is achieved in any one case is measured primarily by the extent to which due process and legislated procedures are followed.

There is mounting evidence that criminal justice agencies and professionals alone cannot prevent and reduce crime, address the needs of crime victims, and ensure effective interventions for offenders. Initiatives aimed at reversing the

**FIGURE 1.10   Consequences of Overdependence on the Criminal
Justice System**

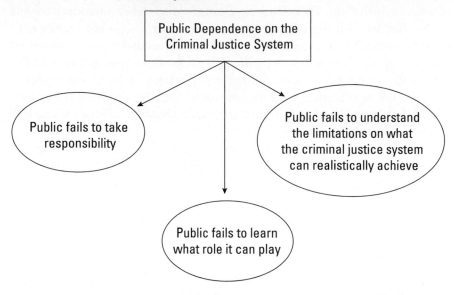

Source: C.G. Nicholl, *Community Policing, Community Justice, and Restorative Justice:
Exploring the Links for the Delivery of a Balanced Approach to Public Safety* (Washington, DC:
Office of Community Oriented Policing Programs, U.S. Department of Justice, 1999), 57.
Reprinted by permission of the U.S. Department of Justice, Office of Community Oriented
Policing Services.

historical trend of excluding the community from efforts to prevent and respond
to crime and criminal offenders include these: the creation of police–community
partnerships within a model of community policing; the use of a wide range of
various restorative justice approaches; and the expansion of community corrections programs. As well, many Aboriginal communities have created reserve-based
police services, community correctional programs, and healing programs.

FIGURE 1.11   Consequences of Unmet Expectations

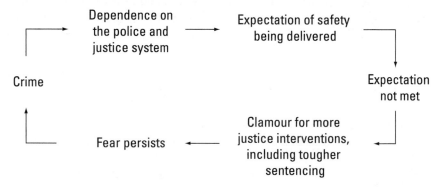

Source: C.G. Nicholl, *Community Policing, Community Justice, and Restorative Justice:
Exploring the Links for the Delivery of a Balanced Approach to Public Safety* (Washington, DC:
Office of Community Oriented Policing Programs, U.S. Department of Justice, 1999), 57.
Reprinted by permission of the U.S. Department of Justice, Office of Community Oriented
Policing Services.

Community residents are also filling thousands of volunteer positions throughout the criminal justice system, at community police stations, victim service offices, and so on. They are working with youth and adult offenders under supervision in the community and in institutions, and they are participating in a wide variety of research activities. Many college and university students do volunteer work in the system in order to gain valuable experience that will help them secure full-time employment on completion of their studies. Volunteers coordinated by various religious groups, service organizations, and private-sector businesses are working in partnership with police services, sponsoring youth programs and camps, and providing assistance to inmates and ex-offenders.

The involvement of community residents as volunteers in the justice system is a positive development; however, there are concerns that governments have taken advantage of this to cut funding for full-time positions. There may also be limitations in the extent to which community residents can function effectively in various capacities in the justice system.

The Emergence of Restorative Justice

The increasing involvement of communities in the prevention of and response to crime has paralleled the emergence of **restorative justice**. Restorative justice is an approach to problem-solving that involves the victim, the offender, their social networks, justice agencies, and the community. The fundamental principle on which restorative justice programs are based is that criminal behaviour injures not only victims but also communities and offenders, and that efforts to address and resolve the problems created by criminal behaviour should involve all of these parties. A widely used definition of restorative justice states that "restorative justice is a process whereby parties with a stake in a specific offence collectively resolve how to deal with the aftermath of the offence and its implications for the future" (Marshall, 1999: 5).

> **restorative justice**
> an approach to justice based on the principle that criminal behaviour injures the victim, the offender, and the community

It is important to note that restorative justice is not a specific practice, but rather a set of principles that provides the basis for a community and the justice system to respond to crime and social disorder. These principles include the following:

- Restorative justice provides for the involvement of offenders, victims, their families, the community, and justice personnel.
- It views crimes in a broad social context.
- It maintains a preventative and problem-solving orientation.
- It is flexible and adaptable. (ibid.)

Figure 1.12 depicts the relationships among the various parties that may be involved in a restorative justice approach.

Restorative justice has six main objectives:

- to fully address the needs of victims of crime;
- to enable offenders to acknowledge and assume responsibility for their behaviour;

FIGURE 1.12   The Relationships of Restorative Justice

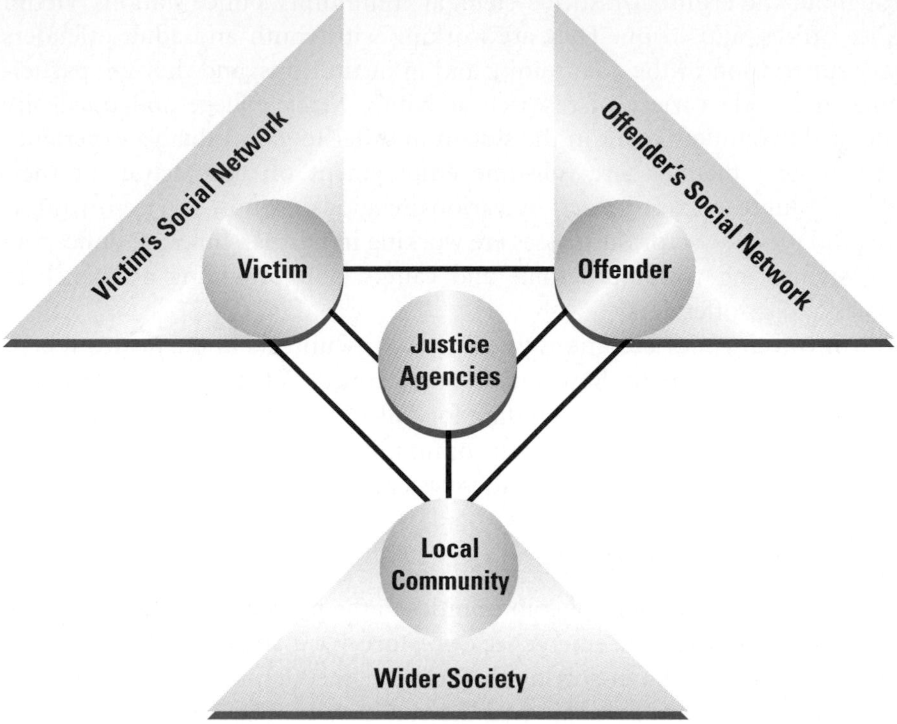

Source: T.F. Marshall, *Restorative Justice: An Overview* (London: Home Office Research and Statistics Directorate, 1999), 5.

- to create a "community" of support and assistance for the victim, the offender, and the long-term interests of the community;
- to provide an alternative to the adversarial system of justice;
- to provide a means of avoiding escalation of legal justice and the associated costs and delays (Marshall, 1999: 6); *and*
- to prevent reoffending by reintegrating offenders back into the community.

As Figure 1.13 illustrates, restorative justice can be utilized at all stages of the justice process.

Accountability of Criminal Justice Personnel and Agencies

Over the past decade there has been a trend toward increasing accountability among criminal justice decision makers and agencies. Justice system personnel may be subject to criminal and/or civil prosecution as well as to both internal and external review bodies. For example, police officers can be held accountable for their actions under civil law, the Criminal Code, provincial statutes, and freedom of information acts. Police boards, complaint commissions, and investigative units both within and outside police services have the authority to

FIGURE 1.13   Restorative Justice: Entry Points in the Criminal Justice System

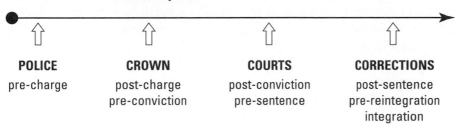

Source: J. Latimer and S. Kleinknecht, *The Effects of Restorative Justice Programming: A Review of the Empirical Research Literature* (Ottawa: Research and Statistics Division, Department of Justice, 2000), 7. Reproduced with the permission of the Minister of Public Works and Government Services Canada.

oversee and review the actions and decisions of police officers. Legislatively, for example, systems of corrections are accountable to

- the Charter of Rights and Freedoms;
- provisions in the Corrections and Conditional Release Act;
- the federal office of the Correctional Investigator;
- provincial correctional ombudsmen;
- the courts;
- coroner's inquests; *and*
- commissions or task forces created to examine specific incidents or actions of corrections personnel.

Criminal justice agencies may also be held accountable by crime victims and offenders through the civil courts. Crime victims may sue to recover damages from justice agencies that did not fulfill their mandate to protect (see Chapter 2). Offenders and suspects may sue to recover damages for actions taken by criminal justice personnel. Examples of such actions include excessive force by police and wrongful conviction by the criminal courts (see Chapter 5).

In an effort to increase the transparency of the justice process, and to increase citizens' access to information, the Ontario Attorney General has posted the entire Crown Policy Manual online so that the public can access it. This manual lists the principles and criteria that guide Crown attorneys in the prosecution of offences. In British Columbia, the provincial court has launched an "Ask the Chief Judge" website (http://www.provincialcourt.bc.ca); through this, the Chief Judge and other members of the provincial court will respond to queries from the general public (other than those asking for legal advice or information on specific cases).

Agencies may also be held accountable by civil suits filed by both offenders and crime victims. Civil litigation brought by crime victims against the justice system is discussed in Chapter 2. For example, in one case a woman who had been raped by a parolee with sixty-three prior criminal offences filed a civil suit against Corrections Canada and settled out of court for $215,000. At the time of the assault the offender was on a twelve-hour day pass from a halfway house. The facility had not yet received from the offender's previous residence (a higher security institution) the case file indicating he was an unsuitable candidate for release into the community (Hall, 2001). Most of these suits are settled out of court; this allows the victim or the victim's family to avoid recounting painful experiences, but it also prevents an examination of the policies and decisions that contributed to the incident.

Increasing Tension Between the Provincial and Federal Governments on Crime and Justice Issues

Although the Constitution Act (1867) sets a clear division of responsibility between the provinces and the federal government with respect to the criminal law and criminal justice, in recent years there has been increasing tension between the two parties. The disagreements have centred on a number of "hot button" issues, including firearms registration, the use of conditional sentences for convicted offenders, the decriminalization of marijuana, and general approaches to responding to criminal offenders. For example, several provinces joined in a suit against the federal government (they lost) to prevent violent offenders from being eligible for conditional sentences, which allow the offender to serve a custodial sentence in the community. Similarly, a number of provincial governments have been extremely vocal in their opposition to the national firearms registry and the decriminalization of marijuana and have, as well, pressured the federal government to amend the Criminal Code to provide for more severe penalties for offenders who are convicted of violent crimes.

Summary

This chapter has provided a general orientation to the structure and operations of the criminal justice system. The federal, provincial/territorial, and municipal governments all play a role in the criminal justice process. The principal components of the criminal justice system are the police, the courts, and correctional agencies. Each of these components has its own mandate, and there are a number of obstacles that hinder information sharing among them. Key themes in criminal justice include the delicate balance between individual rights and public protection, the justice system as a human enterprise, the task environments within which justice personnel carry out their responsibilities, and the role of communities in crime prevention and responses to crime. Various challenges and trends in criminal justice have emerged in recent years.

Key Points Review

1. The police, the courts, and corrections are the basic components of the criminal justice system.
2. Responsibility for criminal justice is shared among the federal, provincial, and municipal governments.

3. The flow and attrition of cases through the criminal justice system can be represented graphically by means of a funnel.
4. There are a number of factors that prevent criminal justice agencies and personnel from operating as a system.
5. Interoperability is the term used to denote the capacity for criminal justice agencies to share information with one another.
6. Criminal law has a number of functions in Canadian society.
7. Explanations vary as to the origin and application of criminal law.
8. A fundamental balance must be struck between protecting the interests of the public and protecting the rights of the individual.
9. The criminal justice system is a human enterprise in that it involves people (criminal justice personnel) making decisions about other people (suspects/offenders).
10. A number of areas present serious challenges to the criminal justice system.

Key Term Questions

1. What is the **Constitution Act, 1867,** and what is its significance to the Canadian criminal justice system?
2. What is meant by the term **interoperability** and what are some examples of this?
3. Why is the **Canadian Charter of Rights and Freedoms** important to any study of the criminal justice system?
4. What are the origins and components of the **rule of law**?
5. What is the definition of **crime,** and what are the two essential ingredients of a crime?
6. Compare and contrast **criminal law** and **civil law**.
7. Compare and contrast the **value consensus model** and the **conflict model** as explanations for the origins and application of criminal law.
8. What is meant by the **due process** and **crime control** orientations of the law and criminal justice?
9. What is a **task environment**, and why is this concept important to any study of the criminal justice system?
10. Discuss the role of **discretion** in the criminal justice system.

References

Auditor General of Canada. (2003). "National Security in Canada—the 2001 Anti-Terrorism Initiative." Retrieved from http://www.oag-bvg.gc.ca

Besserer, S., and J. Tufts. (1999). "Justice Spending in Canada." Catalogue no. 85-002-XPE. *Juristat,* 19(12). Ottawa: Canadian Centre for Justice Statistics, Statistics Canada.

Bolan, K. (2005, August 20). "The Long Arm of Uncle Sam." *Vancouver Sun,* A1, A4.

Canadian Criminal Justice Association. (2000). *Aboriginal Peoples and the Criminal Justice System.* Special issue of the *Bulletin.* Ottawa.

Cherry, P. (2001, February 16). "Anti-Gang Law Convictions Come with a High Price." *National Post*, A5.

Corrections Statistics Committee. (2004). *Corrections and Conditional Release Statistical Overview*. Ottawa: Public Safety and Emergency Preparedness.

CTV News Staff. (2002, December 14). "Native Leader Says Hitler Right to 'Fry' Jews." Retrieved from http://www.CTV.ca

Currie, A., and G. Kiefl. (1994). "Ethnocultural Groups and the Justice System in Canada: A Review of the Issues." Ottawa: Research and Statistics Division, Department of Justice.

Curry, B. (2005, January 5). "Estimated 40% of Country's Foster Children Are Native." *National Post*, A5.

Davis, R.C., and N.J. Henderson. (2003). "Willingness to Report Crimes: The Role of Ethnic Group Membership and Community Efficacy." *Crime & Delinquency*, 49(4), 564–81.

Gomme, I.M. (1993). *The Shadow Line: Deviance and Crime in Canada* Toronto: Harcourt Brace Jovanovich, 158.

GPI Atlantic. (2005). "Crime Costs Nova Scotians $1.2 Billion A Year." Retrieved from http://www.gpiatlantic.org/releases/pr_crime.shtml

Griffiths, C.T. (2004). *Canadian Corrections*, 2nd ed. Scarborough, ON: Thomson Nelson.

Hall, N. (2001, January 9). "Woman Raped by Parolee Gets $215,000.00 from Government." *Vancouver Sun*, A1.

Harding, K. (2005, July 9). "Ahenakew Unapologetic after Conviction." *The Globe and Mail*, A11.

Johnson. S. (2004). "Adult Correctional Services in Canada, 2002/03." Catalogue no. 85-002-XPE, *Juristat* 24(10). Ottawa: Canadian Centre for Justice Statistics, Statistics Canada.

Joseph, R. (1999). *Healing Ways: Aboriginal Health and Service Review*. Vancouver: Vancouver/Richmond Health Board.

Kirkey, S. (2001, June 20). "19% of Children, Teens Sexually Solicited Online." *National Post*, A8.

LaPrairie, C. (2002). "Aboriginal Over-representation in the Criminal Justice System: A Tale of Nine Cities." *Canadian Journal of Criminology*, 44(2):181–208.

Latimer, J., and S. Kleinknecht. (2000). *The Effects of Restorative Justice Programming: A Review of the Empirical Research Literature*. Ottawa: Research and Statistics Division, Department of Justice.

Leslie, K. (2005, September 11). "McGuinty Rejects Shariah Law." Retrieved from http://cnews.canoe.ca

Leung, A. (2004). *The Cost of Pain and Suffering from Crime in Canada*. Ottawa: Research and Statistics Division, Department of Justice.

Li, K. (2004). "Costs of Crime in Canada: An Update." *JustResearch* 12. Ottawa: Research and Statistics Division, Department of Justice Canada. Retrieved from http://canada.justice.gc.ca

Marshall, T.F. (1999). *Restorative Justice: An Overview*. London, England: Research Development and Statistics Directorate, Home Office.

Nicholl, C.G. (1999). *Community Policing, Community Justice, and Restorative Justice: Exploring the Links for the Delivery of a Balanced Approach to Public Safety.* Washington, DC: Office of Community Oriented Policing Program, U.S. Department of Justice.

Ontario Ministry of Finance. (2003). "Census 2001 Highlights. Visible Minorities and Ethnicity in Ontario." *Factsheet 6.* Toronto: Labour and Demographic Analysis Branch, Office of Economic Policy.

Silver, W., K. Mihorean, and A. Taylor-Butts. (2004). "Hate Crime in Canada." Catalogue no. 85-002-XPE. *Juristat,* 24(4). Ottawa: Canadian Centre for Justice Statistics, Statistics Canada.

Statistics Canada. (2003). "Justice Spending, 2000–2001." Retrieved from http://www40.statcan.ca

———. (2005, March 12). "Study: Canada's Visible Minority Population in 2017." *The Daily.* Ottawa: Canadian Centre for Justice Statistics.

Steering Committee on Integrated Justice Information. (1999). *Integrated Justice Information Action Plan.* Ottawa: Strategic Policy and Integrated Justice, Solicitor General Canada, 16.

Taylor-Butts, A. (2004). "Private Security and Public Policing in Canada, 2001." Catalogue no. 85-002-XPE. *Juristat,* 24(7). Ottawa: Canadian Centre for Justice Statistics, Statistics Canada.

Thomas, M. (2004). "Adult Criminal Court Statistics, 2003/04." Catalogue no. 85-002-XPE. *Juristat,* 24(12). Ottawa: Canadian Centre for Justice Statistics, Statistics Canada.

Trevethan, S., and C.J. Rastin. (2004). *A Profile of Visible Minority Offenders in the Federal Correctional System.* Ottawa: Research Branch, Correctional Service of Canada.

WEBSITES

Access to Justice Network
http://www.acjnet.org

A comprehensive website containing extensive materials on law and justice. Information is organized by jurisdiction, subject, and format. This website includes numerous links to other Canadian and American criminal justice sites.

Canadian Criminal Justice Association
http://www.ccja-acjp.ca

A useful site from which to access various Canadian criminology periodicals, journals, and reports. Although this is a membership-based site, you need not be a member to access the information. This site is designed to update members on pending legislation and research and on emerging trends in Canadian criminal justice.

Canadian Legal Information Institute (CanLII)
http://www.canlii.org

This website is sponsored by the Federation of Law Societies of Canada. It contains extensive information on legislative and judicial texts, as well as legal commentaries from federal, provincial, and territorial jurisdictions. An excellent source of information on provincial court systems and decisions, with links to Supreme Court of Canada decisions, provincial court decisions, and other quasi-judicial decisions such as the National Parole Board.

Department of Justice Canada
http://canada.justice.gc.ca

This website contains useful information on the Canadian court system and Canada's laws, a guide to federal Acts and legislation, and other resource materials.

Jurist Canada
http://jurist.law.utoronto.ca

A comprehensive website on Canadian law, legal research, and legal issues. Also contains links to Canadian law schools and information on admissions and programs.

continued

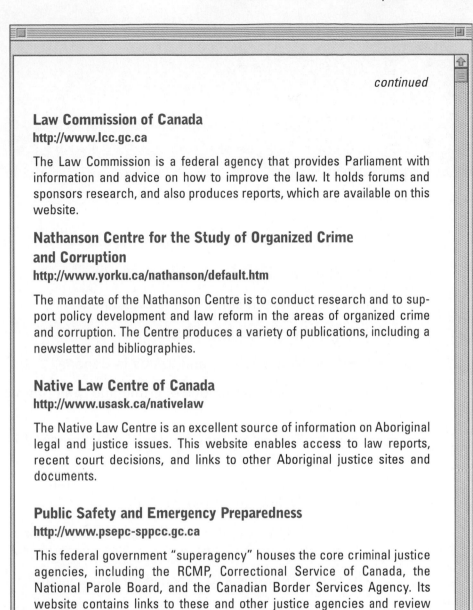

continued

Law Commission of Canada
http://www.lcc.gc.ca

The Law Commission is a federal agency that provides Parliament with information and advice on how to improve the law. It holds forums and sponsors research, and also produces reports, which are available on this website.

Nathanson Centre for the Study of Organized Crime and Corruption
http://www.yorku.ca/nathanson/default.htm

The mandate of the Nathanson Centre is to conduct research and to support policy development and law reform in the areas of organized crime and corruption. The Centre produces a variety of publications, including a newsletter and bibliographies.

Native Law Centre of Canada
http://www.usask.ca/nativelaw

The Native Law Centre is an excellent source of information on Aboriginal legal and justice issues. This website enables access to law reports, recent court decisions, and links to other Aboriginal justice sites and documents.

Public Safety and Emergency Preparedness
http://www.psepc-sppcc.gc.ca

This federal government "superagency" houses the core criminal justice agencies, including the RCMP, Correctional Service of Canada, the National Parole Board, and the Canadian Border Services Agency. Its website contains links to these and other justice agencies and review boards.

chapter

2

Crime, Victimization, and the Canadian Public

learning objectives

After reading this chapter, you should be able to

- Identify the methods used to count crime.

- Describe how the crime rate is calculated.

- Identify the factors related to crime reporting.

- Discuss crime patterns and trends in Canada.

- Discuss public perceptions of crime and the criminal justice system.

- Describe the factors that influence public perceptions of the justice system.

- Discuss the role of the media in creating public images of crime and criminal justice.

- Identify and discuss the factors related to the risk of becoming a crime victim.

- Describe programs and services for crime victims.

- Identify the ways in which victims may seek redress through the courts.

key terms

avoidance behaviours   70

crime rate   51

criminal injury
   compensation   80

dark figure of crime   53

defensive behaviours   70

restitution   80

Much of what is known about crime, victimization, and the attitudes of the Canadian public toward crime and the criminal justice system comes from two sources: the General Social Survey (GSS) and the Uniform Crime Reporting (UCR) system. The GSS is conducted periodically by Statistics Canada and involves using random-digit dialling to contact people fifteen years of age or older in thousands of households in all ten provinces. Respondents are asked about their victimization experiences, their perceptions of crime and the criminal justice system, and related topics. Significantly, the GSS excludes Yukon, Nunavut, and the Northwest Territories, three jurisdictions that tend to have the highest rates of crime in the country. This is a serious limitation to the GSS data in that it prevents a comprehensive assessment of the experiences and perceptions of residents north of the 60th parallel. Also, this omission obscures the very challenging issues related to crime and criminal justice in these regions.

The UCR system is the standardized system by which all incidents reported to the police are recorded and compiled into national statistics. The UCR has always focused on more "traditional" types of crimes, including violent offences and property offences. Beginning in 2005, however, data will be collected on the involvement of organized crime and street gangs in criminal activity as well as on hate-motivated crime, cybercrime and other emerging types of criminal activity. The Aboriginal Peoples Survey provides, among other types of data, information on crime and victimization among Aboriginal people.

CRIME RATES AND PATTERNS

Counting Crime

How many robberies or assaults were committed in Canada last year? The answers to this question and similar ones would seem to be straightforward. However, as Table 2.1 indicates, there is no single, foolproof way to count the number of crimes; each method has its limitations.

Calculating the Crime Rate

The **crime rate** is the number of criminal incidents known to the police expressed in terms of the number of people in the population. It is generally spoken of in terms of the number of incidents for each 100,000 people (adults and children). The steps required for an incident to ultimately be included in the calculation of the crime rate are outlined below.

> **crime rate**
> the number of incidents known to the police expressed in terms of the number of people in the population

Recognizing That a Crime Has Been Committed

Some crimes are not reflected in official statistics because the victims do not realize that the incident in question is against the law. For example, until recently many women who were assaulted by their partners did not call the police, even though such behaviour is a crime.

Similarly, only in recent years has on-ice violence among hockey players been scrutinized by the criminal justice system. In 2004, for example, Todd Bertuzzi, a Vancouver Canucks right winger, pled guilty to assault causing bodily harm for an on-ice attack against Colorado Avalanche forward Steve Moore. Bertuzzi

TABLE 2.1	Methods of Counting Crime		
Technique	**Method**	**Information**	**Limitations**
Self-Report Surveys	Identify a sample and administer survey asking how often they have committed any of the listed crimes and not been detected; generally done via telephone.	Most people have committed minor crimes for which they have not been arrested; a smaller number have committed serious crimes.	Persons without telephones and those who do not speak English/ French excluded; respondents may conceal or exaggerate criminal conduct; often used with small samples which makes it difficult to estimate levels of crime in the general population.
Victimization Surveys	Contact a large, randomly selected group of people to ask who among them has been a victim of crime; respondents interviewed anonymously in their homes or via telephone about victimization experiences during a specific time (e.g., the past year).	Number and types of victimization experiences; reported and not reported to police; relationship between victim and offender; findings indicate that, with the exception of certain offence categories, the majority of crime victims do not report their victimization incident to the police; raises questions about the validity of police statistics.	Respondents may forget, conceal, or exaggerate information; exclusion of Yukon, N.W.T., and Nunavut from the GSS—three jurisdictions with the highest rates of crime and violence.
Police Statistics	Compiled through the Uniform Crime Reporting (UCR) system and sent to the Canadian Centre for Justice Statistics, a division of Statistics Canada.	Detailed information on each criminal incident, including multiple offences; information on location and time of offence, victim/ accused relationship, age and gender of victim and accused, use of weapons.	May undercount crime due to variations in reporting among police services and failure of victims to report criminal incidents.

received a conditional discharge, including one year's probation and eighty hours of community service. If these are successfully completed, he will be deemed not to have a criminal record. This case illustrates that future incidents of violence in sports may be subject to criminal prosecution (see Husa and Thiele, 2002).

Reporting the Crime to the Police

For a variety of reasons, the victims of (or witnesses to) a crime may not report the incident. If the criminal activity is not discovered by the police in some other way, then in terms of official statistics, the crime was never committed. The majority of crimes known to the police are brought to their attention by the victims themselves or by witnesses. Less than 5 percent of crimes are discovered by the police without the assistance of the public or a crime victim.

The spread of wireless technology is having an impact on crime reporting. Millions of calls are made to 9-1-1 emergency numbers every year by people using cell phones. Many of these calls involve traffic accidents and suspected impaired driving.

Findings from the GSS and other studies indicate the following:

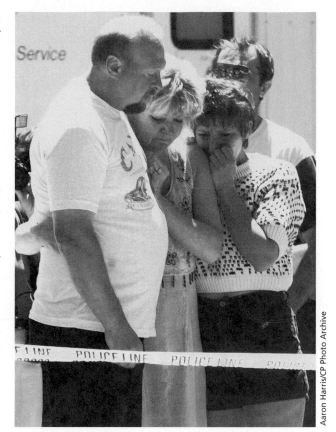

Family and friends gather at the site of a murder-suicide in Kitchener, Ontario.

- There is a **dark figure of crime** that is not reflected in official crime statistics. Rates of reporting to the police have declined in recent years. It is estimated that more than half of all incidents involving victims are never brought to the attention of the police. Fewer than four in ten incidents involving sexual assault, robbery, assault, break and enter, theft of motor vehicles and motor vehicle parts, and vandalism are reported to police. However, these incidents may be reported to insurance companies, which in turn provide information to the police.

> **dark figure of crime**
> the difference between how much crime occurs and how much crime is reported to or discovered by the police

- The likelihood that victims will call the police is not related to the seriousness of the offence.
- Reporting practices vary across different regions of the country.
- There are age and gender differences in reporting by crime victims.
- There is underreporting of crime in many Aboriginal communities. One study of Aboriginal peoples found that even though 52 percent of incidents of victimization involved a violent crime, 74 percent of the victims did not report the crime to the police (LaPrairie, 1994: 34).
- There appears to be a "utilitarian" factor at work in victims' decisions to call the police. Crime victims know that for many types of crimes, there is little the police can do.

For explanations of why crime victims choose to report or not to report their victimization to the police, see Figures 2.1 and 2.2 respectively.

FIGURE 2.1   Victims' Reasons for Reporting to Police

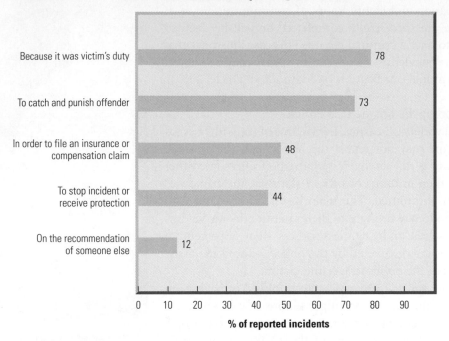

% of reported incidents

Source: Adapted from the Statistics Canada publication, *Juristat*, Criminal Victimization in Canada, 1999, vol. 20, no.10, Catalogue 85-2002, November 2, 2000, p. 12.

FIGURE 2.2   Victims' Reasons for Not Reporting to Police

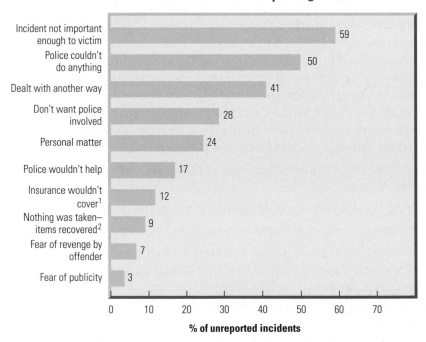

% of unreported incidents

[1]Total exceeds 100 percent owing to multiple responses. Excludes incidents that were not classified by crime type as well as incidents of spousal sexual and physical assault.
[2]Excludes incidents involving sexual assault or a physical attack.

Source: Adapted from the Statistics Canada publication, *Juristat*, Criminal Victimization in Canada, 1999, vol. 20, no.10, Catalogue 85-002, November 2, 2000, p. 12.

Declaring the Case to Be Founded

Not all reported incidents are included in the crime rate. Before a crime is investigated, the police must believe that the incident is founded. In other words, it must be determined that it is a matter appropriate for criminal investigation. Generally, less than 5 percent of reported incidents are declared to be unfounded. Examples include individuals who lost their money claiming they were robbed, and a suspected murder that turns out to have been a suicide. The "unfounded" rate for certain categories of crime may be higher: 15 to 17 percent of sexual assaults are determined, upon investigation, to be unfounded.

Determining That the Incident Is a Criminal Matter

Community residents sometimes call the police for assistance in situations that turn out to be civil rather than criminal matters. Landlord–tenant disputes are a common example. In such cases, the responding officers refer the complainant to the appropriate agency.

CRIME TRENDS

When discussing the incidence and patterns of crime in Canada, it is important to include both statute and Criminal Code offences. Otherwise, some very important crime categories, such as drug offences, will not be included in the totals. In 2004, approximately 2.8 million federal statute incidents, including Criminal Code offences, were reported to the police. These incidents are broken down in Figure 2.3.

FIGURE 2.3 **Number of Federal Statute Incidents, including Criminal Code Offences, Reported to Police by Crime Type, 2004**

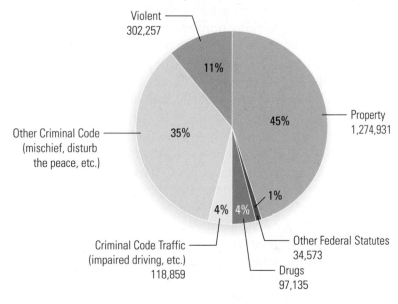

Total Federal Statutes, Including Criminal Code:
2,822,427

Source: Adapted from the Statistics Canada publication, *Juristat,* Crime Statistics in Canada, 2004, vol. 25, no. 5, Catalogue 85-002, July 21, 2005, p. 16.

FIGURE 2.4   Police-Reported Crime Rate, 1984–2004

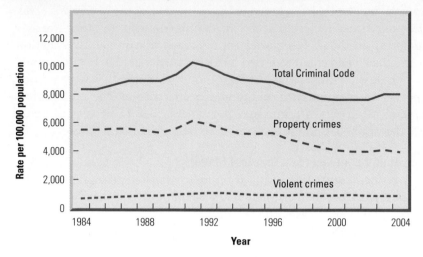

Source: Adapted from the Statistics Canada publication, *Juristat,* Crime Statistics, 2004, Catalogue 11-001, Thursday, July 21, 2005, available at http://www.statcan.ca/Daily/English/ 050721/d050721a.htm.

Focusing just on the Criminal Code incidents (that is, excluding traffic and also excluding federal statute incidents, which include drug offences), there were 8,051 crime incidents per 100,000 population. Twelve percent of these were violent crimes, 50 percent were property crimes, and the remaining 39 percent were other Criminal Code offences such as mischief and disturbing the peace. Following are some highlights of crime statistics in Canada as of 2004:

- *The crime rate reported by police as official statistics is decreasing.* Since peaking in the early 1990s, the crime rate has been falling steadily (see Figure 2.4). This decline is due in part to the shrinking proportion of the population in the crime-prone 15-to-24 age range and to lower crime rates in Ontario.

- *Violent crime continues to decline.* However, homicide rates increased 1 percent during 2003–04, with the highest rates being recorded in the territories and western Canada and in the cities of Winnipeg, Edmonton, and Vancouver.

- *Property crime is continuing to decline.* Prince Edward Island and Ontario reported the lowest break-in rates; the highest rates were in the West. In 2004, British Columbia had the highest rate of theft in the country and Saskatchewan had the highest rate of break-ins. Ontario (−12 percent) and British Columbia (−6 percent) reported the greatest declines in vehicle theft; Newfoundland (+52 percent), Nova Scotia (+24 percent), and Manitoba (+23 percent) recorded the greatest increases.

- *The drug offence rate has increased significantly over the past ten years.* The biggest contributor to this is marijuana possession; incidents of this offence doubled between 1992 and 2002. However, there are significant variations in the rates of drug offences across the provinces and among Canadian cities, mainly owing to differences in police enforcement strategies and resources. The highest rates of drug-related offences per 100,000 population were in British Columbia (544), Saskatchewan (351), and New Brunswick (343), while Alberta (242), Manitoba (232), and Newfoundland (232) reported the lowest rates. Among Canadian cities, Thunder Bay (571) had the highest rate per 100,000, followed by Vancouver (468), Victoria (459), and Trois-Rivières (364), while the cities of St. John's (174), Edmonton (166), and Kitchener (151) had the lowest rates. Can you think of factors other than police activity that might account for the differences in rates among the provinces and among Canadian cities?

There has also been a significant increase in the number of incidents involving marijuana-growing operations ("grow-ops"). Annual production of marijuana in Canada is estimated at over 800 tons. Some grow-ops are "mom and pop" enterprises; many, though, are linked to organized crime groups, including the outlaw biker gangs and Asian crime syndicates.

- *Drugs and alcohol play a significant role in physical and sexual assault.* Half the crime victims surveyed by the GSS believed the assault incident involved alcohol or drug use by the perpetrator. Alcohol plays an especially significant role in domestic violence. Of particular concern is the increasing use of "date rape" drugs, which allow the perpetrator to gain physical control over the victim, leaving that person with little memory of what transpired. These drugs, which are tasteless, odourless, and colourless, are typically placed in the victim's drink in a bar or social setting. It is difficult to determine the incidence of sexual offences associated with date rape drugs, since the victim may have no memory of the assault.

- *The rate of adults charged with criminal offences is declining.* This pattern holds for both men and women.

- *Crime rates tend to increase from east to west, but are highest in the North* (see Figure 2.5). Crime rates also tend to be higher in western cities such as Vancouver, Edmonton, and Winnipeg than in Toronto, Quebec, and Ottawa (Sauve, 2005).

The rates of violent crime in Yukon, the Northwest Territories, and Nunavut were several times higher than the national average and for those of any of the provinces. In Chapter 1 this was identified as a serious challenge to the criminal justice system. There is little evidence that either the justice system or the northern communities themselves are succeeding in reducing these high rates of serious crime (see Figure 2.6).

- *The crime rate among youths between twelve and seventeen continues to decline.*

- *Overall, Canada has a lower crime rate than the United States, although there are notable exceptions.* The crime rates in Yukon, the Northwest Territories, and Nunavut are higher than those in the southern states, which have the highest rates in the United States. As well—and surprisingly, given our general assumptions about crime rates in the two countries—research has found recently that the Prairie provinces have higher rates of crime than adjacent American states to the south (see Box 2.1).

- *In some areas, a handful of offenders may be responsible for committing a large number of reported offences.* These repeat offenders, dubbed "4 percenters" by the Vancouver Police Department, are often drug addicts who commit property offences in order to purchase drugs (Skelton, 2000; Tremblay, 2000).

FIGURE 2.5   Crime Rate per 100,000 Population, Canada and the Provinces/Territories, 2004

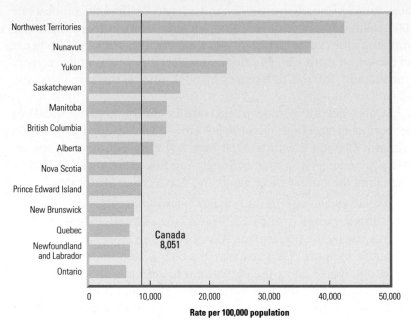

Source: Adapted from the Statistics Canada publication, *Juristat,* Crime Statistics, 2004, Catalogue 11-001, Thursday, July 21, 2005, available at http://www.statcan.ca/Daily/English/050721/d050721a.htm.

FIGURE 2.6   Violent Crime Rate per 100,000 Population, Canada and the Provinces/Territories, 2004

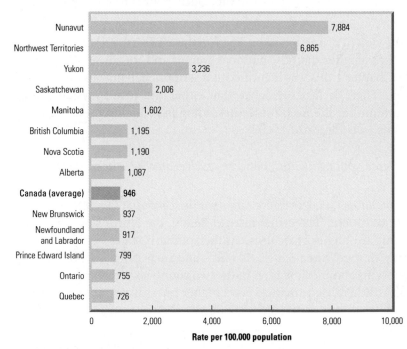

Source: Adapted from the Statistics Canada publication, *Juristat,* Crime Statistics in Canada, 2004, vol. 25, no. 5, Catalogue 85-002, July 21, 2005, p. 18.

focus

BOX 2.1

Crime on the Canadian and American Prairies: A Comparison

Recent studies (Preville and Jackson, 2005; Jackson, 2005) compared crime rates in the Prairie provinces (Manitoba, Saskatchewan, Alberta) with those in the adjacent border states (Minnesota, North Dakota, Montana, Idaho). These provinces and states have similar economies, societies, and populations. After the crime figures for 2003 were adjusted to account for differences in how offences were classified, it was found that both violent and property crime rates were two-thirds higher in the Prairie provinces than in the four border states.

Sources: E. Preville and A. Jackson, *A Comparison of Violent and Firearm Crime Rates in the Canadian Prairie Provinces and Four U.S. Border States, 1961–2003* (Ottawa: Parliamentary Information and Research Service, Library of Parliament, 2005); E. Jackson, *Follow-Up to a Comparison of Violent and Firearm Crime Rates in the United States and Canada* (Ottawa: Parliamentary Information and Research Service, Library of Parliament, 2005).

The specific factors that contribute to these crime trends have not yet been closely examined. It is not clear, for example, why cities in the West have higher crime rates than their counterparts in the East. Differences in the enforcement practices of the police are one possible explanation. (Can you think of others?) One study that sought to explain the differences in homicide rates between Vancouver and Seattle is described in Box 2.2.

Crime in First Nations and Inuit Communities

A major challenge confronting the criminal justice system, and Aboriginal people, is the high rates of crime in many First Nations and Inuit communities—especially in the northern regions of the provinces and in Yukon, the Northwest Territories, and Nunavut. In these places the crime rates are as much as five times higher than for Canada as a whole. The high crime rates in Aboriginal urban communities are often an extension of on-reserve violence, poverty, and addiction.

Among the findings of an in-depth study of Aboriginal people residing in the urban centres of Edmonton, Winnipeg, Toronto, and Montreal were these:

- Nearly two-thirds of the people interviewed had been victims of crime; 84 percent of these victims reported violent victimization.

- Nearly two-thirds of those interviewed had experienced family violence, and nearly half had experienced more than one incident of spousal, child, or sexual abuse.

- The risk of victimization varied in this population. The more marginalized Aboriginal people were in urban areas, the greater the number of incidents of victimization they reported.

What do you think?

What explanations would you offer for the findings of the study in Box 2.1?

The study was used by certain members of Parliament to argue against the federal gun registry—specifically, these MPs made the point that the availability of guns was not related to crime rates. (Guns are much easier to obtain and more prevalent in the United States, where there is no federal gun registry.) Do you agree with this argument?

focus

BOX 2.2

Homicide in Vancouver and Seattle

International reviews of crime rates reveal high homicide rates in the United States relative to other countries. Nearly half of all murders in the United States are committed with handguns. American advocates of stricter controls over the sale and possession of guns often use Canada as an example of a country where legal restrictions on them have been effective in maintaining low rates of violent crime. Pro-gun groups have disputed the argument that strict gun controls are related to low crime rates.

To test this assertion, medical researchers from Washington State compared homicide data from Seattle with data from that city's neighbour to the north, Vancouver. These two cities are similar in geography, climate, demographics, and the penalties for homicide and for crimes committed with firearms. However, it is easier to buy a gun in Seattle, which has a higher rate of gun ownership than Vancouver.

The researchers found that for the years 1980–86, the two cities had similar rates for burglaries, robberies, and assaults. But in Seattle the rate of assault involving firearms was 7.7 times that of Vancouver. They also found that a person in Seattle was almost twice as likely to be murdered as a person in Vancouver. This was linked to the fact that Seattle's rate of homicide by firearm was five times higher than Vancouver's. The availability of firearms was found to be directly related to the levels of risk in the two cities and to the homicide rate as well.

Annual Rates of Homicide in Seattle and Vancouver, 1980–86, according to the Weapon Used

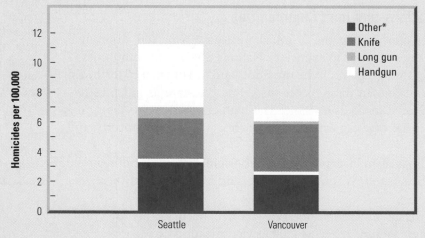

*"Other" includes blunt instruments, other dangerous weapons, and hands, fists, and feet.

Source: J.H. Sloan et al., "Handgun Regulations, Crime, Assaults and Homicide: A Tale of Two Cities," *New England Journal of Medicine* 319 (1988), 1260. Copyright 1988 Massachusetts Medical Society. All rights reserved.

The eastern Arctic community of Pangnirtung, Nunavut

- Aboriginal females residing in inner cities reported the highest rates of victimization (LaPrairie, 1994).

Many First Nations and Inuit communities have high rates of violent crime. In Saskatchewan, for example, assault/homicide and sexual assault involving Aboriginals each account for 26 percent of all major offences. In the Northwest Territories and Nunavut, assault/homicide involving Inuit accounts for 18 percent of all major offences, while sexual assault involving Inuit accounts for 48 percent of all major offences (National Crime Prevention Centre, 2000). In Box 2.3, crime rates in First Nations communities are compared with crime rates in small urban areas and rural areas in the provinces of Ontario and Quebec. These data are a decade old; however, there is nothing to indicate that significant changes have occurred in the high rates of Aboriginal criminality.

Crime patterns vary considerably both among and within First Nations/Inuit reserves and communities. Possible explanations include differences in police enforcement practices, patterns of crime reporting, the capacity of individual communities to address problems of crime and disorder, and the extent to which effective partnerships have been developed between the community and criminal justice agencies.

The high crime rates in First Nations communities have driven the development of a wide range of community-based justice programs, many of which are based on the principles of restorative justice. With a few exceptions, however, the effectiveness of these initiatives has not been assessed, and the search continues for innovative policies and programs that will reduce the high levels of Aboriginal conflict with the law.

What do you think?

The findings of this study (in Box 2.2) appear to contradict those of the cross-border study of crime in the prairie regions of Canada and the United States. How do the two studies differ in terms of the crimes that were examined, and how might this have affected the findings?

focus

BOX 2.3

First Nations Crime in Ontario and Quebec

To gain a better understanding of the patterns of crime in First Nations communities, researchers gathered data from five of these communities in Ontario and twenty in Quebec. They then compared these data with comparable data gathered from small urban areas and rural areas in the two provinces. Their findings (presented below) indicate that the rate of violent crime in Ontario was notably higher in First Nations communities (257 per 10,000 population) than in small urban areas (93 per 10,000) and rural areas (81 per 10,000). Similar data taken from selected reserves in Quebec reveal a violent-crime rate of 224 per 10,000 in First Nations communities, compared with 41

per 10,000 in small urban communities and 40 per 10,000 in rural communities.

The researchers also found that:

- nearly half the female offenders in the surveyed First Nations communities had been charged with violent offences, compared to one-quarter of female offenders in the rural and small urban areas; *and*

- the proportion of youths involved in criminal activity was similar in all three survey areas. However, more youths in First Nations communities had been charged with property offences, relative to their counterparts in the other areas.

Criminal Code Offence Rates, Ontario, 1996

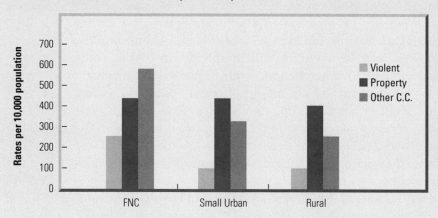

continued

PUBLIC PERCEPTIONS OF CRIME AND THE CRIMINAL JUSTICE SYSTEM

Findings from the GSS (Statistics Canada, 2000, 2004) and reviews of the research by Stein (2001) and Roberts (2004) provide insights into the perceptions of crime and criminal justice held by Canadians and the levels of public confidence in the justice system:

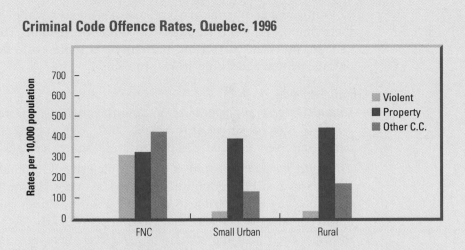

Criminal Code Offence Rates, Quebec, 1996

Source: C. LaPrairie, *Seen But Not Heard: Native People in the Inner City* (Ottawa: Department of Justice, 1994).

Source: S. Lithopoulos, *Police Reported First Nations Crime Statistics, 1996* (Ottawa: Aboriginal Policing Directorate, Solicitor General of Canada, 1999), 6–10. Reproduced with the permission of the Minister of Public Works and Government Services Canada, 2006.

- *Except when it comes to corruption in government, crime and violence typically are not at the top of the list of concerns of most Canadians.* In the 2004 Canada Election Study, Canadians identified health care, corruption in government, taxes, social welfare programs, and the environment as the country's most pressing problems (Gidengil et al., 2005). There has only been one year between 1985 and 2000 in which more than 3 percent of GSS respondents identified crime/violence as the most important problem facing Canada. However, in 2005, there were several high-profile events that may precipitate a shift in attitudes to come: the shooting deaths of four RCMP officers in Mayerthorpe, Alberta, and of a police constable in Laval, Quebec, and of a teenage girl who was caught in the crossfire of rival gangs in downtown Toronto.

- *Younger Canadians (18–29) and older Canadians view law and order differently.* Younger Canadians are more likely to be opposed to the death penalty for persons convicted of murder and to support rehabilitation programs for young offenders (ibid.).

Crime

- *Most Canadians believe that crime rates are stable.* This finding is contradicted by the fact that the rates for most types of crime are actually declining.

- *Canadians tend to think that crime is worse in communities other than their own.* Nearly 60 percent believe that crime in their neighbourhood is lower than elsewhere in Canada, and a majority believe that the levels of crime in their area have remained unchanged over the past five years. It seems that Canadians have a much more accurate view of the crime problem in their own neighbourhoods, where personal experience is likely to outweigh media images.

• *There are regional variations in attitudes toward crime.* For example, British Columbians and Albertans favour harsher sentences, while Quebeckers think there should be more emphasis on crime prevention. The perception is general, however, that crime prevention is more cost-effective than law enforcement in addressing crime problems.

Criminal Justice

• *Canadians generally appear to be dissatisfied with government responses to crime.* In a recent survey (Ipsos-Reid, 2002), only 6 percent of the respondents were "very satisfied"; 54 percent were "somewhat satisfied."

• *The police have a high approval rating.* A majority of respondents to the GSS felt that the police were doing a "good job" (as opposed to an "average job") in terms of being approachable (65 percent), ensuring citizen safety (61 percent), enforcing laws (59 percent), and treating people fairly (59 percent). Within these generally high approval ratings, however, there were variations among cities. With respect to people's perceptions of how well the police were enforcing the law, the approval ratings were highest in Quebec—including Chicoutimi–Jonquière (74 percent), Trois-Rivières (68 percent), and Quebec City (68 percent)—and in Victoria, B.C. (68 percent). In several cities on the Prairies, fewer residents felt that the police were doing a "good job" at enforcing the law (Winnipeg, 53 percent; Regina, 49 percent; Saskatoon, 40 percent). A similar pattern was evident in these cities with respect to the perceptions of how well the police were doing with respect to ensuring public safety.

What factors might be contributing to these variations among cities?

• *The public is less satisfied with the performance of the criminal courts.* The courts received low marks in terms of doing a "good job" on a number of counts, including determining whether an accused is guilty (27 percent), assisting crime victims (20 percent), and dispensing justice in a timely manner (15 percent). As well, the courts were often viewed as too lenient on convicted offenders. Again, in the context of these low overall ratings, there are differences among cities. For example, 27 percent of the respondents in Chicoutimi–Jonquière felt that the courts were doing a "good job" in helping crime victims, but only 14 percent of the respondents in St. Catharines–Niagara agreed. Interestingly, visible minorities and immigrants were less critical of the courts than other members of the Canadian public.

• *The prison system receives low marks from the public.* The prison system was awarded an approval rating of just over 31 percent with respect to its ability to supervise and control offenders. Even fewer respondents (18 percent) felt that the prison system was effective at helping offenders become law-abiding citizens. It is noteworthy that the respondents were less sure of themselves when it came to rating the prison system than when it came to rating the police and the courts. This suggests that corrections systems need to develop better methods for communicating with the general public and for informing the public about their activities and outcomes. And perhaps they should also improve their methods for treating offenders (more on this in Chapter 7).

• *The parole system is the least respected part of the entire criminal justice system.* Only 17 percent of the respondents perceived that parole boards were doing a "good job" in releasing offenders who were not likely to commit another crime; even fewer (15 percent) felt that community corrections did a "good job" in supervising parolees. The lowest ratings for the parole system were recorded in the West.

The extremely low ratings for systems of release and community supervision are grounds for concern and suggest that it may be time to revisit the systems presently in place (see Chapter 8).

- *Support for the death penalty is strong but declining.* In a 2001 poll, a slim majority of Canadians (52 percent) supported the reinstatement of capital punishment. The declining support for the death penalty in recent years is partly attributable to several widely publicized cases of persons who were wrongly convicted of crimes (Besserer and Trainor, 2000; Cobb, 2001; Tufts, 2000; and see Chapter 6). Interestingly, a Gallup poll in 1998 found that 75 percent of those who favoured the death penalty would continue to do so even if it were proven that capital punishment did not deter crime. This suggests that some Canadians consider the death penalty appropriate for certain serious crimes, whether it is a deterrent or not (Stein, 2001).

Public Confidence in the Criminal Justice System

The above findings and other research studies point to high levels of distrust of Canada's criminal justice system (Roberts, 2004). Indeed, the Canadian public seems to have less confidence in the justice system than in other public-sector institutions such as education. However, Roberts (2004: 4) notes that a similar situation exists in other Western countries such as the United States and England. The public *does* seem to trust the police; 73 percent say they do. Compare this with the trust rating for nurses (89 percent), doctors (79 percent), and teachers (74 percent). In contrast, only 34 percent of respondents in the same survey (Ekos Research Associates, 2000) expressed a high level of trust in lawyers. The rates of public trust in the justice system in Canada are, however, higher than those in England and slightly higher than the average for fourteen European countries (Page, Wake, and Ames, 2004; Roberts, 2004).

Factors Affecting Public Attitudes Toward, and Trust in, the Criminal Justice System

Research findings suggest that trust in the criminal justice system is influenced by a number of factors:

- *Type of contact with the system.* Less positive views of the police are held by those who have been arrested by police or who have had contact with them as victims of crime (particularly violent crimes), as traffic violators, or as witnesses to a crime. Individuals whose contact with the police has been restricted to non-enforcement situations, such as public information sessions, hold more positive attitudes toward the police. People who have had contact with the criminal courts are more likely to rate the courts as doing a poor job of providing justice quickly; but they also tend to view the courts favourably in terms of ensuring a fair trial for the accused.

- *Feelings of personal safety.* Individuals who are more satisfied with their personal safety tend to have more positive attitudes toward the police.
- *Gender.* Women tend to rate the police more favourably, while men have more positive attitudes toward the criminal courts and the prison and parole systems. The specific reasons for these gender differences remain to be explored.
- *Age.* Older Canadians tend to be more satisfied with the performance of the police, while younger ones tend to hold more positive attitudes toward the criminal courts and the prison and parole systems.
- *Level of education.* The most positive views of the police, the criminal courts, and the prison and parole systems are held by those with lower levels of education (high school or less).
- *The visibility of the criminal justice professionals.* Roberts (2004: 20) makes this important observation: the police enjoy high approval ratings in part because they are the most visible component of the justice system. Officers are on the street, and of all the criminal justice professionals, it is they whom the general public is likely to have contact with in either an enforcement or some other context. The high visibility of the police, however, could also result in more negative experiences and attitudes among the general public. In contrast, the general public has little contact with probation officers, judges, criminal lawyers, correctional officers, parole boards, or parole officers, unless, of course, they are convicted of a criminal offence. So it is hard for Canadians to develop an understanding of what these people do.

The public's main source of information about criminal justice professionals is the media. As noted below, the incidents that are reported and featured in the media tend to be those with negative outcomes. When was the last time you heard a "good news" story about offenders released from prison on parole—as in, "As is our custom on the final Friday of each month on this newscast, we congratulate all of the offenders who successfully completed parole this month!" You are far more likely to hear a news reader announce: "Smith, on parole from Collins Bay penitentiary, was arrested for a series of armed robberies in the Kingston area."

This raises the possibility that the public's responses to the GSS and other surveys are based not on factual knowledge about various criminal justice professionals and the components of the justice system in which they work, but rather on media reports about crime rates and the justice system's responses. Roberts (2004: 21) notes, for example, that Canadians consistently:

- overestimate the levels of crime and the rates of recidivism;
- underestimate the severity of punishment given to offenders (see the discussion of the sentencing of impaired drivers in Chapter 6);
- tend to believe that the criminal justice system is biased in favour of defendants; *and*
- overestimate both the rates of parole release and the reoffending rates of parolees.

All of this suggests that we carefully examine the role of the media in relation to crime and criminal justice.

THE MEDIA, CRIME, AND CRIMINAL JUSTICE

> The ways in which the news media collect, sort, and contextualize crime reports help to shape public consciousness regarding which conditions need to be seen as urgent problems, what kinds of problems they represent, and by implication, how they should be resolved. — Sacco (1995: 141)

For most Canadians, television and newspapers are the primary sources of information about crime and the criminal justice system. The media's tendency to sensationalize heinous crimes—which are statistically rare and yet cause considerable harm to victims, their families, and communities—skews people's perceptions of crime and justice as well as their responses to survey questions such as those asked on the GSS. It is widely believed that the media have a significant impact on Canadians' perceptions of crime and criminal justice, although in one survey (cited in Stein, 2001), respondents rejected the notion that media reporting affected their attitudes toward crime. Recall that crime is not listed among the top concerns of Canadians. This suggests that more research is required in order to determine exactly how crime reporting and the depictions of crime and criminal justice in the popular media affect perceptions of crime and expectations of, and attitudes toward, the criminal justice system.

Research on how the news media cover crime has found that:

- media reporting of crime does not reflect the actual patterns of crime;

- violent offences are overrepresented in media coverage;

- there is a disproportionate focus on sensational events;

- crime stories rarely explore the causes of the crime, particularly the role of various criminogenic factors such as poverty and physical and/or sexual abuse experienced by the offender; *and*

- most coverage is at the pre-arrest stage—there is little subsequent information (Dowler, 2004).

Front page, *Toronto Star,* April 26, 2005

Reprinted with permission of the *Toronto Star*

An obvious question: Why does the media do this? The most obvious answer: ratings, which translate into advertising dollars and revenue for the newspaper, radio or television station. The public seems to have an insatiable appetite for crime and chaos; witness the success of the television drama *CSI* and its various spin-offs. Crime and police shows produced in Canada and the United States consistently attract the largest viewing audiences. There are other reasons, which are highlighted in a recent study comparing the reporting of crime stories on local TV newscasts in Canada and the United States (see Box 2.4).

One trend is the increasing use of the Internet as a source of news. It remains to be seen how this source of information will affect public perceptions of crime. Many youth do not access traditional sources of information; they are more likely to watch satirical "news" programs such as the *Daily Show* with Jon Stuart, which, needless to say, provides a unique perspective on current events, including crime and criminal justice.

focus

BOX 2.4

Local Television Crime Coverage in Canada and the United States: A Comparison

As any Canadian who has watched the local news via cable (or illegal satellite dish) from New York, Detroit, Chicago, or Seattle knows, local TV newscasts place a strong emphasis on the "crime(s) of the day," often making these the lead stories. Are there are any differences in the crime reporting on local TV newscasts between Canada and the United States? A larger question is why the news media emphasize certain facets of crime and ignore others.

A recent study by Kenneth Dowler of Wilfrid Laurier University–Brantford explored these questions. He gathered 400 thirty-minute local TV newscasts from Detroit, Michigan; Toledo, Ohio; and Toronto and Kitchener in Ontario. He then conducted a content analysis of these broadcasts, which included a total of 1,042 crime stories. Information was recorded on the characteristics of the story (lead story? live story? interview? and so on), the length of the story, the stage of the justice system at which the crime was reported, latent variables (fear presented? sensationalism presented? and so on), and the type of crime reported (homicide, sexual assault, and so on). The results of this content analysis may surprise you; see Table 2.2.

The content analysis of the news stories indicated that with a few exceptions, there was no difference in the types of crimes presented on American and Canadian TV newscasts. Examine the table and note where differences did arise. What might account for these differences between the two countries?

Dowler (2004: 591) concluded that crime news is reported in similar ways (in his terms, "news is constructed selectively") in order to attract viewers and advertisers and to maintain the status quo. Note well, though, that his study focused on coverage of local crime. Had the study included national crime stories in the two countries, the results probably would have revealed a considerable amount of media attention on the conviction, sentencing, custodial time, and post-release life of white-collar offenders such as lifestyle guru Martha Stewart and on the sentencing of Enron's former executives. In Canada, media attention nationally would have focused on the sponsorship scandal, the Air India trial, and other high-profile national stories. That said, local crime stories may well exert a greater influence on the attitudes and perceptions of the general public with regard to crime and criminal justice.

TABLE 2.2   Descriptive Characteristics of Canadian and U.S. Local Broadcast Crime Stories

	Canadian Stories (%)	American Stories (%)
Story Characteristics		
Lead Story	10.7	8.3
Live Story**	7.3	19.8
Local Story**	72.8	81.2
Interview	41.3	37.0
Firearm Reported**	19.1	31.1
Motive Presented	18.4	19.1
Length of Story (mean, in seconds)*	55.06	50.18
Stage of Crime**		
Pre-arrest	45.8	53.9
Arrest	24.2	27.5
Court	22.5	16.0
Disposition	7.5	2.6
Latent Variables		
Fear presented	12.8	13.4
Outrage or sympathy presented	28.7	25.4
Sensationalism presented**	32.1	40.9
Pro-active police response**	44.2	26.9
Type of Crime		
Homicide	33.6	33.6
Sexual assault	11.6	9.4
Assault	8.1	8.7
Drug crime	4.5	4.3
Theft or robbery	12.6	15.0
Threats or harassment	7.5	5.0
White collar crime	4.9	5.0
Police misconduct	3.9	2.8
Traffic violations	5.8	6.3

*p < 0.05
**p < 0.01

Source: K. Dowler, "Comparing American and Canadian Local Television Crime Stories: A Content Analysis," *Canadian Journal of Criminal Justice*, October 2004, *46*(5), p. 584. Reprinted by permission of the Canadian Criminal Justice Association.

THE FEAR OF CRIME AND THE RISKS OF VICTIMIZATION

Findings from the GSS, surveys of crime victimization, and other Canadian surveys indicate the following:

- *The large majority of Canadians (94 percent) are satisfied with their personal safety.* This includes immigrants, visible minorities, and people who live in communities with high crime rates. The fear of crime is generally higher among women than among men, but these differences have narrowed in recent years. Interestingly, fear of crime among women appears to *decrease* with age, while among men it remains stable or increases. The sources of these differences are not readily apparent. It has often been assumed that elderly people in society are more fearful of crime than the young.

- *The large majority of Canadians (95 percent) are not concerned about becoming the victim of a crime because of their race, ethnicity, language and/or religion.* The figure is lower (89 percent) when only visible minority responses are considered (Silver, Mihorean, and Taylor-Butts, 2004). As well, the vast majority of Canadians (93 percent) indicate that they have never or almost never experienced discrimination or unfair treatment because of their ethnocultural attributes. However, 20 percent of visible minority people indicated that they *had* experienced discrimination (Statistics Canada, 2003).

- *There are high rates of victimization among Aboriginal people.* Eight percent of women and 7 percent of men have been victims of violence at the hands of a spouse or common-law partner. The rates are much higher among Aboriginal people: 25 percent of Aboriginal women and 13 percent of Aboriginal men reported that they experienced violence from a current or past partner in the previous five years. Also, eight times more Aboriginal men, and eighteen times more Aboriginal women, die at the hands of their spouses (Lane, Bopp, and Bopp, 2003: 26).

- *Most Aboriginal people feel safe in their communities, even when those communities face high levels of violence.* One possible reason why is that for many Aboriginal people, violence is a "normal," unavoidable feature of everyday life (National Crime Prevention Centre, 2000). Note that, again, statistics of victimization are not available for Yukon, the Northwest Territories, and Nunavut—the jurisdictions with the highest rates of crime and violent crime in Canada. The continued unavailability of these data—for reasons that are not publicly specified by Statistics Canada—makes it difficult for us to understand the issues surrounding fear of crime and victimization in Canada. It also hinders the development of policies and programs to address the needs of crime victims in northern communities.

Precautionary Measures

Fear of crime may cause people to take the following steps to protect themselves:

avoidance behaviours
lifestyle changes designed to reduce one's risk of becoming a victim of crime

defensive behaviours
specific anticrime measures designed to reduce one's risk of becoming a victim of crime

- *Avoidance behaviours.* **Avoidance behaviours** are lifestyle choices designed to reduce one's risk of becoming a victim of crime. Examples include moving to a safer neighbourhood and avoiding high-crime neighbourhoods.

- *Defensive behaviours.* **Defensive behaviours** are steps taken to reduce one's risk of becoming a victim of crime. Examples include keeping a dog, carrying Mace (pepper spray), asking neighbours to pick up the mail and "keep an eye out" during absences, and installing a household alarm. University campuses across the country have established Safe Walk programs; these provide, on request, security patrol escorts to people who are walking across campus or to their vehicle.

Risk Factors

A number of factors (many of them interrelated) are linked to the risk of becoming a victim:

- *Age.* Age is one of the most reliable correlates of criminal victimization. People in the 15–24 age range report the highest levels of victimization; those over sixty-five report the lowest levels. A person in the former group is twenty-one times more likely than a person in the latter group to be the victim of a violent offence. Nearly 60 percent of the victims of sexual assault, for example, are under eighteen. The risk of victimization decreases with age for both men and women. The greater risk for younger people may have something to do with the link between participating in evening activities outside the home and increased risk of victimization.

- *Gender.* Men and women face roughly the same overall risk of becoming the victim of a crime against the person. However, men are more likely to be the victims of assault or robbery, while women are more likely to be the victims of sexual assault.

- *Marital status.* Single men and women face the highest rates of victimization by assault, robbery, and other crimes against the person. Women who are separated from their partner are five times more likely than other women to be killed by their partner.

- *Residence.* People who live in urban areas and in the West are at greater risk of being victimized. Given the high rates of violent crime in the North, the residents of communities in these regions are at the highest risk of becoming crime victims.

- *Income.* People with annual household incomes under $15,000 are at greater risk, with the rates of violent victimization twice those of higher income households.

- *Season.* Not surprisingly, the highest number of criminal victimizations occur during the summer months (June, July, August).

- *Relationships.* Ironically, the street may be safer than one's own home. Nearly 50 percent of victimizations occur in or around the victim's home. This is because people are usually victimized by people they know (Besserer and Trainor, 2000; McGillivray and Comaskey, 1999; National Crime Prevention Centre, 2000; Ontario Native Women's Association, 1989).

In many regions of the country, Aboriginal people face a high risk of being victimized. Aboriginal communities have rates of violent death three times higher than non-Aboriginal communities; these rates are even higher in some Inuit communities in the Eastern Arctic. Compounding the problem is the lack of resources, including victim services, in northern and remote communities.

Police records and victimization surveys indicate that Aboriginal people are victimized mainly in their own communities, and that such incidents most often involve Aboriginals as both the offender and the victim. A study by the Ontario Native Women's Association found that 80 percent of the women surveyed had been victims of violence; of these, 87 percent had been physically abused and 57 percent had been sexually abused (Ontario Native Women's Association, 1989; see also McGillivray and Comaskey, 1999).

<u>VICTIMS OF CRIME</u>

> I am politely reminded that the court tries to balance the rights of the accused and the victim. I hear less talk from the legal professionals and the justice system about the plight of victims who can be treated as intruders even though it is victims who live with the long-term effects of the crime.
> —a victim/survivor of homicide, in Magnussen (2005)

Over the past two decades, a wide range of programs and services have been developed to address the physical, emotional, and financial consequences of victimization, as well as to provide victims with information about the justice process and the progress of cases through the justice system. There have also been increased efforts to involve victims in the criminal justice process, mainly through the presentation of victim impact statements in the courts and through victims' attendance at parole board hearings. Some observers have argued that empowering crime victims by involving them in the criminal justice system injects undue emotionalism and increases the punitiveness of the system; however, such resistance has diminished in recent years.

Federal and provincial legislation sets out the rights of victims and defines their role in the justice process. Procedures have been established to provide victims with information on the status of cases and to notify victims if the offender is being released or has escaped from custody. In addition, in several jurisdictions the definition of "victim" has been expanded to include the immediate family members of murder victims. These developments have been driven at least in part by the recognition that the criminal justice system has for too long focused more on the rights of offenders than on the rights of crime victims. There is also evidence that the psychological harm that criminal offenders inflict on victims and communities has never been addressed adequately by the criminal justice system.

Several observers have argued that the Charter of Rights and Freedoms should be amended to include victims' rights. In rebuttal, the Canadian legal scholar Kent Roach (2005) has argued that such an amendment would not significantly benefit crime victims. Such an amendment might well diminish the role of governments in promoting victims' rights and shift the onus for protecting those rights onto the courts.

The role of crime victims in the criminal justice process will be highlighted throughout this book. Victims often initiate the criminal justice process by reporting the incident to the police (Chapter 2). They may also be involved in subsequent stages of the justice process, from the decision of the Crown counsel to lay a charge (Chapter 5), to the presentation of a victim impact statement in court (Chapter 6), to attendance at a parole board hearing in jurisdictions where this is permitted (Chapter 8). Crime victims are also involved in various restorative justice programs, including victim–offender mediation and circle sentencing.

In 2003 the federal, provincial, and territorial ministers responsible for justice set out a new Canadian Statement of Basic Principles of Justice for Victims of Crime. This was an update of the statement of principles that was first issued nearly two decades ago. These principles provide the framework within which policies and programs and legislation for crime victims are to be developed;

focus

BOX 2.5

Basic Principles of Justice for Victims of Crime, 2003

1. Victims of crime should be treated with courtesy, compassion, and respect.
2. The privacy of victims should be considered and respected to the greatest extent possible.
3. All reasonable measures should be taken to minimize inconvenience to victims.
4. The safety and security of victims should be considered at all stages of the criminal justice process and appropriate measures should be taken when necessary to protect victims from intimidation and retaliation.
5. Information should be provided to victims about the criminal justice system and the victim's role and opportunities to participate in criminal justice processes.
6. Victims should be given information, in accordance with prevailing law, policies, and procedures, about the status of the investigation; the scheduling, progress and final outcome of the proceedings; and the status of the offender in the correctional system.
7. Information should be provided to victims about available victim assistance services, other programs and assistance available to them, and means of obtaining financial reparation.
8. The views, concerns and representations of victims are an important consideration in criminal justice processes and should be considered in accordance with prevailing law, policies and procedures.
9. The needs, concerns and diversity of victims should be considered in the development and delivery of programs and services, and in related education and training.
10. Information should be provided to victims about available options to raise their concerns when they believe that these principles have not been followed.

Source: Basic Principles of Justice for Victims of Crime, 2003, http://www.canada.justice.gc.ca/en/ps/voc/ publications/03/basic_prin.html. Department of Justice. Reproduced with the permission of the Minister of Public Works and Government Services Canada.

they also provide the foundation for the work of Policy Centre for Victim Issues (http://canada.justice.gc.ca/en/ps/voc/index.html and see Box 2.5).

The first efforts to assist crime victims focused on the needs of those affected by crime. For example, the grassroots work and political advocacy of the women's movement was instrumental in creating services for abused women, such as shelters and crisis centres. A parallel development was the emergence of the victims' rights movement, which was based on the view that the rights of victims had been unfairly eclipsed by those of offenders. Most jurisdictions have enacted legislation setting out the rights of crime victims, and the federal government has amended the Criminal Code to ensure that crime victims are involved in the justice process.

Ontario has been highly proactive in developing policies and programs that establish victims' rights and that address the needs of crime victims

focus

BOX 2.6

An Act Respecting Victims of Crime—Victims' Bill of Rights, 1995 (Ontario)

This act is designed to support the needs and rights of crime victims. It includes provisions that do the following:

- Establish a set of principles to support crime victims throughout the criminal justice process. This includes providing crime victims with information on sources of support and assistance, the progress of the case through the justice system, any plea negotiations and pretrial arrangements, and the release of the offender from custody.

- Create a framework within which victims of crime can file civil suits against their perpetrators for emotional distress caused by the offence and, if successful, recover legal costs from the defendant.

- Enshrine the Victims' Justice Fund in the Victims' Bill of Rights to ensure that monies collected under the victim fine surcharge are used to provide services for victims. This surcharge is attached to fines under the Provincial Offences Act (except parking violations) and to federal offences as well. Among the programs and services developed with these monies have been the Victim Support Line; the Victim–Witness Assistance Program (V/WAP), the Victim Crisis Assistance and Referral Service (VCARS), and the Community Victims' Initiatives Program (CVIP), which provides funding to community groups involved in assisting the victims of crime.

Source: Attorney General of Ontario (http://www.e-laws.gov.on.ca/DBLaws/Statutes/English/95v06_e.htm). © Queen's Printer for Ontario, 1995. Reproduced with permission.

(see Box 2.6). Specific measures designed to improve the services available to victims and to protect the victims of specific types of crimes include the following:

- The Victims' Bill of Rights (1995)(see Box 2.6).
- Creation of the Office for Victims of Crime (OVC), staffed by victims and criminal justice personnel, to oversee the implementation of the Victims' Bill of Rights and to implement standards of service for crime victims.
- A Victim Support Line and the SupportLink program. The support line is a provincewide, toll-free information line that provides information on victim services, the criminal justice system, and the status and scheduled release date of provincially sentenced offenders. The SupportLink program provides women who are at risk of domestic violence with comprehensive personal-safety planning and resources, including preprogrammed cell phones for use in the event of a personal emergency.
- Domestic violence courts that provide early intervention counselling as well as the coordinated prosecution of domestic violence cases (see Chapter 5).

- The Protecting Children from Sexual Exploitation Act (2000), designed to give the police and child welfare workers the power to remove children (with or without a warrant) from sexually exploitive situations such as prostitution, and to pursue people who sexually exploit children. Similar legislation has been passed in other jurisdictions, including British Columbia.

- The Prohibiting Profiting from Recounting Crimes Act (S.C., 2000), which prevents offenders from profiting from their crimes by providing for the seizure of any monetary proceeds that offenders might receive as a consequence of telling about their crimes in books or other media. These monies would be placed in a fund for crime victims.

Manitoba's Victims' Bill of Rights includes a statutory requirement that Crown attorneys consult with crime victims or the victims' families during the prosecution of the case. Police and Crown counsel must keep victims apprised of court proceedings, including decisions to stay criminal charges or to plea bargain, sentencing, and release from custody.

Victim Complaints about the Justice System

I was so confused by it all . . . It is a tough world trying to understand the judicial system. — Crime victim (Magnussen, 2005)

After the initial trauma of the crime, victims can be made to feel worse by the actions of criminal justice officials. The result has been termed "revictimization." Although victims are a heterogeneous group with diverse needs, perceptions, and experiences, there are common themes in their complaints and concerns. Victims often express the following criticisms:

- They do not receive sufficient information about developments in the case.

- They do not have input into key decisions.

- They often feel that the court process is too long (that is, it involves too many adjournments and delays).

- Personnel are not adequately trained to be sensitive to victims' needs.

- Procedures are not explained using plain language.

- Victims are not adequately protected from retaliation by the offender.

In interviews with crime victims, Roberts and Roach (2004) found that many crime victims find the justice system complex and confusing and have difficulty understanding the decisions it makes and the sanctions it imposes on offenders. For example, they often confuse probation with a conditional sentence, and they often do not understand how the presiding judge weighed the various factors involved in the crime when deciding on the sentence.

Recall the findings from the GSS presented earlier in this chapter. The courts receive very low marks from the community in terms of how they address the needs of crime victims. This suggests that despite the various initiatives in this area over the past twenty years, much more work needs to be done, especially in the criminal courts.

Note that most crime victims have had little previous involvement with the criminal justice system. For most victims, this involvement begins and ends with the call to the police. Only a small number of offenders identified by the police are charged, and a much smaller number are convicted of criminal offences.

Programs and Services for Crime Victims

Crime affects victims physically, psychologically, emotionally, financially, and socially. The programs and services operating across the country are designed to help victims deal with these impacts. Programs and services for crime victims and witnesses to crime are operated by governments and by private, for-profit and not-for-profit agencies; many of these rely heavily on volunteer staff. Although programs are offered across Canada, their availability is uneven. Resources are especially limited for crime victims in rural and northern regions, where such resources may be required the most. The gaps in the delivery of services to crime victims are often due to a lack of integration between the program and the broader justice process.

One program that has overcome this problem is found in Prince Edward Island (see Box 2.7). Crisis intervention centres are generally operated by community agencies (see Box 2.8). In Ontario, the Aboriginal Healing and Wellness Strategy, delivered under the auspices of the Ministry of Community and Social Services, supports a broad range of programs and services in First Nations communities; these focus on reducing family violence and improving health. For example, they offer shelters for abused women and their children and healing lodges.

Many victim service programs are affiliated with police departments. They are generally staffed by non-police personnel or volunteers, with referrals made by police officers. Police-based programs for victims usually offer preliminary counselling for crime victims, refer victims to other services in the community, and provide victims with information on the progress of case investigations. Many police departments have developed specialized services for the victims of specific types of crime such as robbery and sexual assault. Court-based victim programs assist victims whose cases result in charges and possible prosecution. These programs are usually located in courthouses and may be affiliated with Crown attorneys' offices. Services to victims include notifying them of court dates and adjournments, explaining the legal process, offering courthouse tours before the day of court, providing separate waiting areas on the day of court (so that they don't encounter the defendant), and extending emotional support during the often traumatic task of testifying. A primary objective of these programs is to ensure that revictimization does not occur.

A number of services that are not part of the justice system assist crime victims with their medical, therapeutic, spiritual, and housing needs. These include crisis lines, sexual assault crisis centres, hospitals, mental health agencies, translation and cultural interpretation services, child protection agencies, shelters for battered women and children, and second-stage housing programs. Self-help groups have been created by groups of former victims in all parts of the country, and several of these organizations both help victims and lobby for victim-oriented reforms. As noted in Chapter 1, these include the Canadian

focus

BOX 2.7

The Victim Services Program, Prince Edward Island

Prince Edward Island's Victims Services Program is operated by the Attorney General and is staffed by social workers and other criminal justice workers who have been trained to work with crime victims. Unlike programs that are affiliated with police services or criminal courts, the Island's program is integrated with the justice system as a whole.

Under the program, one staff member is assigned to each victim at the beginning of the process (when the police first get involved) and remains involved with the victim until the criminal case is concluded. The assigned staff member can:

- explain the process and the legal terminology;
- provide information on case developments;
- perform a referral function by connecting the victim with community and legal services;
- listen and offer emotional support;
- assist those who are called to court as witnesses;
- assist in the preparation and filing of victim impact statements;
- organize criminal injury applications and investigate claims; *and*
- explain after-court considerations such as parole release.

This assistance is accompanied by coordination of and liaison with all components of the justice system. The program's innovative features include the following:

- Services are available *and* evenly accessible to people across the province, including those in rural areas.
- Services are offered to victims on a proactive basis, either after police referral to the program or because of ongoing review of police occurrence reports.
- One agency helps victims at all stages of the process; in this way, the victim is not passed along from service to service as the case passes through the system.
- Victims can qualify for services even if no charge is laid.

The services are free and confidential. Priority is given to the victims of domestic violence, sexual offences, child abuse, and other cases of personal injury.

Source: http://www.gov.pe.ca. Reprinted by permission of the Office of the Attorney General, Prince Edward Island.

Resource Centre for the Victims of Crime and Mothers Against Drunk Driving (MADD).

Information Services

Information services are designed to protect the rights of victims while at the same time ensuring their participation in the criminal justice process. Provincial/territorial legislation generally requires that crime victims be provided with information on the status of the police investigation, the role of the victim in the justice process, the opportunity to make victim impact statements, and the status of the offender vis-à-vis sentencing, parole eligibility, release from custody, escape from custody, and so on.

focus

BOX 2.8

Ontario's Victim Crisis Assistance and Referral Service (VCARS)

VCARS is a community-based crisis intervention program that receives referrals from the police. It is designed to respond to the short-term needs of "victims of crime and tragic circumstance." In each VCARS program, a full-time coordinator is aided by a team of trained volunteers. Services are available around the clock, seven days a week. Volunteers work in pairs and are on the scene within thirty minutes of a request from the police. Once there, they may help clean up after a burglary, make tea for a distraught assault victim, or sit with parents whose teenager has been killed in a car accident. Most cases involve family violence. Follow-up is undertaken the next day, when victims can be referred to appropriate agencies and services in the community. VCARS is funded through the Victim's Justice Fund from surcharges on fines assessed on convicted offenders in court. Currently, VCARS is managed by community boards composed of volunteers and operates in thirty-eight sites across the province.

Source: http://www.attorneygeneral.jus.gov.on.ca
© Queen's Printer for Ontario. Reproduced with permission.

Reprinted by permission of MADD

A "postcard" from the Mothers Against Drunk Driving website

In most regions, victims can access information through a toll-free number. In Ontario, for example, people affected by crime can call the Victim Support Line (VSL) for information on the release status of any adult offender in the provincial system; on support services for victims; and on ways that victims can play an active role in the criminal justice system. In British Columbia, a twenty-four-hour telephone service immediately notifies victims when offenders are being transferred, released on day parole, or released from jail entirely. Within thirty minutes of any changes in an offender's status, the automated system calls the victim. Victims may also call the system twenty-four hours a day from anywhere in Canada to find out about a provincial offender's jail or parole status. These notification systems track only provincial offenders; there is no national system in place yet for federal offenders.

Victim Utilization of Victim-Focused Programs and Services

Although there has been a proliferation of programs and services for crime victims in recent decades, there is some question as to the extent to which these resources are accessed. Findings on victimization from the GSS, for example, indicate the following:

- Less than four in ten victims reported their victimization to the police.
- Few victims made use of formal victim services, although services were more frequently used by victims of spousal violence.
- In the large majority (93 percent) of violent incidents that were reported to the police, the victim did not contact or use a police or court-based victim service (Kong, 2004).

Immigrant and visible minority women may find it especially difficult to access victim programs and services. These difficulties generally relate to the inability to speak or understand either English or French; a lack of knowledge about the legal and social service system and the resources that are available; fear of deportation; and pressure from extended family members to remain silent (Smith, 2004). Policies and programs need to be developed that are specifically geared to the needs of immigrant and visible minority women.

These figures on the non-utilization of victim services are significant and disturbing. At present, there is no research that addresses why crime victims do not contact victim services or which factors encourage crime victims to make use of the programs and services that do exist (or discourage them from doing so). Clearly, this area requires more attention.

FINANCIAL COMPENSATION FOR CRIME VICTIMS

Crime victims can seek financial redress for the harm caused by the victimization. There are a number of ways.

Compensation for Property Offences

People whose property has been stolen or damaged have several possible avenues of redress. The police may recover the stolen property, or the offender may agree to pay restitution to the victim. Such outcomes are rare, however. The most common way crime victims are compensated for property losses is through private insurance.

Criminal Injury Compensation

Victims of personal injury offences can apply for financial compensation from the provincial government to cover expenses and damages directly related to the crime. **Criminal injury compensation** programs operate in all provinces. The key features of these programs are highlighted in Box 2.9. Note that the Northwest Territories, Nunavut, and Yukon do not have compensation programs. Given the high rates of crime and victimization in these jurisdictions, this is of concern.

Each system is unique, but all share the following features:

- A person can qualify for compensation if he or she is a victim of a violent offence such as assault, sexual assault, or robbery. Injuries caused by motor vehicle accidents are generally not covered.

- To qualify, victims must report the offence to the police; cooperate with the investigation; and apply for compensation promptly (usually within one year). This time limit may be waived if the case involves a sexual offence, childhood victimization, and domestic violence.

- It does not matter whether charges are laid or not, whether an offender is never identified, or whether the person charged with the crime is found not guilty in court. Applicants need only prove on the balance of probabilities (rather than "beyond a reasonable doubt," the standard used in criminal court) that they have been victimized.

- Surviving relatives of murder victims may qualify for compensation, including support payments if the victim was a family breadwinner.

- "Good Samaritans"—people injured while preventing a crime or assisting a police officer—may also qualify for compensation.

- If victims contributed to their own injuries, they may receive a lower award or be disqualified from compensation altogether.

- Those not satisfied with the decision have a right of appeal.

See Box 2.10 for a summary of the types of expenses covered by the various compensation programs.

Victims may be compensated for out-of-pocket expenses such as lost wages, ambulance trips, damaged eyeglasses, dental bills, medical supplies, and prescription drugs. In a few provinces, including British Columbia, Ontario, New Brunswick, and Prince Edward Island, victims can be compensated for "pain and suffering" caused by the offence, referred to as *general damages*.

Restitution

Offenders may be required to pay **restitution** to crime victims as part of the sentence ordered by the court. Restitution orders are most commonly used for property-related offences such as fraud and theft. They typically involve cash payments by the offender to the victim; or they compensate for stolen or

criminal injury compensation
financial remuneration paid to crime victims

restitution
a court-ordered payment that the offender makes to the victim to compensate for loss of or damage to property

focus

BOX 2.9

Summary of Criminal Injuries Compensation Programs

Prov	Include secondary victim?	Time limit[1]	Interim awards?[2]	Police report required (cooperation)?	Reduction or denial if victim culpable	Max. awards	Periodic payments	Offender notified?
BC	Yes	1 year	Yes	Report is not required, cooperation is	Yes	No maximum provided, some maximums provided for different types of offences.	Yes	No
AB	No	2 years	Yes	Yes	Yes	$110,000	Yes	No
SK	Yes	1 year	Yes	Yes	Potentially	$25,000	Yes	No
MB	Yes	1 year	Yes	Yes	Yes	$100,000 (equivalent of worker's compensation)	Yes	No
ON	Yes	2 years	Yes	No, but it would be taken into consideration	Yes, denial	$25,000 or $1,000 / month	Yes	Yes, if there is no conviction
PQ	Yes	1 year	Yes	Yes	Yes	No maximum, except in relation to salary replacement: up to 90%, maximum $53,500.	Yes	No
NB	Yes	1 year	Yes	Yes	Yes	$5,000	No	No
NS[3]	Yes	1 year	Yes	Yes	Yes	$2,000	Yes	No
NF[4]	—	—	—	—	—		—	—
PE	Yes	1 year	Yes	Yes	Yes	Depends on claim		No
NU[5]	—	—	—	—	—	—	—	—
YT[6]	—	—	—	—	—	—	—	—
NW[5]	—	—	—	—	—	—	—	—

Notes:
1. In most provinces, the board may extend the time limit for filing an application. 2. The board may consider awarding interim payments if the victim is in actual financial need and if it appears to the board that it will probably grant compensation to the applicant. 3. The Criminal Compensation program in Nova Scotia no longer provides monetary compensation for lost wages, or for medical/dental and/or funeral services, to a person harmed as a result of a violent crime. The program will now provide only counselling for victims. 4. The Crimes Compensation program in Newfoundland was abolished in 1992. 5. The Crimes Compensation program in the Northwest Territories was abolished in 1996. Nunavut has the same Act. 6. The Crimes Compensation program in Yukon was abolished in 1993.

Source: Canadian Resource Centre for Victims of Crime, online at http://www.crcvc.ca/en/resources/com_summary1.php. Reprinted by permission of the Canadian Resource Centre for Victims of Crime.

focus

BOX 2.10

Summary of Types of Compensation Offered to Crime Victims, by Province

Type of Compensation	BC	AB	MB	SK	ON	PQ	PE	NS	NB
Expenses reasonably and actually incurred as a result of the victim's injury or death	Yes	Yes	Yes	Yes	Yes	Yes	Yes	No	Yes
Loss of earnings due to victim's total or partial disability to work	Yes	No	Yes	Yes	Yes	Yes	Yes	No	No
Funeral or transportation costs	Yes	No	Yes	Yes	Yes	Yes	Yes	No	Yes
Pain and suffering	No	No	No	No	Yes	No	Yes	No	Yes
Support of a child born out of rape	Yes	No	No	No	Yes	Yes	Yes	No	No
Other pecuniary loss that, in the opinion of the Board, the victim is likely to incur	Yes	No	Yes	Yes	Yes	Yes	Yes	No	Yes

Source: www.crcvc.ca/en/resources/comp_summary2.php. Reprinted by permission of the Canadian Resource Centre for Victims of Crime.

damaged property or for property that has been seized from a person who innocently purchased it without knowing that it had been stolen.

Restitution is also a key component of restorative justice. Across the country there are victim–offender mediation (VOM) programs that are designed to address the needs of crime victims and to ensure that offenders are held accountable for their offending behaviour (see Chapter 6).

Civil Litigation

Victims of crime can seek redress for their victimization through civil litigation directed toward the alleged offender, a third party, or the justice system itself.

Suing the Alleged Offender A person acquitted in court can still be sued for damages, as the following case illustrates:

A B.C. woman was awarded $50,000 in a civil suit against a man who was found not guilty of sexual assault in three separate criminal trials. The first trial ended in a hung jury, the second in a mistrial, and the third in an acquittal by a jury. Nevertheless, in the civil suit, a B.C. Supreme Court judge found that the evidence strongly supported the plaintiff's accusations. The award included $35,000 in general damages and $15,000 in punitive damages; in addition, the woman's common-law spouse, who cared for her following the incident, was awarded $7,000 for lost wages (Hall, 2000: A1).

Suing an offender is a lengthy and risky process. If the suit fails, the victim–plaintiff may have to pay the defendant's court costs. On the other hand, victim–plaintiffs have more control over and involvement in civil actions than they have as complainants in criminal proceedings. In civil suits the focus is on the harm done to the victim rather than solely on the defendant's culpability. The civil law burden of proof (that is, balance of probabilities) is less onerous than the "beyond a reasonable doubt" standard of the Criminal Code. And the court awards in civil suits are generally higher than those provided by criminal injury compensation boards.

As the following case illustrates, an alleged offender can be sued in civil court even if no criminal charges are laid:

A woman in Maple Ridge, B.C., was awarded $200,000 for the sexual abuse her stepfather had inflicted on her two decades earlier. The offences had been reported to the police several years previously but no criminal charges had been laid. The award in court included $85,000 for lost wages, $50,000 for loss of future opportunities, $50,000 in general damages, $15,000 for the cost of future therapy, and $10,000 in punitive damages. The victim–plaintiff was also awarded interest and the costs of bringing the suit. The stepfather was liable for the full cost of the award (Hall, 1998: B5).

Suing a Third Party In some cases the victim may file a lawsuit against an institution or third party that facilitated (even if inadvertently) the offender's criminal behaviour. Perhaps the most prominent case in Canada is the class action and individual civil suits being brought against the federal government by Aboriginal people who were forced into residential schools in the last century. As of 2004, around 12,000 Aboriginal people were seeking compensation for the physical and psychological mistreatment they had suffered in government- and church-operated residential schools. It is estimated that 90,000 Aboriginal people living today are survivors of the residential schools. To date, most of them have not been compensated, although there have been settlements in some of the civil suits brought against the four mainstream Christian churches (Anglican, Roman Catholic, Presbyterian, and United) that operated residential schools until 1970. Some settlements have pushed local parishes and dioceses into bankruptcy. A number of civil actions are still pending against the churches. In addition, criminal charges have been brought against individual teachers and church officials in the residential schools; a number of these people have been convicted of sexual assault.

Also, in recent years bar owners have been sued successfully for continuing to serve alcohol to patrons who subsequently drove impaired and caused an accident resulting in the death of another driver or a pedestrian.

Suing the Justice System As noted in Chapter 1, criminal justice agencies and personnel are increasingly being held accountable for their actions. Criminal justice agencies can be sued in the civil courts by crime victims. Recent years have seen a significant increase in the number of civil suits filed against systems of corrections in relation to heinous crimes committed by offenders on parole from correctional institutions.

In 2005, for example, the Government of Saskatchewan paid $142,000 to the family of a North Battleford woman who had been stabbed to death eight years earlier by two teenage girls who were in custody at her community group home. The girls were later convicted of second-degree murder. In the suit the family argued that the province had knowingly put the deceased woman in grave danger when it placed the two girls in her home without telling her about the girls' backgrounds or providing sufficient training. One of the girls was serving a sentence for murder at the time. She had been diagnosed with fetal alcohol spectrum disorder (FASD) at age fourteen, when she was convicted of drowning a child (CBC News, 2005).

As well, the police have been sued by crime victims. One of the more high-profile cases is set out in Box 2.11.

What do you think?

Do you agree with the decision and reasoning of the B.C. Supreme Court (in Box 2.11)? Explain.

focus

BOX 2.11

Victim Sues RCMP

In April 1996 a woman in Prince George, B.C., barely escaped with her life when her common-law husband went on a violent rampage in her home. In the attack, the woman's daughter was seriously wounded and her best friend was killed. Her ex-husband attempted to set the home on fire before ending his own life. Several months earlier he had served three weeks in jail for assaulting his common-law wife. Following his release, he had tried to run her down with his Jeep. Both incidents were reported to the RCMP.

The woman filed a civil suit seeking more than $400,000 in damages from the federal and provincial governments and the RCMP for failing to prevent the attack on her home. Specifically, the suit alleged that the RCMP had not trained its officers adequately to respond to cases of domestic violence; moreover, the police had failed to protect her, failed to prevent the death of her best friend, and failed to prevent the injury to her daughter.

In June 2001 a B.C. Supreme Court justice dismissed the suit. In his ruling he stated that although there was some evidence that the

RCMP officer about whom the complaints were made had not fulfilled his duties in their entirety, the police could not be expected to provide absolute protection to the victim: "It can reasonably be expected that the police are guardians, not guarantors, of public well-being." In July 2004, in a two-to-one decision, the B.C. Court of Appeal dismissed the lawsuit. The dissenting judge argued that there were sufficient grounds for an appeal. A subsequent application to the Supreme Court of Canada for leave to appeal was dismissed in 2005. For a case in which the police were found to be negligent in not alerting women about a serial rapist operating in their neighbourhood, one of whom was subsequently raped, see *Jane Doe v. Board of Commissioners of Police of Toronto* [1998] O.J. No. 2681 (Gen. Div.).

Sources: J. Sandler, "Battered Woman Can't Sue Over Ex-Husband's Deadly Rampage," *Vancouver Sun* (2001, June 6), A2; A.L. Choo, "Court Lets RCMP Off the Hook," *Vancouver Sun* (2004, July 23), A1, A2. Reprinted by permission of the *Vancouver Sun*.

Summary

The discussion in this chapter has centred on crime, the perceptions Canadians have of crime and the criminal justice system, and the victims of crime. Overall, the rates of crime in Canada are declining, although crime rates remain high in the territories and are higher in the West than in the East. Surveys of public perceptions of crime and criminal justice indicate that with the exception of corruption in government, crime is not at the top of the list of concerns of most Canadians. This may have changed with the death of a teenage girl who was caught in the crossfire between two rival gangs in downtown Toronto on Boxing Day, 2005. Although the long-term impact of this and other high-profile crime events remains to be seen. At the same time, however, there is general dissatisfaction with governments' responses to crime. The police continue to enjoy high approval ratings; the courts, prison systems, and parole boards do not. It is difficult to determine precisely how the media influence public attitudes toward, and perceptions of, crime and criminal justice; there is some evidence, though, that the Canadian and American media take very similar approaches to crime reporting.

Most Canadians do not seem too concerned about their personal safety nor are they afraid of becoming victims of crime. Factors associated with the risk of becoming a crime victim include age, gender, marital status, residence, income, ethnicity, season of the year, and the interpersonal relationships between offenders and victims. Attempts to address the needs of crime victims over the past two decades have included provincial and federal legislation enshrining victims' rights and the development of a wide range of programs and services. Crime victims can seek redress through civil litigation and criminal injury compensation programs.

Key Points Review

1. There are several key steps involved in calculating the crime rate.
2. The police-reported crime rate in Canada is declining, and crime rates tend to increase from east to west (but are highest in the North).
3. Many Aboriginal communities have high rates of crime and violence, and there are high rates of violence among urban Aboriginal people.
4. Crime and violence are not a high priority for most Canadians, although several high-profile events in 2005 may change this.
5. The public has generally positive attitudes toward the police but are less favourably disposed toward the performance of the courts and the systems of corrections.
6. Attitudes toward the criminal justice system are influenced by such factors as a person's type of contact with the system, feelings of personal safety, gender, age, and level of education.
7. It is difficult to determine the influence of the media on Canadians' perceptions of crime and criminal justice.
8. The risk of becoming a crime victim depends on such factors as age, gender, marital status, residence, income, being Aboriginal, season, and interpersonal relationships.
9. The federal and provincial/territorial governments have given increasing attention to the needs of crime victims through legislation, programs, and services.

10. Programs and services for crime victims include crisis intervention services, programs that define the rights of victims and ensure their participation in the criminal justice system, and compensation programs.
11. There is some evidence that many crime victims do not access programs and services.
12. Crime victims may seek redress through civil litigation.

Key Term Questions

1. What is the **crime rate** and how is it calculated?
2. What is meant by the **dark figure of crime**?
3. Define **avoidance behaviours** and **defensive behaviours** and give examples of each.
4. Discuss the role of **criminal injury compensation** programs and provide examples of the types of claims that are made by crime victims.
5. What is **restitution** and how is it related to the needs of crime victims?

References

Besserer S., and C. Trainor. (2000). "Criminal Victimization in Canada, 1999." *Juristat*, 20(10). Ottawa: Canadian Centre for Justice Statistics, Statistics Canada.

CBC News. (2005, March 14). "Province Pays $142K to Family of Slain Group Home Operator." Retrieved from http://sask.cbc.ca/regional/servlet/View?filename=montgomery050314

Choo, A.L. (2004, July 23). "Court Lets RCMP Off the Hook." *Vancouver Sun*, A1, A2.

Cobb, C. (2001, March 12). "Canadians Prefer Prevention over Prisons." *Vancouver Sun*, A3.

Dowler, K. (2004). "Comparing American and Canadian Local Television Crime Stories: A Content Analysis." *Canadian Journal of Criminology and Criminal Justice*, 46(5):573–96.

Ekos Research Associates. (2000). *Rethinking Government*. Ottawa: Department of Justice.

Gidengil, E., A. Blais, J. Everitt, P. Fournier, and N. Nevitte. (2005, January). "Missing the Message: Young Adults and Election Issues." *Electoral Insight*. Ottawa: Elections Canada. Retrieved from http://www.elections.ca/eca/eim/article_search/article.asp?id=122&lang=e&frmPageSize=&textonly=false

Hall, N. (1998, April 8). "Judge Awards $200,000 for Abuse." *Vancouver Sun*, B5.

———. (2000, February 25). "B.C. Woman Awarded $50,000.00 in Civil Suit for Rape." *Vancouver Sun*, A1.

Husa, A., and S. Thiele. (2002). "In the Name of the Game: Hockey Violence and the Criminal Justice System." *Criminal Law Quarterly*, 45(4):509–28.

Ipsos-Reid. (2002). *Public Views on Information Sharing in the Criminal Justice System*. Final Report. Ottawa: Solicitor General of Canada.

Jackson, E. (2005). *Follow-Up to a Comparison of Violent and Firearm Crime Rates in the United States and Canada.* Ottawa: Parliamentary Information and Research Service, Library of Parliament.

Kong, R. (2004). "Victim Services in Canada, 2002/03." Catalogue no. 85-002-XPE *Juristat,* 24(11). Ottawa: Canadian Centre for Justice Statistics, Statistics Canada.

LaPrairie, C. (1994). *Seen but Not Heard: Native People in the Inner City.* Ottawa: Department of Justice.

Lane, P., J. Bopp, and M. Bopp. (2003). *Aboriginal Domestic Violence in Canada.* Ottawa: Aboriginal Healing Foundation.

Lithopoulos, S. (1999). *Police Reported First Nations Crime Statistics, 1996.* Ottawa: Aboriginal Policing Directorate, Solicitor General of Canada, 6–10.

Magnussen, H.J. (2005). "Future Directions for Victim Assistance: Victim Advocates or Guardians of the System?" *Info Papers: A Victim's Perspective.* Ottawa: Canadian Resource Centre for the Victims of Crime. Retrieved from http://www.crcvc.ca/en/

McGillivray, A., and B. Comaskey. (1999). *Black Eyes All of the Time: Intimate Violence, Aboriginal Women, and the Justice System.* Toronto: University of Toronto Press.

National Crime Prevention Centre. (2000). *Aboriginal Canadians: Violence, Victimization, and Prevention.* Ottawa: Department of Justice.

Ontario Native Women's Association. (1989). *Breaking Free: A Proposal for Change to Aboriginal Family Violence.* Thunder Bay, ON: Author.

Page, B., R. Wake, and A. Ames. (2004). *Public Confidence in the Criminal Justice System. Research Findings,* no. 221. London, ON: Research and Statistics Directorate, Home Office.

Preville, E., and A. Jackson. (2005). *A Comparison of Violent and Firearm Crime Rates in the Canadian Prairie Provinces and Four U.S. Border States, 1961–2003.* Ottawa: Parliamentary Information and Research Service, Library of Parliament.

Roach, K. (2005). "Victims' Rights and the Charter." *Criminal Law Quarterly,* 49(4):474–516.

Roberts, J.V. (2004). *Public Confidence in Criminal Justice: A Review of Recent Trends.* Ottawa: Public Safety and Emergency Preparedness Canada.

Roberts, J.V. and K. Roach. (2004). *Community-Based Sentencing: The Perspectives of Crime Victims.* Ottawa: Research and Statistics, Department of Justice.

Sacco, V.F. (1995). "Media Constructions of Crime." *Annals of the Academy of Political and Social Science,* 539:141–54.

Sandler, J. (2001, June 6). "Battered Woman Can't Sue over Ex-Husband's Deadly Rampage." *Vancouver Sun,* A2.

Sauve, J. (2005). "Crime Statistics in Canada." Catalogue no. 85-002-XPE. *Juristat,* 25(5). Ottawa: Canadian Centre for Justice Statistics, Statistics Canada.

Silver, W., K. Mihorean, and A. Taylor-Butts. (2004). "Hate Crime in Canada." Catalogue no. 85-002-XPE. *Juristat,* 24(4). Ottawa: Canadian Centre for Justice Statistics, Statistics Canada.

Skelton, C. (2000, August 19). "The 'Four-Per-Centers': Our Real Crime Problem." *Vancouver Sun*, A1, A16–A17.

Sloan, J.H., A.L. Kellerman, D.T. Reay, J.A. Ferris, T. Koepsell, F.P. Rivara, C. Rice, L. Gray, and J. LoGerfo. (1988). "Handgun Regulations, Crime, Assaults and Homicide: A Tale of Two Cities." *New England Journal of Medicine*, 319:1256–62.

Smith, E. (2004). *Nowhere to Turn? Responding to Partner Violence against Immigrant and Visible Minority Women*. Ottawa: Canadian Council on Social Development.

Statistics Canada. (2000). *General Social Survey on Victimization. Cycle 13, 1999*. Ottawa: Statistics Canada.

———. (2003). *Ethnic Diversity Survey: Portrait of a Multicultural Society*. Ottawa: Ottawa: Statistics Canada.

———. (2005). *General Social Survey, Cycle 18. Overview: Safety and Perceptions of the Criminal Justice System*. Ottawa: Social and Aboriginal Statistics Division.

Stein, K. (2001). *Public Perceptions of Crime and Justice in Canada: A Review of Opinion Polls*. Ottawa: Research and Statistics, Department of Justice.

Tremblay, S. (2000). "Crime Statistics in Canada." Catalogue no. 85-002-XPE. *Juristat*, 20(5). Ottawa: Canadian Centre for Justice Statistics, Statistics Canada.

Tufts, J. (2000). "Public Attitudes toward the Criminal Justice System." Catalogue no. 85-002-XPE. *Juristat*, 20(12). Ottawa: Canadian Centre for Justice Statistics, Statistics Canada.

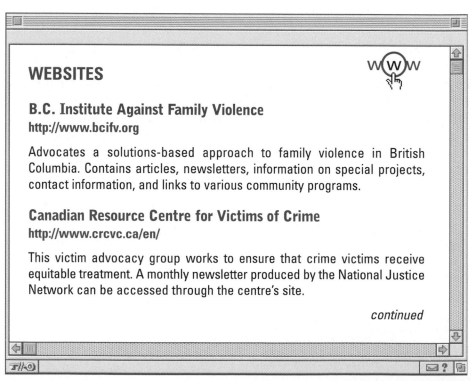

WEBSITES

B.C. Institute Against Family Violence
http://www.bcifv.org

Advocates a solutions-based approach to family violence in British Columbia. Contains articles, newsletters, information on special projects, contact information, and links to various community programs.

Canadian Resource Centre for Victims of Crime
http://www.crcvc.ca/en/

This victim advocacy group works to ensure that crime victims receive equitable treatment. A monthly newsletter produced by the National Justice Network can be accessed through the centre's site.

continued

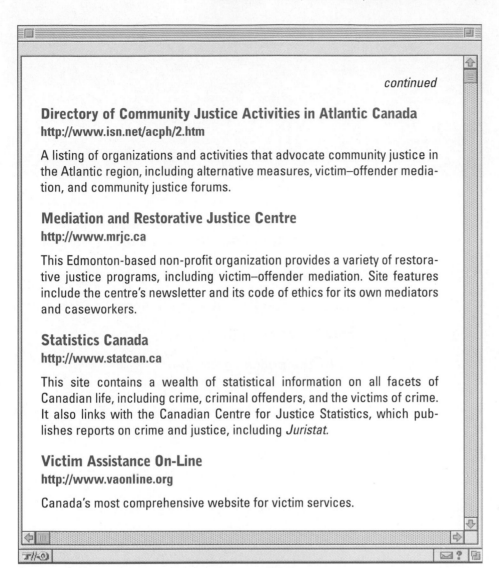

continued

Directory of Community Justice Activities in Atlantic Canada
http://www.isn.net/acph/2.htm

A listing of organizations and activities that advocate community justice in the Atlantic region, including alternative measures, victim–offender mediation, and community justice forums.

Mediation and Restorative Justice Centre
http://www.mrjc.ca

This Edmonton-based non-profit organization provides a variety of restorative justice programs, including victim–offender mediation. Site features include the centre's newsletter and its code of ethics for its own mediators and caseworkers.

Statistics Canada
http://www.statcan.ca

This site contains a wealth of statistical information on all facets of Canadian life, including crime, criminal offenders, and the victims of crime. It also links with the Canadian Centre for Justice Statistics, which publishes reports on crime and justice, including *Juristat.*

Victim Assistance On-Line
http://www.vaonline.org

Canada's most comprehensive website for victim services.

chapter

3

The Police

learning objectives

After reading this chapter, you should be able to

- Discuss the structure of policing in Canada.

- Discuss the recruitment and training of police officers.

- Identify the roles of the police.

- Describe the police occupation.

- Describe the powers of the police with respect to detention and arrest, search and seizure, entrapment, use of force, and interrogation of crime suspects.

- Discuss the influence of the Charter of Rights and Freedoms on police powers.

- Discuss the structures of police accountability.

key terms

Perhaps no component of the criminal justice system has been the subject of more books, television series, and films, or been more involved in key events in Canadian history, than the police. The police respond to a wide variety of demands and situations—many unrelated to crime and to the maintenance of public order—and carry out their duties in settings ranging from megacities (such as Montreal, Toronto, and Vancouver) to rural communities and hamlets in the remote North.

Being on the front lines, the police have more contact with the general public than other criminal justice personnel. Many police are highly visible; by contrast, other professionals in the criminal justice system (judges, probation officers, correctional officers) operate in relative obscurity. Criminal Court judges, for example, make decisions within the safe confines of a courthouse and have the luxury of considering the case facts and carefully weighing the interests of the community, the victim, and the offender before issuing a judgment; although their decisions may be controversial, they are in large measure immune from the scrutiny that police officers receive in carrying out their day-to-day tasks.

A BRIEF HISTORY OF POLICING

In the early days, before Canada existed as the country it is today, laws were enforced on an informal basis by community residents. In Halifax, for example, tavern owners were charged with maintaining order. For many years, policing remained closely tied to local communities. As settlements grew and the demands of law and order increased, however, this arrangement lost its effectiveness. The first police constables appeared on the streets of Quebec City in the mid-1600s and in Upper Canada (now the Province of Ontario) in the early 1800s. These early police constables performed a number of duties—for example, they were night watchmen, tax collectors, bailiffs, and inspectors of chimneys and buildings (Griffiths, Whitelaw, and Parent, 1999). As communities grew, policing became more formalized and organized police services were formed and expanded.

Many of the jurisdictions that ultimately became provinces after Confederation in 1867 originally had their own police forces. Most often, these were established in response to the disorder associated with gold strikes (for example, in British Columbia and Ontario). The earliest police force of this type was founded in 1858 in British Columbia (then a colony); that force continued to police the province until 1950, when policing services were contracted out to the RCMP. The police forces that had been established in Alberta, Saskatchewan, and Manitoba suffered from poor leadership and a lack of qualified officers. Between 1917 and 1950 the RCMP assumed provincial policing responsibilities in all provinces except Ontario, Quebec, and parts of Newfoundland. To this day, those are the only three provinces with provincial police forces.

The North-West Mounted Police (now the RCMP) was founded in 1873 to maintain law and order in, and to ensure the orderly settlement of, the previously unpoliced and sparsely settled North-West Territories (in rough terms, present-day Alberta and Saskatchewan). During its early years, the force was beset by internal difficulties and resented by both settlers and federal legislators. The historical record points to high rates of desertion, resignation, and improper conduct, due in part to the harsh conditions of the frontier. Attempts during the 1920s to phase out the force were driven by resistance in many regions to its expansion into provincial policing. It was anticipated that as these areas became more populated, the responsibility for policing would shift to local communities; for a variety of reasons, this did not happen. The emergence of the RCMP as a national police force involved in policing provinces and municipalities was, in fact, more an accident of history than part of a master plan (Griffiths, Whitelaw, and Parent, 1999).

Today there are about 60,000 police officers in Canada, or one police officer for every 529 Canadians. This is 22 percent lower than in the United States and Australia and 28 percent lower than in England and Wales. Some other attributes of the Canadian police:

• The number of police officers has been increasing since 1998.
• Five police services—the RCMP, the Toronto Police Service, the Ontario Provincial Police (OPP), the Sûreté du Québec (SQ), and the Service de police de la Communauté urbaine de Montréal (SPCUM)—account for just over 60 percent of all police officers in Canada.
• Policing is the largest component of the justice system and receives the biggest slice of the funding pie (see Figure 1.6).
• The proportion of female police officers has been steadily increasing and women now comprise 17 percent of the total number of officers.
• The Peel Regional Police in Ontario is the largest regional police force; the Toronto Police Service is the largest municipal police department (Sauvé and Reitano, 2005).

The shoulder patches of several Canadian police services are presented below.

Reprinted by permission of Halifax Regional Police

Reprinted by permission of Nishnawbe-Aski Police Services

Reprinted by permission of Fredericton Police Force

Reprinted by permission of Durham Regional Police

Reprinted by permission of Vancouver Police Department

Reprinted by permission of the Royal Canadian Mounted Police

THE STRUCTURE OF POLICING

Policing in Canada is carried out at four levels: federal, provincial, municipal, and First Nations. Other police services include the Canadian Pacific Railway Police and the Canadian Pacific Investigation Service (which fulfill policing roles for their respective organizations). Also, in major urban centres such as Montreal, Toronto, and Vancouver, transit police forces provide security and protection for property and passengers on public transportation systems. In 2005 the officers of the Greater Vancouver Transportation Authority Police Service were given full powers as peace officers and became the first transit officers in Canada to carry handguns.

The arrangements for delivering police services across Canada are quite complex. In Ontario, for example, the London Police—an independent municipal police service—is responsible for policing within the city boundaries, while the London detachment of the Ontario Provincial Police (OPP) has jurisdiction in the rural areas outside the city. The RCMP has its provincial headquarters in London and operates as a federal police force in the areas policed by the London Police and the OPP.

In Greater Vancouver there is a checkered pattern of municipal police departments and RCMP detachments. The municipalities of West Vancouver, Vancouver, New Westminster, Delta, Port Moody, and Abbotsford are policed by independent municipal police departments, whereas North Vancouver, Burnaby, Surrey, Coquitlam, and Richmond are policed by RCMP detachments under contract with the municipal government. And the Royal Newfoundland Constabulary—a provincial police force—is responsible for providing policing services to three areas of Newfoundland and Labrador: St. John's, Mount Pearl, and the surrounding communities referred to as Northeast Avalon; Corner Brook; and Labrador West, which includes Labrador City, Wabush, and Churchill Falls. The rest of the province is policed, under contract, by the RCMP.

Federal Police: The Royal Canadian Mounted Police

It is often said that Canada is the only country in the world with a police force as its primary symbol. The RCMP has long played a pivotal role in maintaining the Canadian identity and in fostering national unity (Griffiths, Whitelaw, and Parent, 1999: 16). The mythology that surrounds the RCMP has been the subject of countless books, movies, and academic studies.

The RCMP, which operates under the umbrella of Public Safety and Emergency Preparedness Canada, is organized into sixteen divisions, fourteen of which are operational divisions. These are organized into the following four regions, each of which is under the direction of a deputy commissioner:

- *Pacific Region:* E Division—British Columbia; M Division—Yukon.
- *Northwest Region:* D Division—Manitoba; F Division—Saskatchewan; G Division—Northwest Territories; V Division—Nunavut; K Division—Alberta; Depot Division—Saskatchewan.
- *Central Region:* O Division—Ontario; C Division—Quebec; A Division—National Capital Region.
- *Atlantic Region:* B Division—Newfoundland and Labrador; H Division—Nova Scotia; J Division—New Brunswick; L Division—Prince Edward Island.

Royal Canadian Mounted Police Act
federal legislation that provides the framework for the operation of the RCMP

Generally, these divisions are headquartered in the provincial and territorial capitals and are under the supervision of a commanding officer.

The **Royal Canadian Mounted Police Act** (R.S. 1985, c. R-10) provides the framework for the force's operations. As the federal police force in all provinces and territories, the RCMP enforces most federal statutes, the Controlled Drugs and Substances Act, the Securities Act, and lesser known statutes such as the Canada Shipping Act and the Student Loans Act. When acting in the capacity of a federal police service, the RCMP does not normally enforce the Criminal

Glenbow Archives, NA-52-1

North-West Mounted Police officers, Fort Walsh, Saskatchewan (ca 1874–1880)

Code, except when it receives a request from a federal government department to investigate allegations of fraud connected with the use of public funds or when it lays a conspiracy charge in relation to a drug offence under the Controlled Drugs and Substances Act.

Policing Provinces and Municipalities under Contract

The RCMP is a federal police force, yet about 60 percent of its personnel are involved in **contract policing**—that is, they serve as provincial and municipal police officers under agreements between the RCMP and the provinces/territories. (Note again that Ontario and Quebec have their own provincial police forces and that the Royal Newfoundland Constabulary serves parts of Newfoundland and Labrador.) These RCMP officers are not subject to the provincial police acts in the jurisdictions they police. They are accountable only to the RCMP Act and to other directives issued by RCMP headquarters in Ottawa. At the RCMP's national headquarters in Ottawa, there is a deputy commissioner who is in charge of the various directorates located there.

The RCMP does not act in isolation, even when operating as a federal police force. It operates a number of specialized branches—for example, the Economic Crime Branch and the National Crime Intelligence Branch—that focus on specific types of crimes or activities. It also serves as an information clearing-house for all of Canada's police departments. In addition, it operates crime laboratories and computerized information systems (including CPIC—the Canadian Police Information Centre) as well as the Canadian Police College, which offers a range of education and training programs.

contract policing
an arrangement whereby the RCMP and provincial police forces provide provincial and municipal policing services

The RCMP's involvement in policing at the municipal level (under contract with municipalities), the provincial level (under contract with provinces), and the federal level makes it a truly national police force. Indeed, the Mountie in red serge astride a horse is a Canadian icon. The RCMP's reach extends to the international level. It has posted a number of liaison officers in various countries in Asia, Europe, and the Americas. This reflects the increasing globalization of crime and criminal justice (see Chapter 1). RCMP liaison officers provide a bridge between foreign police forces and their Canadian counterparts; they also assist in cross-national investigations.

Peacekeeping

The international profile of the RCMP has been significantly enhanced by its involvement in a number of peacekeeping missions under the auspices of the United Nations. The force's Civilian Police Peacekeeping Operations coordinates and manages the participation of Canadian police officers in international peacekeeping activities. In 2005, RCMP officers—along with provincial and municipal police officers—were serving as peacekeepers in Sierra Leone, Afghanistan, Jordan and Iraq, Ivory Coast, the Democratic Republic of Congo, Sudan, and Haiti. In those countries they function mainly as technical advisers and instruct police forces in new policing strategies. Police officers from provincial forces in Quebec and Ontario, and from municipal forces across the country, have also participated in these peacekeeping activities. The RCMP has extensive experience in peacekeeping; even so, the effectiveness of these deployments is open to debate, given that there have been few evaluations of the impact that

Courtesy of UNPOL Media Relations Unit, MINUSTAH, Port-au-Prince, Haiti

RCMP peacekeeper demonstrates self-defence techniques to cadets at the police academy in Haiti.

RCMP officers and their provincial and municipal counterparts have had in the countries to which they have been deployed. Among the difficulties that have been identified are the lack of predeployment training for officers being sent on peacekeeping missions and the fact that Canadian officers are often part of a multinational force of police officers, among whom there is wide disparity in both skills and level of professionalism (Donais, 2004).

Organizational Features of the RCMP

As a national police force, the RCMP has several distinctive organizational characteristics.

Accountability Municipal police departments are subject to local police boards and elected municipal councils; in contrast, RCMP detachments are not legally accountable to the municipalities they police under contract. In those municipalities which the RCMP polices under contract, there are no police boards and the local mayor and council have no mandate to oversee their work. Nor are RCMP officers subject to provincial police acts; rather, they are governed by the RCMP Act and by two national review commissions. Some observers thus argue that the RCMP is "in" but not "of" the communities they police, and that it is often difficult to ensure that RCMP detachments are responsive to the community's priorities and requirements.

Nationwide Recruiting and Centralized Training Officers are recruited from across the country and trained in Regina, Saskatchewan, at a central facility known as the Training Academy, informally referred to as "Depot."

Policing Diverse Task Environments RCMP members carry out their duties in a variety of environments across the country, from small coastal villages in Newfoundland and British Columbia, to Aboriginal communities in the North, to large suburban communities. The three largest RCMP detachments— each of which has several hundred officers—are all in British Columbia (Surrey, Burnaby, Coquitlam). At the other extreme are two-officer detachments, such as the one in the Inuit community of Resolute Bay (population 150), in Nunavut.

Transfer Policy Quite unlike their municipal police counterparts, RCMP officers have traditionally been rotated among detachments every two years or so. Owing to fiscal restraints, transfers have declined in recent years in both frequency and number, except in northern and remote areas of the country, where officers are generally moved every two or three years.

Nonunion Unlike their provincial and municipal counterparts across the country, RCMP officers are prohibited by legislation from forming a union (see *Delisle v. Canada (Deputy Attorney General)* [1999] 2 S.C.R. 989). Rather, their interests are represented through the **Division Staff Relations Representative (DivRep) Program.** Full-time DivReps are elected from each of the geographic divisions of the RCMP; it is mainly through them that officers express their concerns about employment issues.

Division Staff Relations Representative (DivRep) Program program that provides RCMP officers with a way to express their concerns to management

Broad Mandate The RCMP is involved in a broad range of policing activities, from peacekeeping to policing national parks. RCMP officers guard dignitaries, staff drug enforcement offices in foreign countries, and cooperate with other countries in the fight against organized crime. In recent years, fiscal restraint in Ottawa has led some observers to question whether the RCMP has sufficient resources to deliver policing services effectively on all of these fronts (Palango, 1998).

Provincial Police

There are three provincial police forces in Canada: the Ontario Provincial Police (OPP; http://www.opp.ca), the Sûreté du Québec (SQ; http://www.suretequebec.gouv.qc.ca), and the Royal Newfoundland Constabulary (RNC; http://www.justice.gov.nl.ca/rnc). Provincial police forces are responsible for policing rural areas and the areas outside municipalities and cities. They enforce provincial laws as well as the Criminal Code. Some municipalities in Ontario are policed under contract by the OPP. Outside Ontario and Quebec and certain parts of Newfoundland, the RCMP provides provincial policing under contract with provincial governments. When the RCMP acts as a provincial police force, it has full jurisdiction over the Criminal Code as well as provincial laws.

The OPP is organized into six regions. Its responsibilities include:

- providing policing services to communities that do not have municipal police services;
- with a few exceptions, policing all domestic waterways, trails, and roadways;
- maintaining the province's ViCLAS (Violent Crime Linkages Analysis System) and the provincial Sex Offender Registries; *and*
- providing policing services to a number of First Nations communities that have not exercised the option to have a First Nations police service.

The OPP also operates a number of innovative community policing initiatives (see Chapter 4).

Regional Police Services

A notable trend in Canada has been the creation of regional police services. This generally involves amalgamating several independent police departments to form one large organization. Regional police services have been a feature of policing in Ontario for many years. Today a number of regional police services, including the Peel Regional Police (the largest regional police force in Canada) and the Halton Regional Police, are providing policing services to more than 50 percent of Ontarians. In Quebec, the Service de police de la Communauté urbaine de Montréal provides policing services to the City of Montreal as well as several surrounding municipalities. There are only two regional police forces west of Ontario: the Dakota Ojibway Police Service in Manitoba, and the Lesser Slave Lake Regional Police, an Aboriginal police force that provides services to six First Nations territories in northern Alberta.

Proponents of regional policing contend that it is more effective at providing a full range of policing services to communities and is less expensive than having a number of independent municipal departments. Critics of regional policing argue that a regional police service is too centralized and does not offer the opportunity for effective community policing. The trend toward regionalization will continue to be driven by fiscal considerations and by the growing need for police services to maintain interoperability (see Chapter 1).

Municipal Police

As the name suggests, municipal police services have jurisdiction within a city's boundaries. Municipal police officers enforce the Criminal Code, provincial statutes, and municipal bylaws, as well as certain federal statutes such as the Controlled Drugs and Substances Act. Most police work is performed by services operating at this level.

A municipality can provide police services in one of three ways: by creating its own independent police service; by joining with another municipality's existing police force, which often means involving itself with a regional police force; or by contracting with a provincial police force—the OPP in Ontario, the RCMP in the rest of Canada (except Quebec).

Municipal police officers constitute the largest body of police personnel in the country, if you include both police employed by municipal departments and those who have been contracted through the RCMP or the OPP. There is no provision under Quebec provincial law for the Sûreté du Québec to contract out municipal policing services. The Toronto Police Service has more than 5,000 officers; at the other end of the spectrum, some remote communities are policed by detachments of only one or two officers.

Municipalities with their own policing services generally assume the costs of those services, which are sometimes underwritten by the provincial government. A notable trend in Ontario has been a decline in the number of independent municipal police services in favour of contracting with the OPP. Today, about 60 municipalities operate their own independent police services; just over 300 are policed by the OPP.

First Nations Police

In Chapter 1 it was noted that across Canada, Aboriginal peoples are becoming increasingly involved in the creation and control of justice programs. It is in the area of policing that they have assumed the greatest control over the delivery of justice services. This is perhaps appropriate, given the conflicts that have arisen between the police and Aboriginal peoples both in the past and in the present day.

Within the framework of the federal First Nations Policing Policy, the federal government, the provincial and territorial governments, and First Nations communities can negotiate agreements for police services that best meet the needs of First Nations communities. These communities have the option of developing an autonomous, reserve-based police force or using First Nations officers from the RCMP or the OPP in Ontario. Funding for Aboriginal police forces is split between the province and the federal government (Sauvé and Reitano, 2005).

Autonomous Aboriginal police services are now operating in all provinces and territories except Newfoundland, the Northwest Territories, Nunavut, and Prince Edward Island. Among the larger Aboriginal police forces—which are involved in policing multiple reserve communities—are the Ontario First Nations Constable Program, the Six Nations Tribal Police in Ontario, the Amerindian Police in Quebec, and the Dakota Ojibway Police Service in Manitoba. There are smaller Aboriginal police forces in other provinces. Aboriginal police officers generally have full powers to enforce on reserve lands the Criminal Code, federal and provincial statutes, and band bylaws. In a recent decision, the Supreme Court of Canada held that First Nations police constables in Ontario were empowered to operate a R.I.D.E. (roadside alcohol check program) off reserve and that the "territorial jurisdiction" of the members of the Anishinabek Police Service was not confined to the territorial boundaries of the reserve (*R. v DeCorte* [2005] 1 S.C.R. 133).

The activities of First Nations police forces are overseen by reserve-based police commissions or by the local band council. Aboriginal police forces often work closely with the OPP, the SQ, and the RCMP.

The Challenges of First Nations Policing

First Nations police services confront a number of challenges, including the demands of policing in communities that are often afflicted with poverty and high levels of violent crime. A review of First Nations Police Services in Alberta (Cardinal, 1998) noted:

- a lack of qualified on-reserve police candidates;
- intense competition for qualified candidates;
- low recruitment standards, and political interference in the selection process; *and*
- a lack of pre-employment training and upgrading programs to prepare potential First Nations recruits for a career in policing.

The same review also found low levels of satisfaction with the police in some of the communities, as well as a reliance on the RCMP to assist in providing policing services. Furthermore, members of First Nations Police Commissions often lacked an understanding of their role. A more recent audit of First Nations policing agreements in Ontario (Moore, 2003) found that Aboriginal police forces are still finding it difficult to attract qualified candidates and that a number of First Nations constables had not completed training at the Ontario Police College. Officers are now required to complete this training program and to meet minimum educational requirements.

PRIVATE SECURITY SERVICES

Recent years have seen exponential growth in private security, which is now providing services previously performed by provincial and municipal police services. There are two main types of private security: security firms that sell their services to businesses, industries, private residences, and neighbourhoods; and alternatively, companies that employ their own in-house security officers.

focus   BOX 3.1

Private Police: Coming to a Neighbourhood Near You?

In June 2005 a private security firm began a twenty-four-hour security patrol of a twenty-five-block area on Vancouver's affluent West Side. The firm is absorbing the $150,000 cost of operating the program for one year; during that time it will deploy GPS- and video camera-equipped vehicles and guarantee a five-minute response time to alarm calls in the area. The company's goal is to generate customers for its alarm-monitoring business and to sustain the patrols beyond the year-long trial period.

Source: M. Bridge (2005, June 15), "Security Firm Starts Street Patrol," *Vancouver Sun*, B1, B6. Reprinted by permission of the *Vancouver Sun*.

Private security officers outnumber police officers in Canada and are engaged in a wide range of activities, including crowd control, protecting businesses and property (including shopping malls and college and university campuses), and conducting investigations for individuals and businesses. In some venues, such as sporting events and concerts, private security officers and police officers may work in collaboration. At least one private security firm has been hired by a neighbourhood's residents for preventive patrol—a development that raises a number of questions (see Box 3.1). In 2005, Manitoba Justice was developing a plan that would allow private security agencies to supplement the policing services provided by the RCMP—for example, by enforcing the Highway Traffic Act and the provincial liquor act (Canadian Association of Police Boards, 2004a).

Some private security firms specialize in forensic accounting and other investigative activities. These firms are often staffed by ex-police officers and have an international clientele. Notable examples include KPMG and Forensic Investigative Associates Inc.

The expansion of the activities of private security firms into areas traditionally serviced by public police organizations has often resulted in uncertainty about the powers and authority of private security officers. Generally, private security personnel have no more legal authority than ordinary citizens to enforce the law or protect property. However, private security officers can arrest and detain people who commit crimes on private property. Recent court cases suggest that private security personnel must adhere to the provisions in the Charter of Rights and Freedoms only when making an arrest. The rapid growth of the private security industry has prompted calls for standardized training and accreditation for private security officers. The private security industry is gradually being professionalized through provincial legislation, community college programs, and various other training programs.

Several observers have raised questions about the transformation of private security officers into "parapolice" through the extension of their activities beyond loss prevention and the protection of property to encompass order maintenance and enforcement (see McLeod, 2002; Rigakos, 2003). Other

What do you think?

What do you think of the Manitoba proposal to extend the activities of private security firms into the enforcement arena? Also, what concerns, if any, do you have about the one-year trial program in Vancouver?

observers have expressed the concern that although public police are account-able to oversight commissions and—in the case of municipal and provincial police forces—to elected community officials, no similar systems of governance are in place for private security officers (Burbidge, 2005). So it will be impor-tant to follow the proposed plan in Manitoba, the neighbourhood private patrol experiment in Vancouver, and other initiatives involving private security firms in other provinces across the country.

POLICE BOARDS AND COMMISSIONS

In many jurisdictions, municipal police boards and provincial police commis-sions play a major role in overseeing the activities of police services. A munic-ipal police board is composed of community members and city council members and is usually chaired by the mayor. Activities of the board typically include hiring the chief constable, preparing and overseeing the police budget, and authorizing increases in police personnel. Provincial police commissions are involved in developing policing standards, promoting research, and pro-viding training programs for municipal and provincial officers. The authority of police boards is derived from provincial police acts.

POLICE RECRUITMENT

Recruiting qualified people is one of the greatest challenges facing Canadian police services in the early twenty-first century. A "retirement bulge" in recent years has led to the departure of hundreds of police officers, and this has only heightened the competition among police services for qualified applicants. Several large urban police services in Canada now find themselves having to attract hundreds of new recruits just to replace those officers who are retiring. An additional challenge is to increase the numbers of female, visible minority, and Aboriginal recruits.

basic qualifications
(for police candidates)
the minimum requirements for candidates applying for employment in policing

preferred qualifica-tions (for police candidates)
requirements that increase the competitive-ness of applicants seeking employment in policing

People who are interested in a career in policing must have both **basic qualifi-cations** and **preferred qualifications**. The basic qualifications include Canadian citizenship (although some departments consider landed immigrants), a min-imum age of nineteen (the average age of police recruits in many departments is over twenty-five), physical fitness, and a grade twelve education. Also, the appli-cant cannot have any prior criminal convictions or pending charges, and must exhibit common sense and good judgment. Preferred qualifications—which are highly prized by police services—include knowledge of a second language or culture, related volunteer experience, postsecondary education, and work/life experience. Ontario has standardized the criteria for assessing prospective appli-cants through the Constable Selection System, which is used by most of the province's police services. Prospective recruits file one application, which is then vetted through this system. This has done away with multiple applications to several police services and consequent duplication of the assessment effort.

Over the past decade, police recruiting has undergone significant changes because of the increasing pressure on police services to reflect the gender and cultural and ethnic diversity of the communities they police. Police services are actively seeking female and visible minority applicants. Although the numbers of women in policing have increased significantly over the past decade, women still comprise only 17 percent of police officers (Taylor-Butts, 2004). As well,

focus

BOX 3.2

Profile of Ontario Police College Recruits, 1998–2003

Following is the demographic profile of the more than 9,700 recruits from municipal police forces and the OPP who attended the Ontario Police College between 1998 and 2003:

- Nineteen percent of the recruits were women.

- The recruits ranged in age from twenty to fifty, with an average age of just under twenty-eight.

- The percentage of police recruits over thirty has steadily increased.

- Just under 11 percent of the recruits reported being a visible minority or Aboriginal (not including recruits from First Nations police services).

- About 22 percent of the recruits indicated that they spoke French.

- Ninety-five percent of the recruits had completed at least some college or university, nearly half had completed a college program, and one-third had earned a university degree.

Source: R. Morris, "Ontario Police College Recruit Profile: September 1998 to September 2003," *Canadian Review of Policing Research,* 1 (2004), 205–19.

women are severely underrepresented at the senior administrative levels of police services; they are less than 5 percent of senior officers. With respect to diversity in Canadian police services:

- Visible minorities (excluding Aboriginal people) are 13 percent of the national population over the age of fifteen but are only 4 percent of police officers.

- Visible minorities make up just over 30 percent of the populations of Vancouver and Toronto but less than 10 percent of police officers in those cities.

- Visible minorities are just over 21 percent of British Columbia's population but only 3 percent of RCMP officers in that province (Canadian Association of Police Boards, 2004b).

The profile of recruits attending the Ontario Police College between 1998 and 2003 is set out in Box 3.2.

On a more positive note, Aboriginal people make up approximately 3 percent of the national population and 3 percent of police officers, owing in large measure to the creation and growth of First Nations police services. Also, female police officers have begun to break through the so-called glass ceiling, and some are serving as members of specialty units such as SWAT and dive teams.

Interestingly, the profiles of private security officers indicate that Aboriginal people are equally represented (3 percent) in private security firms. There are, however, more women in private security than in public policing: women are 25 percent of private investigators and 23 percent of security guards but only

focus

BOX 3.3

Point/Counterpoint: Modifying Standards to Recruit Women and Visible Minorities

Point

- Police departments have been largely unsuccessful in recruiting sufficient numbers of women and visible minorities.

- Current recruitment standards, especially physical requirements and certain elements of written examinations, discriminate against certain groups of individuals.

- The first priority should be to ensure that the composition of police forces mirrors that of the general community.

Counterpoint

- Police recruiting standards reflect the physical and mental abilities required to carry out the demands of the job.

- Modifying recruiting standards for certain groups of individuals constitutes reverse discrimination.

- Lowering entrance requirements places the recruit and other police members at risk.

- There are many reasons, besides recruiting standards, why women and visible minorities are underrepresented in police departments.

- Differential standards result in women and visible minority officers being viewed as less qualified than other police officers.

What do you think?

Do you support or oppose the modification of recruiting standards (including fitness requirements and entrance examination scores) in order to attract more women and visible minorities into policing? Explain.

17 percent of police officers. Similarly, there was a higher percentage of visible minorities in private security—11 percent of private investigators and 16 percent of security guards, compared to 4 percent of police (Taylor-Butts, 2004). This suggests that there may be considerable interest among Aboriginal people and visible minorities in a career in policing, but that these persons are not applying to police services, do not meet the basic qualifications, or encounter bias in the recruiting process. Programs such as the RCMP's Cadet Development Program (discussed below) have assumed even greater importance as strategies for increasing the number of visible minorities and Aboriginal people in policing.

A subject of considerable debate is whether entry and training requirements (including physical fitness and entrance examination standards) should be modified in an attempt to increase the numbers of women and visible minorities in policing (see Box 3.3). Although police services have generally resisted this course of action, the RCMP has lowered its physical fitness entrance requirements for female applicants and rewritten portions of its entrance examination in response to concerns that some Aboriginal applicants and members of visible minorities score lower than white, male applicants.

Recruiting Visible Minorities and Aboriginal People

Visible minority and Aboriginal police officers are much more common than they used to be. Many police services have developed special initiatives and programs to attract qualified visible minority and Aboriginal recruits. For example, the following programs are operated by the OPP:

- *PEACE* (Police Ethnic and Cultural Exchange). Visible minority students participate in a police-sponsored summer employment program.
- *Asian Experience.* Potential recruits of Asian background accompany in-service officers for several days.
- *OPPBound.* Over several days, potential recruits (specifically, women and visible minorities) participate in a variety of activities with in-service officers and at the Provincial Police Academy.

The RCMP operates a number of programs across the country to attract Aboriginal people and visible minorities to a policing career:

- The Aboriginal Youth Training Program provides selected Aboriginal youth with seventeen weeks of summer employment. After three weeks at the Training Depot in Regina, students return to a detachment near their home community, where they work under the supervision of an RCMP officer.
- The Aboriginal Cadet Development Program helps Aboriginal applicants upgrade their skills and education so that they can meet the force's entrance requirements.
- In Nova Scotia, the Diversity and Career Development Program helps visible minority candidates upgrade their qualifications to meet entrance requirements. As well, the Visible Minority Summer Student Program offers twelve weeks' summer employment to Nova Scotian youth of African-Canadian background. These young people participate in a number of activities designed to foster an understanding of, and positive attitudes toward, the criminal justice system.

Competition for Previously Experienced Officers (PEOs)

The need for qualified recruits has also spawned increased competition between municipal and provincial police forces and the RCMP for qualified in-service police officers (also commonly referred to as previously experienced officers, or PEOs) who may be amenable for a variety of personal and professional reasons to leaving their current police service. A number of police services have special recruiting teams that target PEOs, and it is not unusual for the police services to "raid" officers from one another. Most of the larger police services have a specific section on their websites devoted to in-service officers—for example, the Calgary Police Service has its Experienced Police Officer Campaign, the Edmonton Police Service has an Experienced Officer Program, and the Winnipeg Police Service operates a Lateral Entry Candidate web page. These pages provide information to prospective in-service candidates on the department's lateral entry policy and entrance requirements and on the criteria for determining salary levels.

Inuit RCMP member, Baker Lake, Nunavut

The recruiting of PEOs has increased the mobility of police officers (often referred to as "gypsy officers") between police services to levels heretofore unseen in Canadian policing. Officers are often able to move to police services closer to their residence or to their home province; or they can take advantage of career opportunities that may exist in another police service. Some of them may choose to leave a department in order to avoid discipline or termination.

This unprecedented movement of officers presents challenges to police departments not only to attract personnel but also to retain them. It can no longer be assumed that a police officer will spend an entire career with the same police department. Significantly, the RCMP is also trying to attract in-service municipal and provincial police officers; historically, it has preferred to recruit and train its own members and has generally not been amenable to hiring officers from municipal or provincial police services.

With regard to this now widespread practice, Middleton-Hope (2004) has raised a word of caution. In a preliminary study of one police service, he found that as a group, gypsy officers may have more complaints filed against them than other officers. So it is important that the recruiting department screen in-service applicants carefully, making sure that the previous department discloses all relevant information.

POLICE TRAINING

Just as important as recruiting qualified people to become police officers is training them well. There are several different models of police training in Canada.

RCMP recruits receive basic training in the twenty-four-week Cadet Training Program at the Training Academy ("Depot") in Regina. This rigorous course has both physical and academic components. The cadets are trained in the law, community policing, operating police vehicles, handling firearms, self-defence, and tactics. Issues relating to law enforcement and public security are also part of the curriculum. Unlike recruits to municipal and provincial police forces across the country, RCMP recruits in training are on a pre-employment contract and are not yet members of the force. Although recruits who complete the training program in Regina are generally hired by the RCMP, employment is not guaranteed. Even after they are hired, the new members must still complete a six-month Field Coaching Program in a field training detachment. The new member is assigned to an officer who serves as mentor for the training period.

Several provinces and regions operate police academies to standardize the training of municipal police officers:

- The Atlantic Police Academy, in Charlottetown, serves Nova Scotia, Prince Edward Island, New Brunswick, and Newfoundland.
- The Institut de police du Québec provides basic training, in-service training, and specialized training for both provincial and municipal police officers in Quebec.
- The Ontario Police College in Aylmer provides training for all police services in Ontario, including the OPP.
- The Justice Institute of British Columbia Police Academy is the training centre for all police recruits in that province's independent municipal police services.

In those provinces which do not have provincial academies, individual police services operate their own training centres. In 2005, Alberta was preparing to establish a central training facility for all municipal police services in that province.

Police recruits in Canada generally receive instruction in the law, community relations, methods of patrol and investigation, and firearms handling. They are also provided with driver training and physical training. Having completed this training, the recruits are usually assigned to general patrol duties for three to five years. Thereafter, others are eligible to apply to specialty units.

Municipal recruits in British Columbia are trained at the Justice Institute of British Columbia Police Academy (http://www.jibc.ca). This training has four blocks:

- Block I (thirteen weeks), at the Police Academy. The emphasis is on police skills including driving, handling firearms, arrest and control procedures, and investigation and patrol techniques. Recruits also study the law and improve their physical fitness.
- Block II (thirteen to seventeen weeks), at the recruit's home police department. This involves field training under the guidance of a field trainer, with opportunities to apply knowledge gained in Block I in an operational setting.
- Block III (eight weeks), at the Police Academy. The emphasis is on developing additional knowledge and skills.
- Block IV. The recruit begins general patrol duties. After twelve to eighteen months, on the recommendation of a supervising officer, the constable is designated a "Certified Municipal Constable."

Reprinted by permission of the Vancouver Police Department

Police recruits struggle up a hill during physical training at the British Columbia Police Academy near Vancouver.

Training programs are constantly evolving to reflect changes in the demands placed on police services. In 2005, for example, a mock crystal meth lab was created at the Ontario Police College to help train recruits and working police on the dangers associated with these labs and on techniques for dismantling them (Erwin, 2005). For more detailed information on training programs, see the websites for individual police services.

There are also pre-employment programs for people interested in a policing career. The most notable of these is the Police Foundations Program in Ontario, which is offered through several community colleges. The program includes courses in criminal justice, diversity, community policing, criminal law, and case investigation as well as more general courses in human relations. Students who complete the two-year program and who pass a provincial examination are eligible for employment by any police service in Ontario (with the exception of the RCMP). Successful applicants to police services receive further training at the Ontario Police College.

THE ROLES OF THE POLICE

In the discussion of the role of media in criminal justice in Chapter 1, it was noted that the public seems to have an insatiable interest in the police, especially in those activities which involve the traditional police role of "catching crooks." This is illustrated by the ongoing popularity in Canada of American television programs such as *COPS*. But besides presenting a one-dimensional portrait of the police as crime fighters, these depictions on American TV programs and in the print media have led many Canadians to assume that policing in Canada is similar to policing in the United States. In fact, there are many differences between the two systems. For example, while country sheriffs are elected by popular vote in the U.S., there are no elected police officials (or criminal justice officials) in Canada.

In considering the myriad roles of the police in Canadian society, the following questions can be posed: What does the community expect of the police? How do the police view their role? What demands does the public place on the police? How can the police become more effective?

Police Duties and Activities

The duties of police officers are set out in provincial statutes. Ontario's Police Services Act, for example, sets out the following duties of police officers:

- preserving the peace;
- preventing crimes and other offences;
- assisting victims of crime;
- apprehending criminals;
- laying charges and participating in prosecutions;
- executing warrants;
- performing the lawful duties that the chief of police assigns; *and*
- completing the prescribed training.

Over the years, the responsibility for maintaining order, preventing crime, and responding to crime has become increasingly centralized in the agencies of the criminal justice system, including the police. This centralization of authority has been accompanied by a reduction in community involvement to the point where the Canadian public has come to rely—some would say rely too heavily—on the police to respond to and solve a wide variety of problems and situations. Often, the police on their own lack the capacity to address these problems and situations effectively.

The many activities that police officers become involved in can be divided into three major categories:

- *Crime control.* Responding to and investigating crimes, and patrolling the streets to prevent offences from occurring.
- *Order maintenance.* Preventing and controlling behaviour that disturbs the public peace, including quieting loud parties, responding to (and often mediating) domestic and neighbourhood disputes, and intervening in conflicts that arise between citizens.
- *Crime prevention and service.* Providing a wide range of services to the community, often as a consequence of the twenty-four-hour availability of the police, including responding to traffic accidents, assisting in searches for missing persons, and acting as an information/referral agency for victims of crime and for the general public.

**FIGURE 3.1   Calls to Communication and Dispatched Calls, Calgary
Police Service**

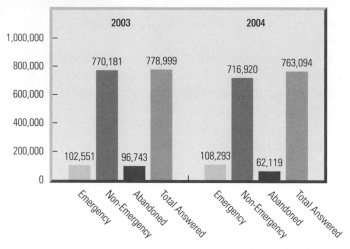

Calls to Communication

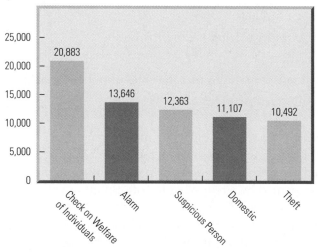

Dispatched Calls

Source: Calgary Police Service, *2004 Annual Report* (2004), 17. Retrieved from
http://www.calgarypolice.ca/facts/2004_annual_report.pdf. Reprinted by permission of the
Calgary Police Service.

Although both police officers and the public have long viewed crime control as
the key role of the police, research studies indicate that these activities generally
occupy less than 25 percent of police officers' time. (For most officers this figure is
considerably lower.) Most of the calls made to the police are for information, order
maintenance, and other service-related activities. Figure 3.1 presents information
on calls for service and dispatched calls for the Calgary Police Service.

As Kappeler and colleagues (1996: 212) note: "If television were to create a
program that realistically depicted police work, it would soon go off the air due
to poor ratings. It would offer little in the way of 'action' and would quickly be
tuned out by bored viewers."

The police are often called on to do the "dirty work" of society. They are often
the first to encounter social problems in the community, whether these involve

street people, labour strikes, conflicts between ethnic groups, violence within or outside the home, or drug and alcohol abuse. Police officers are often the first to arrive on the scene of fires, accidents, suicides, personal disputes and disagreements, and, of course, crime. In fact, police officers are involved in a wide range of tasks, many of which are not directly related to law enforcement.

The disorders and crimes to which police officers respond are often only symptoms of deeper personal and societal ills, and it is unrealistic to expect the police on their own to have either the resources or the knowledge to address the causes of these problems. Police services employ a wide range of strategies to prevent and respond to crime and social disorder; many of these involve partnerships with the community, the private sector, and various justice and social service agencies. These are considered in Chapter 4. It is widely recognized that merely "calling the cops"—that is, relying solely on the police to solve problems—is not an especially effective strategy.

Factors Influencing the Role and Activities of the Police

Several factors can significantly affect the demands on police, the role they play in a community, and their ability to respond to community needs and expectations. These include, but certainly are not limited to, the following:

Christina Laforce/CP Photo Archive

An RCMP Emergency Response Team member rushes a baby to a waiting car following the peaceful conclusion to a lengthy standoff in Halifax, Nova Scotia.

- *Legislation.* New laws and amendments to existing legislation can have a sharp impact on police powers, on the demands placed on police services, and on how police services set (and try to achieve) their operational priorities. Literally overnight, behaviour that was once criminal can become legal, and behaviour that was once legal can become criminalized. As well, as noted in Chapter 1, increasing concern about public security and safety in the post-9/11 world and about the activities of organized criminal groups has often strained police resources. The Anti-Terrorism Act, for example, gives police expanded powers to deal with individuals identified as posing a threat to safety and security; it has also established a new crime— "terrorist activity." The expectation is that the police have the capacity to respond to the multitude of demands that are placed upon them.
- *Geography and demographics.* Canada is a huge but sparsely populated country. RCMP, OPP, and SQ officers can be posted to rural and remote settings, often in small detachments of five or less. These settings—many of which include Aboriginal reserves—place unique demands on the officers and require them to develop strategies for delivering effective police services.
- *The ethnic and cultural diversity of Canadian society.* Many of Canada's communities are rich in ethnic and cultural diversity. Fifteen percent of this country's population is "foreign-born," and that percentage is expected to rise

in the coming years. Two-thirds of arriving immigrants settle in the megacities of Vancouver, Toronto, and Montreal, and many of them have had negative experiences with, or hold less than favourable attitudes toward the police in their countries of origin. Urban centres are also attracting more and more Aboriginal people from rural and remote areas. Police services are being pressured to adapt their recruitment, training, and delivery practices to reflect the diverse needs of multicultural communities and neighbourhoods.

- *Police funding.* As public-sector organizations, police services depend for their funding on municipal councils or provincial legislatures (in the case of the RCMP, on federal funding). Many factors influence the budgets of police services, including economic downturns and efforts to reduce the public debt and the reluctance of politicians to increase taxes in order to increase public services. Police services, like many other government-funded agencies and organizations, are under increasing pressure to adopt private-sector practices such as zero-based budgeting and performance review systems.

- *Economic, political, and cultural trends.* As noted in Chapter 1, the complexity of criminal cases has increased along with the "globalization" of crime. Conflicts over environmental issues such as logging, ongoing protests against globalization as represented by the World Trade Organization (WTO), unresolved Aboriginal land claims, and political terrorism are presenting new challenges to police organizations. International crimes such as drug trafficking, Internet crime, human smuggling, sex tourism, and genocide require the police in many countries to coordinate their activities.

THE POLICE OCCUPATION

Clearly, police officers are required to play a multifaceted role: counsellor, psychologist, enforcer, mediator, and listener. They must be able to understand and empathize with the feelings and frustrations of crime victims; at the same time, they must develop strategies to cope with the dark side of human behaviour, which they encounter every day. Officers must walk a fine line between carrying out their enforcement role and ensuring that the rights of law-abiding citizens and suspects are protected.

The "Working Personality" of Police Officers

working personality of the police

a set of attitudinal and behavioural attributes that develops as a consequence of the unique role and activities of police officers

The various pressures and demands placed on police officers contribute to what researchers have called the **working personality of the police.** This concept is used to explain how the police view their role and the world around them. It was first identified and defined by Jerome Skolnick (1966: 4): "The police, as a result of the combined features of their social situation, tend to develop ways of looking at the world distinctive to themselves, cognitive lenses through which to see situations and events."

Among the features of the working personality are a preoccupation with danger, excessive suspiciousness of people and activities, a protective cynicism, and difficulties exercising authority in a manner that balances the rights of citizens with the need to maintain order. It is argued that as a consequence of these "personality" attributes, many police officers:

- tend to view policing as a career and a way of life, rather than merely a nine-to-five job;

- value secrecy and practise a code of silence to protect fellow officers;

- exhibit strong in-group solidarity—often referred to as "the blue wall"—owing to job-related stresses, shift work, and an "us versus them" division between police and non-police (Goldsmith, 1990);

- tend to hold conservative political and moral views; *and*

- exhibit attitudes, often referred to as "the blue light syndrome," that emphasize the high-risk, high-action component of police work.

In the more than thirty years since Skolnick first proposed the notion of a "working personality," there have been many changes in the activities and strategies of police officers and in the diversity of the officers themselves. One of the most significant developments has been the emergence of community policing, a model of policing centred on police–community partnerships that bring officers into close contact with community residents in a wide range of crime prevention and response activities. Various community policing strategies can succeed in reducing the distance (and distrust) between the police and the communities they serve. This has led to the suggestion that the "us [police] versus them [public]" dichotomy is much too general and that it fails to account for the wide variety of relationships which exist between the public and the police, as well as the differences among police officers themselves with respect to how the police role is carried out (Herbert, 1998; Paoline, 2004).

That a police subculture exists has many positive implications. For example, it encourages camaraderie and trust among police officers, helps individual officers cope with the more stressful aspects of police work, and is a source of general support. A more negative view of the police subculture is warranted, however, if and when the group solidarity it generates comes at the expense of positive police–community relations and openness to new strategies and models of policing.

POLICE POWERS

One of the key issues in the study of policing is how much power they should have, and what type. How can society extend the police sufficient authority to ensure order and pursue criminals, while at the same time protecting the rights of citizens? To imagine what life would be like in a "police state," you need only look to countries where the police have no limits on their power. A police force with unlimited power might be more effective, but it would also interfere with the freedoms Canadians enjoy. Police powers in Canada were greatly expanded when the Anti-Terrorism Act was enacted in December 2001. Since that time, though, the police have rarely exercised the powers given to them under this legislation.

Charter Rights and Police Powers

One of the most significant developments in terms of police powers was the Canadian Charter of Rights and Freedoms, enacted in 1982. Besides entrenching constitutional rights for persons accused of crimes, the Charter gave those accused the right to challenge the actions of the police in situations where those rights might have been violated. Charter rights combine with pre-existing legal rules to prevent the unlimited use of police power. These safeguards include the following:

- The police cannot use certain investigative techniques (such as electronic surveillance) without prior judicial authorization.
- If the police gather evidence illegally, it may be excluded from a trial if its use would bring the administration of justice into disrepute.
- A defendant who feels that police officers or prosecutors have used unfair tactics can plead not guilty and cite "abuse of process" as a defence.
- A judge can remedy a violation of a defendant's rights by ordering a stay of proceedings or by ordering the Crown attorney to pay some or all of the defendant's legal fees.

Lest this sound one-sided, note that the police have retained significant powers (see below). As well, the courts have so far upheld the provisions of the Anti-Terrorism Act, which have greatly expanded the powers of the police.

In interpreting the Charter in the context of specific cases, Canadian courts have restricted, extended, or better defined the Charter rights of citizens. Consider the Nova Scotia case of *R. v. Borden* ([1994] 3 S.C.R. 145), which was heard by the Supreme Court of Canada. Borden had been convicted and sentenced to six years for the rape of a sixty-nine-year-old woman in a senior citizens' home in 1989. DNA testing proved conclusively that Borden was the per-petrator. But Borden had volunteered the blood sample after being arrested for assaulting an exotic dancer. He had assumed the test was related to that investigation. On appeal, the court ruled that because he had not been informed of the true reason he was asked for the blood sample, the police had violated his Charter right against unreasonable search or seizure. His conviction was overturned.

Predictably, Borden's lawyer applauded the decision: "It sends a message to the police and to the prosecution that people's rights have to be respected—that the Charter of Rights has meaning" (Canadian Press, 1994). The prosecutor's comments reflected a contrary view: "The balance between the rights of the accused person and rights of the community to be protected from crime is shifting very much in the wrong direction." Indeed, before the Charter was enacted, the test results would have been used and the conviction never questioned. As it was, Borden served four years for sexually assaulting the exotic dancer. Amendments to the Criminal Code now permit the police to obtain DNA warrants from judges for the collection of tissue samples. This example illustrates how the balance between police powers and the rights of citizens is constantly shifting and, in this case, how the exercise of discretion by judges is involved as well.

Do you know what your rights are under the Charter? Some of them have been reproduced in Box 3.4.

Vast amounts have been written about police powers in Canada. The limits of those powers are established through the decisions of the courts, which means that the process is ongoing. This makes it difficult to spell out precisely which powers the police have and do not have in any particular situation at

focus

BOX 3.4

Excerpts from the Canadian Charter of Rights and Freedoms

1. The Canadian Charter of Rights and Freedoms guarantees the rights and freedoms set out in it subject only to such reasonable limits prescribed by law as can be demonstrably justified in a free and democratic society.

Fundamental Freedoms

2. Everyone has the following fundamental freedoms:
   (a) freedom of conscience and religion;
   (b) freedom of thought, belief, opinion and expression, including freedom of the press and other media of communication;
   (c) freedom of peaceful assembly; and
   (d) freedom of association.

Legal Rights

7. Everyone has the right to life, liberty and security of the person and the right not to be deprived thereof except in accordance with the principles of fundamental justice.
8. Everyone has the right to be secure against unreasonable search or seizure.
9. Everyone has the right not to be arbitrarily detained or imprisoned.
10. Everyone has the right on arrest or detention
    (a) to be informed promptly of the reasons therefore;
    (b) to retain and instruct counsel without delay and to be informed of that right; and
    (c) to have the validity of the detention determined by way of habeas corpus and to be released if the detention is not lawful.

11. Any person charged with an offence has the right
    (a) to be informed without unreasonable delay of the specific offence;
    (b) to be tried within a reasonable time;
    (c) not to be compelled to be a witness in proceedings against that person in respect of the offence;
    (d) to be presumed innocent until proven guilty according to law in a fair and public hearing by an independent and impartial tribunal;
    (e) not to be denied reasonable bail without cause;
    (f) except in the case of an offence under military law tried before a military tribunal, to the benefit of trial by jury where the maximum punishment for the offence is imprisonment for five years or a more severe punishment;
    (g) not to be found guilty on account of any act or omission unless, at the time of the act or omission, it constituted an offence under Canadian or international law or was criminal according to the general principles of law recognized by the community of nations;
    (h) if finally acquitted of the offence, not to be tried for it again and, if finally found guilty and punished for the offence, not to be tried or punished for it again; and
    (i) if found guilty of the offence and if the punishment for the offence has been varied between the time of the commission and the time of sentencing, to the benefit of the lesser punishment.

12. Everyone has the right not to be subjected to any cruel and unusual treatment or punishment.

13. A witness who testified in any proceedings has the right not to have any incriminating evidence so given used to incriminate that witness in any other proceedings, except in a prosecution for perjury or for the giving of contradictory evidence.

14. A party or witness in any proceedings who does not understand or speak the language in which the proceedings are conducted or who is deaf has the right to the assistance of an interpreter.

Equality Rights

15. (1) Every individual is equal before and under the law and has the right to the equal protection and equal benefit of the law without discrimination and, in particular, without discrimination based on race, national or ethnic origin, colour, religion, sex, age or mental or physical disability.

    (2) Subsection (1) does not preclude any law, program or activity that has as its object the amelioration of conditions of disadvantaged individuals or groups including those that are disadvantaged because of race, national or ethnic origin, colour, religion, sex, age or mental or physical disability.

Enforcement

24. (1) Anyone whose rights or freedoms, as guaranteed by this Charter, have been infringed or denied may apply to a court of competent jurisdiction to obtain such remedy as the court considers appropriate and just in the circumstances.

    (2) Where, in proceedings under subsection (1), a court concludes that evidence was obtained in a manner that infringed or denied any rights or freedoms guaranteed by this Charter, the evidence shall be excluded if it is established that, having regard to all the circumstances, the admission of it in the proceedings would bring the administration of justice into disrepute.

Source: Department of Canadian Heritage, excerpts from Canadian Charter of Rights and Freedoms. Reproduced with the permission of the Minister of Canadian Heritage and the Minister of Public Works and Government Services. The Charter is available online at http://laws.justice.gc.ca/en/charter.

any one point in time. (For summaries of recent Charter cases, see http://www.mapleleafweb.com). The following issues are discussed below: the power to detain and arrest; search and seizure; entrapment; the use of force by police; and the interrogation of crime suspects.

The Power to Detain and Arrest

When most people think of police powers, they think automatically of arrest. Over the years, considerable confusion has surrounded the process of arrest. Many citizens do not know when the police have the right to make an arrest; nor do they know what their rights are in an arrest situation.

The power to arrest is provided by the Criminal Code and other federal statutes as well as by provincial legislation such as motor vehicle statutes. An arrest can be made to prevent a crime from being committed, to terminate a breach of the peace, or to compel an accused person to attend trial (Griffiths and Verdun-Jones, 1994: 107).

Contrary to popular belief, only a handful of criminal suspects are formally "arrested" when they are charged with an offence. Most are issued an *appearance notice* by the police officer or are summoned to court by a justice of the peace (JP). A criminal suspect who is arrested will generally be released from custody as soon as possible, on the authority of the arresting officer, the officer in charge of the police lockup, or a JP. Further information on the arrest process is provided in Chapter 5. Here the power of the police to detain and arrest suspects is considered.

If an arrest is warranted, and if there is time to do so, a police officer can seek an **arrest warrant** by swearing an **information** in front of a JP. If the JP agrees that there are "reasonable grounds to believe that it is necessary in the public interest," a warrant will be issued directing the local police to arrest the person. Accessing a JP can pose difficulties in rural areas. Several provinces (including British Columbia, Ontario, Manitoba, and Alberta) have developed "telewarrant" programs that provide twenty-four-hour access to JPs. Police officers can apply for and receive warrants by fax or telephone instead of having to appear in person before a JP.

Sometimes the police must act quickly and have no time to secure a warrant from a JP. Police officers can arrest a suspect *without* an arrest warrant in the following circumstances:

- they have caught a person in the act of committing an offence.
- they believe, on reasonable grounds, that a person has committed an **indictable offence.**
- they believe, on reasonable grounds, that a person is about to commit an indictable offence.

Two additional conditions apply to making an arrest. First, the officer must not make an arrest if he or she has "no reasonable grounds" to believe that the person will fail to appear in court. Second, the officer must believe on "reasonable grounds" that an arrest is "necessary in the public interest." This is defined specifically as the need to:

- establish the identity of the person;
- secure or preserve evidence of or relating to the offence; *and/or*
- prevent the continuation or repetition of the offence or the commission of another offence.

However, provisions in the Anti-Terrorism Act give the police the power of preventative arrest. This allows them to arrest persons without a warrant on "reasonable suspicion" (rather than the standard "reasonable grounds") if it is believed that the arrest will prevent a terrorist activity. The person need not have committed any crime and can be detained for up to seventy-two hours.

In practice, arrests are usually made only in the case of indictable offences. For minor crimes, called **summary conviction offences**, arrest is legal only if the police find someone actually committing the offence or if there is an outstanding arrest warrant or *warrant of committal* (a document issued by a judge directing prison authorities to accept a person into custody upon his or her

arrest warrant
a document that permits a police officer to arrest a specific person for a specified reason

information
a written statement sworn by an informant alleging that a person has committed a specific criminal offence

indictable offence
a serious criminal offence that often carries a maximum prison sentence of fourteen years or more

summary conviction offence
a less serious criminal offence, which is generally heard before a justice of the peace or provincial court judge

sentencing, a *bench warrant* for failure to appear at a court process, or a document issued by a parole board to revoke an offender's conditional release).

In some circumstances, an arrest can be unlawful. To make an arrest without a warrant, the officer must have "reasonable grounds" (formerly "reasonable and probable grounds") to believe that a person has committed an offence. An officer who makes an arrest without reasonable grounds risks being sued civilly for assault or false imprisonment. Moreover, a person who resists an unlawful arrest is not guilty of resisting a police officer in the execution of his or her duty. To make a *lawful* arrest, "a police officer should identify himself or herself, tell the suspect that he or she is being arrested, inform the suspect of the reason for the arrest or show the suspect the warrant if there is one, and, where feasible, touch the suspect on the shoulder as a physical indication of the confinement" (Bolton, 1991: 24).

What is the difference between arrest and detention? An officer can detain a person without arrest. The Supreme Court of Canada has held that a detention occurs when a police officer "assumes control over the movement of a person by a demand or direction that may have significant legal consequence and that prevents or impedes access to [legal] counsel" (*R. v. Schmautz* [1990] 1 S.C.R. 398). In contrast, the primary purpose of an arrest is to compel an accused to appear at trial.

Whether the person has been arrested or detained, an important threshold in the criminal process has been crossed. According to Section 10 of the Charter, anyone who has been arrested or detained has the right to be informed promptly of the reason for the arrest or detention. That person also has the right to retain and instruct counsel without delay, and furthermore, must be told about that right without delay. However, the suspect can choose to exercise that right or not. Also, a suspect who is interviewed by Canadian police officers in the United States must be informed of the right to counsel (*R. v. Cook* [1998] 2 S.C.R. 597). The Charter-based warning read by police officers in independent municipal police services in British Columbia is reproduced in Box 3.5. The wording of this communication of Charter rights may vary from police service to police service depending on the jurisdiction.

Suspects have a right to retain counsel but do not have an absolute right to have that counsel paid for by the state. Moreover, Section 10 of the Charter does not impose a duty on provincial governments to provide free legal representation to everyone who cannot afford it. In many provinces, free preliminary legal advice is available through a toll-free number on a twenty-four-hour basis. When an arrested or detained person does not have or know a lawyer, police must inform that person of this number and hold off on further questioning to give the suspect an opportunity to access this advice. After that, however, to get free legal representation the suspect must qualify for legal aid (see Chapter 5).

Failure to advise a person in a timely manner of the right to counsel upon arrest is an infringement of his or her Charter rights. In addition, the Supreme Court of Canada has held that a person's Charter rights are violated when the police:

- refuse to hold off and continue to question an arrested person despite repeated statements that he or she will say nothing without consulting a lawyer;

focus

BOX 3.5

Communicating Charter Rights upon Arrest or Detention

Sec. 10(a) I am arresting/detaining you for _____ (State reason for arrest/detention, including the offence and provide known information about the offence, including date and place.)

Sec. 10(b) It is my duty to inform you that you have the right to retain and instruct counsel in private without delay. You may call any lawyer you want.

There is a 24-hour telephone service available which provides a legal aid duty lawyer who can give you legal advice in private. This advice is given without charge and the lawyer can explain the legal aid plan to you.

If you wish to contact a legal aid lawyer I can provide you with a telephone number.

Do you understand?

Do you want to call a lawyer?

Supplementary Charter Warning: (If an arrested or detained person initially indicated that he or she wished to contact legal counsel and then subsequently indicates that he or she no longer wishes to exercise the right to counsel, read the following additional charter warning.)

You have the right to a reasonable opportunity to contact counsel. I am not obliged to take a statement from you or ask you to participate in any process which could provide incriminating evidence until you are certain about whether you want to exercise this right.

Do you understand?

What do you wish to do?

Secondary Warning: (Name), you are detained with respect to (reason for detainment). If you have spoken to any police officer (including myself) with respect to this matter, who has offered to you any hope of advantage or suggested any fear of prejudice should you speak or refuse to speak with me (us) at this time, it is my duty to warn you that no such offer or suggestion can be of any effect and must not influence you or make you feel compelled to say anything to me (us) for any reason, but anything you do say may be used in evidence.

Written Statement Caution: I have been advised by (investigating officer) that I am not obliged to say anything, but anything I do say may be given in evidence. I understand the meaning of the foregoing and I choose to make the following statement.

Approved Screening Device (ASD) Demand: In accordance with the provisions of the Criminal Code, I hereby demand that you provide a sample of your breath, forthwith, suitable for analysis using an approved screening device.

Breath Demand: I have reasonable and probable grounds to believe that you are committing, or within the preceding three hours have, as a result of the consumption of alcohol, committed an offence under Section 253 of the Criminal Code, and I hereby demand that you provide now, or as soon as is practicable, such samples of your breath as are necessary to enable a proper analysis to be made to determine the concentration, if any, of alcohol in your blood and to accompany me for the purpose of enabling such samples to be taken.

Blood Demand: I have reasonable and probable grounds to believe that you are committing, or within the preceding three hours have, as a result of the consumption of alcohol, committed an offence under Section 253 of the Criminal Code, and I hereby demand that you provide now, or as soon as is practicable, such samples of your blood as are necessary to enable a proper analysis to be made to determine the concentration, if any, of alcohol in your blood.

Samples of your blood will be taken by, or under the direction of, a qualified medical practitioner who is satisfied that the taking of those samples will not endanger you or your health.

MVA Section 90.3—12-Hour Licence Suspension: I have reasonable and probable grounds to believe

(1) you have alcohol in your body
    or
(2) you have failed or refused to comply with the demand to provide a sample of your breath that is necessary to enable a proper analysis of your breath to be made by means of an approved screening device.

I therefore direct you to surrender your driver's licence. Your licence to drive is now suspended for a period of 12 hours from this time and date.

If you produce, to a Peace Officer having charge of this matter, a certificate of a medical practitioner signed after this suspension is issued stating that your blood alcohol level does not exceed 3 milligrams of alcohol in 100 millilitres of blood at the time the certificate was signed, the suspension is terminated.

MVA Section 215—24-Hour Roadside Prohibition: I have reasonable and probable grounds to believe that your ability to drive a motor vehicle is affected by alcohol (or by drug), and I therefore direct you to surrender your driver's licence.

You are now prohibited from driving a motor vehicle for a period of 24 hours from this time and date.

(for alcohol) However, if you do not accept this prohibition, you have a right to either request a breath test or obtain a certificate from a medical practitioner. In the event that your blood alcohol level is shown not to exceed 50 milligrams of alcohol in 100 millilitres of blood by the test or certificate, this prohibition from driving is terminated.

(for drug) However, if you do not accept this prohibition, you have a right to attempt to satisfy a peace officer having charge of this matter that your ability to drive a motor vehicle is not affected by a drug other than alcohol, and if the peace officer is so satisfied this prohibition from driving is terminated.

OFFICIAL WARNING: You are not obliged to say anything, but anything you do say may be given in evidence.

- belittle the person's lawyer with the express goal or intent of undermining the person's relationship with that lawyer; *or*
- pressure the person to accept a "deal" without first affording the person the option to consult with a lawyer.

Search and Seizure

The power of the police to search people and places and to seize evidence also illustrates the fine balance that must be maintained between crime control and due process. Historically, under the common law, the manner in which evidence was gathered did not affect its admissibility in a criminal trial. That all changed with the Charter, Section 8 of which protects all citizens against "unreasonable" search or seizure. Evidence obtained during an illegal search

focus

BOX 3.6

R v. Mann—a Case of Detainment, Search, and Seizure

As two police officers approached the scene of a reported break and enter, they observed M., who matched the description of the suspect, walking casually along the sidewalk. They stopped him. M. identified himself and complied with a pat-down search of his person for concealed weapons. During the search, one officer felt a soft object in M.'s pocket. He reached into the pocket and found a small plastic bag containing marijuana. He also found a number of small plastic baggies in another pocket. M. was arrested and charged with possession of marijuana for the purpose of trafficking. The trial judge found that the search of M.'s pocket contravened Section 8 of the Canadian Charter of Rights and Freedoms. He held that the police officer was justified in his search of M. for security reasons, but that there was no basis to infer that it was reasonable to look inside M.'s pocket for security reasons. The evidence was excluded under s. 24(2) of the Charter, as its admission would interfere with the fairness of the trial, and the accused was acquitted. The Court of Appeal set aside the acquittal and ordered a new trial, finding that the detention and the pat-down search were authorized by law and were reasonable in the circumstances.

The Supreme Court subsequently decided that the acquittal should be restored. The majority of the Justices found that the police were entitled to detain M. for investigative purposes and to conduct a pat-down search to ensure their safety, but that the search of M.'s pockets was unjustified and that the evidence discovered in them must be excluded. The Court found that the police officers had reasonable grounds to detain M. and to conduct a protective search, but no reasonable basis for reaching into M.'s pocket. This more intrusive part of the search was an unreasonable violation of M.'s reasonable expectation of privacy with respect to the contents of his pockets.

Source: Canadian Legal Information Institute. Online at http://www.canlii.org/ca/cas/scc/2004/2004scc52.html. Reprinted by permission of the Canadian Legal Information Institute.

---

may be excluded from trial if, as indicated in Section 24 of the Charter, its use would bring the justice system into disrepute.

The Supreme Court of Canada has held *R v. S.A.B.* ([2003] 2 S.C.R. 678) that for a search to be reasonable, (a) it must be authorized by law, (b) the law itself must be reasonable, and (c) the manner in which the search was carried out must be reasonable. This is illustrated in the case of *R v. Mann* ([2004] 3 S.C.R.) presented in Box 3.6.

There is considerable room for interpretation by the courts as to what constitutes an unreasonable search in any particular case and when admission of evidence would bring the administration of justice into disrepute. Since the passage of the Charter in 1982 there have been hundreds of court cases and numerous books and legal articles dealing with this issue; the same two decades have seen an ongoing debate about what constitutes a reasonable search. As a result,

What do you think?

Do you agree with the decision of the Supreme Court in this case? Explain. Does this decision place too many restrictions on the powers of the police? What if the officers had found a handgun rather than marijuana?

search warrant

a document that permits the police to search a specific location and take items that might be evidence of a crime

conditions and requirements have emerged regarding prior authorization for a search. Generally, a **search warrant** must be issued. The Supreme Court of Canada has decided that warrants are required in the following situations:

- where there is to be secret recording of conversations by state agents;
- in cases involving video surveillance;
- for perimeter searches of residential premises;
- before the installation of tracking devices to monitor people's movements; *and*
- for searches of automobiles.

Search warrants are generally issued by JPs. Before a warrant can be issued, an information must be sworn under oath before a JP to convince him or her that there are reasonable and probable grounds that there is, in a building or place, (1) evidence relating to an act in violation of the Criminal Code or other federal statute, (2) evidence that might exist in relation to such a violation, or (3) evidence intended to be used to commit an offence against a person for which an individual may be arrested without a warrant.

The following scenario illustrates the principle of reasonable and probable grounds. Your neighbours feel that you match the description of a crime suspect in a bank robbery reenacted on a televised Crime Stopper program. They telephone the police and anonymously provide your name and address. Can this tip be used to establish reasonable and probable grounds for a search of your home? The answer is no. Although a possible starting point for a police investigation, anonymous tips do not provide reasonable and probable grounds. A concern in establishing reasonable and probable grounds is the source of the information, the credibility of which is likely to be questioned if it is anonymous.

A search without a warrant will generally be illegal, except in two types of situations:

1. While arresting a person, the officer may search the person and the immediate surroundings for self-protection (that is, to seize weapons) or to prevent the destruction of evidence (for example, to stop the person from swallowing drugs).
2. In an emergency situation where an officer believes that an offence is being, or is likely to be, committed or that someone in the premises is in danger of injury, a premise may be entered. In *R. v. Godoy* ([1999] 1 S.C.R. 311), for example, the Supreme Court of Canada held that the forced entry of police officers into a residence from which a disconnected 9-1-1 call had been made, and the subsequent arrest of a suspect who had physically abused his common-law partner, was justifiable.

Ultimately it is the courts that decide whether a search warrant has been properly obtained and executed or whether a warrantless search was legal. The passage in 2001 of Bill C-36, the Anti-Terrorism Act, expanded the authority of the police to search property associated with terrorist groups and/ or activity.

Entrapment

Entrapment means just what it sounds like: a person ends up committing an offence that he or she would not otherwise have committed, largely as a result of pressure or cunning on the part of the police. In these situations the police are most often operating undercover. The following are controversial examples of police practice:

- An expensive car is left with the keys in the ignition, observed by concealed officers waiting to arrest anyone who steals it.
- A police officer poses as a young girl while trolling websites frequented by pedophiles.
- An undercover officer poses as an intoxicated subway passenger, wearing expensive jewellery and a Rolex watch. Anyone who mugs him is arrested.
- An undercover officer poses as a potential client to arrest a prostitute who offers his or her sexual services.

Proactive techniques like these can be an effective and cost-efficient use of personnel. They can help prevent crime in "victimless" offences (such as prostitution and drug possession) of the sort that are unlikely to generate citizen complaints. The controversy stems from the fact that there is a line between catching those habitually involved in lawbreaking and creating situational criminals. The concern is that in some situations, typically law-abiding people could be enticed into committing a crime.

The courts have determined that the line is crossed when a person is persistently harassed into committing an offence that he or she would not have committed had it not been for the actions of the police. People cannot be targeted at random. Rather, there should be a reasonable suspicion that the person is already engaged in criminal activity. For example, in the prostitution example above, the actions of the police do not constitute entrapment because such a reasonable suspicion exists. One of the more important cases on entrapment is *R. v. Mack* ([1988] 2 S.C.R. 903), presented in Box 3.7.

Canadian courts have generally not allowed the defence of entrapment, which requires there to have been a clear abuse of process. In *R. v. Pearson* ([1998] 3 S.C.R. 620), the Court made a clear distinction between the issue of entrapment and innocence: "Entrapment is completely separate from the issue of guilt or innocence. It is concerned with the conduct of the police and is dealt with at a separate proceeding from the trial on the merits" (see also *R. v. Campbell,* [1998] 3 S.C.R. 533).

The Use of Force

The legal authority for the police to use force is found in the Criminal Code, which sets out the following principles:

- Officers exercising force must be performing a duty they are required or authorized to do.
- They must act on reasonable grounds.

focus

BOX 3.7

The Case of the Reluctant Drug Trafficker

The defendant was charged with drug trafficking. At the close of his defence, he brought an application for a stay of proceedings on the basis of entrapment. His testimony indicated that he had persistently refused the approaches of a police informer over the course of six months and that he was only persuaded to sell him drugs because of the informer's persistence, his use of threats, and the inducement of a large amount of money. He also testified that he had previously been addicted to drugs but that he had given up his use of narcotics. The application for a stay of proceedings was refused, and he was convicted of drug trafficking. The Court of Appeal dismissed an appeal from that conviction.

The central issue for the Supreme Court of Canada was whether the defendant had been entrapped into committing the offence of drug trafficking. The Court held that the police in this case were not interrupting an ongoing criminal enterprise; the offence was clearly brought about by their conduct and would not have occurred without their involvement. The Court stated that the persistence of the police requests and the equally persistent refusals, and the length of time needed to secure the defendant's participation in the offence, indicated that the police had tried to make the appellant take up his former lifestyle and had gone further than merely providing him with the opportunity. For the Court, the most important and determinative factor was that the defendant had been threatened and had been told to get his act together when he did not provide the requested drugs. This conduct was unacceptable and went beyond providing the appellant with an opportunity. The Court found that the average person in the appellant's position might also have committed the offence, if only to finally satisfy this threatening informer and end all further contact. The Court ruled that the trial judge should have entered a stay of proceedings.

Source: Canadian Legal Information Institute. Online at http://www.canlii.org/ca/cas/scc/1988/1988scc100.html. Reprinted by permission of the Canadian Legal Information Institute.

- They may use only so much force as is necessary under the circumstances.
- They are responsible for any excessive use of force.

Provisions governing the use of force are also contained in provincial police statutes.

The use of force is intended to gain control and compliance—for example, during an arrest or while breaking up an altercation. Degrees of force can be placed on a continuum from officer presence and verbal commands through to deadly force. Police are trained to match the degree of force to the immediate requirements of the situation. The degree of force is to be escalated only in response to escalating risk from the person they are subduing. The generally accepted use-of-force standard is "**one plus one**," meaning that police officers have the authority to use one higher level of force than that with which they are confronted. The use of force in excess of what is necessary can leave the officer criminally or civilly liable for assault.

one-plus-one use of force standard

police officers have the authority to use one higher level of force than that with which they are confronted

focus

BOX 3.8

The Taser: Less than Lethal or Lethal Weapon?

Conducted-energy devices (more commonly referred to as Tasers or "stun guns") were adopted by Canadian police services in the late 1990s. A Taser "gun" fires two metal darts, which are attached to wires and enter the subject's skin, providing a shock of up to 50,000 volts (as compared to 110 or 220 volts for a North American electrical wall receptacle). Tasers have been credited with reducing both the number of deaths resulting from police confrontations and the number of officers injured in the course of their duties. However, there has been widespread concern about the use of Tasers on those who are in a state of "excited delirium," which can result from heavy drug use (often cocaine or crystal meth), mental illness episode, or other causes. People in this state are often incoherent, violent, and non-compliant; the concern is that for someone in this condition, an electric shock can cause a heart attack.

Two recent reviews of the research on Tasers, prompted by a number of deaths of people who were subdued by Taser guns, did not find any link between the use of Tasers and death; nor did they find any increased risk of cardiac harm to people who are shot with a Taser. A recent review (2005) of the use of Tasers in British Columbia recommended that:

- Tasers should be used only against people who are actively resisting arrest and/or who pose a risk to others;
- they should not be used on persons who are "passively" resisting the police; *and*
- they should not be used to shock a person multiple times.

It was also recommended that police services in the province standardize Taser training and that officers be required to report every instance in which a Taser was used.

Sources: British Columbia Police Complaints Commissioner, 2005; Canadian Police Research Centre, 2005.

Police officers are aware of the high risks associated with using force, so in the vast majority of cases they make every attempt to avoid physical conflict with suspects. In many encounters, tactical communication (dialogue with the suspect, often referred to as "verbal judo") is effective in de-escalating the emotions and behaviour of the individuals involved in the conflict. If a higher level of response is required, officers may use less-than-lethal force options, which include the Taser (an electronic stun gun), Mace, pepper spray, the baton, and tear gas. The use of the Taser has been the subject of considerable controversy (see Box 3.8).

The decision to use deadly force is the most critical one any police officer can take. The decision is often made in a split second in circumstances involving fear, a rush of adrenaline, and confusion. Only a very small percentage of police–suspect encounters involve the police use of deadly force. Generally, officers are permitted to use guns only to protect themselves or others from serious injury or to stop a fleeing felon whose escape is likely to result in serious injury or death. In victim-precipitated incidents—more commonly known as "suicide by cop"—a suicidal individual confronts the police with a loaded firearm, a lethal instrument such as a knife, or discharges a weapon at the police in order to compel police use of deadly force.

In recent years there has been an increased emphasis on programs to ensure that police officers are trained in the appropriate use of force. The force options approach is the foundation of most police training in Canada. Traditional training techniques include role playing as well as practice in the use of pepper spray and empty-hand control (which includes various methods of restraint, such as lateral neck restraint or the locking suspect's arm behind the back). Computer simulations of police calls are also used. A scenario, such as an arrest or a domestic disturbance call, is projected onto a large screen. Trainers then vary key aspects of the scenario in response to the officer's actions and reactions. Debriefing and discussion following the session can help even experienced officers hone their skills for these potentially dangerous encounter situations.

Police officers, their managers and their police organizations are liable both criminally and civilly for the decisions they make to deploy force. Most civil suits that are filed against the police, either by the target of a police action or by the survivors of a deceased person who had an encounter with the police, are settled out of court. Out-of-court settlements avoid the costs and emotional pain associated with a public trial, but they also impede the development of insights into the dynamics of situations that culminate in the use of force—in some instances, deadly force. This information would be invaluable for training purposes and possibly also for police policy and practice.

Interrogation of Crime Suspects

The questioning of suspects is one of the most important yet least studied aspects of police operations. The information that is gathered through an interrogation can be a vital element of a criminal case (Williams, 2000).

Under Canadian law, police officers do not have any formal powers to compel crime suspects to answer their questions. Suspects have a right to remain silent, and police officers must inform them of that right. There are some exceptions to this. The right to remain silent does not extend to situations where it would permit a citizen to obstruct a police officer from carrying out his or her duties. For example, if you ride your bike through a red light and a police officer wants to issue you a traffic citation, you must produce identification. (And, in practice, remaining silent may only make things worse: a person who refuses to answer some general questions asked by the officer may raise suspicions that result in an arrest.) Police officers must also inform suspects that any statements they do make may be used against them in a criminal trial.

Other issues concern the rights of citizens in interrogation situations. In the case of confessions, for example, the rule of law is that statements made by crime suspects to persons in positions of authority outside of the court are not admissible unless it can be proven to the court that the statements were made freely and voluntarily.

The courts have also taken a dim view of the use of trickery by police to obtain confessions. The classic case is when an undercover police officer is placed in a cell with a crime suspect and then attempts to encourage the suspect to make incriminating statements. The Supreme Court of Canada has held that there are strict limits on the extent to which police can use this tactic to obtain a confession from a suspect who has refused to make a formal statement to the police. Voluntary statements made by a suspect to a cellmate (who may be an undercover police officer) may not violate the suspect's right to remain silent and may be admissible at trial if such admission does not bring the administration of justice into disrepute.

Final Thoughts on the Charter and Police Powers

Perhaps no other single piece of legislation has had a greater impact on the powers and activities of the police than the Charter of Rights and Freedoms. Defining the limits of police power is an ongoing process. As a result of decisions in many cases, police officers now have the authority to:

- use a warrant to obtain DNA from a suspect, by force if necessary;
- obtain a variety of warrants to intercept private audio and video communications;
- run "reverse stings" (for example, sell drugs as part of an undercover operation and then seize both the money and the drugs); *and*
- obtain foot, palm, and teeth impressions from a suspect.

The powers and authority of the police have also been restricted by the decisions of judges in Charter cases:

- All relevant information gathered during a case investigation must be disclosed to the defence attorney. This has increased investigative time and expense.
- The common-law "knock notice rule," which allowed police officers to knock and give notice before entering a dwelling where a person subject to arrest was known to be hiding, has been overturned and replaced by a warrant procedure that makes it difficult for police to apprehend a person hiding in a house (especially after 9 p.m.).
- Warrantless searches have been deemed unreasonable. Police officers must now clearly articulate why they have conducted a search without a warrant.
- Severe restrictions have been placed on the investigative strategy of placing an undercover officer in a jail cell to elicit evidence from a criminal suspect (Griffiths, Whitelaw, and Parent, 1999: 385–88).

The provisions of the Anti-Terrorism Act have expanded the powers of the police in some areas; however, it remains to be seen to what extent these powers will be employed and how the courts will respond to the litigation that may result from the exercise of these powers.

POLICE ACCOUNTABILITY

Police officers are accountable for their actions to the law, to federal and provincial governments, to local police boards, and ultimately to the citizenry (Murray, 2001). Although Canadians generally hold positive attitudes toward the police, incidents do occur as a result of which citizens take issue with police attitudes and behaviours or their failure to take action and exercise their discretion appropriately. Police services have traditionally resisted attempts to establish processes for external review. However, systems for citizens to file complaints against police officers are one way to ensure police accountability and to maintain the public's trust and confidence. There is considerable diversity across Canada in the arrangements for filing complaints against the police and in the review and investigative mechanisms in place. For example, in Nova Scotia a citizen has thirty days to file a complaint, whereas in Newfoundland the time limit is three months. Complaints must typically be made in writing.

Historically, people in the community who had complaints about the behaviour of police officers were required to file their grievances with the officer's department, which then conducted an investigation. This was an intimidating process and probably deterred many potential complainants. Today, police activities are overseen by a number of commissions, boards, and agencies established under provisions in provincial police acts (for RCMP officers, in the RCMP Act). In addition, there are units within police services that investigate alleged misconduct by officers.

Ontario's Civilian Commission on Police Services (OCCPS) receives appeals from municipalities concerning the alleged misconduct of police officers; reviews public complaints against police officers not resolved to the satisfaction of citizen-complainants at the local level; and conducts its own inquiries relating to complaints and to the disposition of complaints. The Special Investigations Unit (SIU), which operates under the province's attorney general, investigates cases involving serious injury, sexual assault, or death that may have been the result of criminal offences committed by municipal, regional, or provincial officers in the province.

There are two external boards of review that oversee the activities of RCMP officers: the External Review Committee and the Commission for Public Complaints against the RCMP. The former hears appeals from RCMP members who have been disciplined for an infraction of force regulations. The latter is an independent federal agency that receives and reviews complaints made by citizens about the conduct of RCMP officers who are policing under contract (that is, who are serving as provincial or municipal police officers; see http://www.cpc-cpp. gc.ca). These complaints are initially referred to the RCMP for investigation and disposition. If the complainant is not satisfied with the outcome of the RCMP investigation, the commission may conduct an independent review and recommend appropriate actions to the RCMP Commissioner.

Roughly one-tenth of the 2,500 or so complaints that are made against the RCMP each year are reviewed by the commission at the request of complainants. In most cases, the findings of the commission support the RCMP's disposition of the complaint. In many cases the commission uses alternative dispute resolution (ADR); this process brings together the complainant and the

focus

BOX 3.9

Disposition of Citizen Complaints against the Police— Selected Cases

Case 1: Complaint against an RCMP Officer Resolved by ADR

An eighteen-year-old woman was attacked while at school. Although the RCMP attended and took a report, she was concerned that charges would not be laid. A commission analyst contacted the local watch commander, who met with the complainant. The complainant explained that she felt she had been lost in the system. The watch commander assured her that this was not the case and that someone would certainly be available to assist her. He also referred her to Victim Services. The complainant was satisfied and withdrew her complaint.

Case 2: Complaint against a Member of the OPP for Inappropriate Use of Police Information Systems

A police officer's daughter asked to stay overnight at the home of a girlfriend. The officer used CPIC and an internal police records system to conduct a background check on the girlfriend's parents. He learned in this way that one of the parents had a criminal record; out of concern for his daughter, he would not allow her to stay overnight at the girlfriend's home. The girlfriend and her parents learned about this. The police service determined that there was misconduct of a minor nature. The girlfriend's parents requested a review, feeling that the penalty was insufficient. On review, the commission panel determined that there was misconduct of a serious nature and referred the case back to the police service for a hearing. The officer pled guilty at the hearing. After all parties made submissions, the penalty agreed on was three days' forfeiture of days off. The officer apologized in writing and in person to the complainant and gave a written apology to her family.

Case 3: Complaint of Abuse of Authority by British Columbia Police Officer

The complainant and his friend were at a house party. The complainant's friend left the house and walked outside when the respondent officer approached the friend and asked how much he had had to drink. The friend replied that he had had a lot to drink. The complainant's friend was placed in the back of the officer's vehicle. The complainant then walked outside looking for his friend because he had called a cab. The police officer advised him that his friend had been placed in the back of the police vehicle, and he was asked how much he had had to drink. The complainant replied that he had had two beers, and he was escorted into the back of the vehicle. The complainant felt that the officer used excessive force and that he and his friend should not have been placed in the drunk tank when they were waiting to take a cab home.

The complainant filed a Record of Complaint with the Office of the Police Complaint Commissioner, and the complaint was classified as one of public trust. The respondent officer and the complainant both agreed to meet and attempt to informally resolve the complaint. The parties had an opportunity to explain why they had each reacted in the manner they did, and the complaint was informally resolved. The Office of the Police Complaint Commissioner reviewed the file and agreed that informal resolution was an appropriate disposition to the complaint.

Source: Commission for Public Complaints Against the RCMP, *Annual Report, 2003–2004;* Ontario Civilian Commission on Police Services, 2003; British Columbia Office of the Police Complaint Commissioner, *Annual Report, 2004.*

RCMP member(s) as soon as possible after the incident in an attempt to resolve the outstanding issues informally.

The most frequent complaints against police officers involve abuses of authority, the attitudes of officers, and the quality of service provided. Complaints are sometimes brought against the police for very serious charges, including excessive use of force and death. Box 3.9 presents a selection of cases that illustrate the large majority of the types of complaints heard by police review commissions. Note that the vast majority of complaints are resolved informally at the department or detachment level and are not forwarded to complaints commissions.

In recent years the chair of the commission has criticized the RCMP for being less than forthcoming with documents and information required to investigate citizens' complaints (Commission for Public Complaints Against the RCMP, 2005). This has led some observers and politicians to call for increased powers for the commission, including legislation that would force the RCMP to respond to requests made by the chair (Gordon, 2005).

Summary

This chapter has examined the structure and operations of police services in Canada. The mix of municipal, provincial, and federal police services, and the involvement of the RCMP at all three levels of policing, contribute to an often confusing array of policing services across the country. Police services play a number of roles in carrying out their mandate. Many of their activities are only peripherally related to controlling crime and apprehending criminal suspects.

Perhaps at no other stage in the criminal justice process is the tension between due process and crime control more evident than in policing. In carrying out their tasks, police officers must constantly balance the rights of the suspect with the need to protect the community. The parameters of police powers are continually being defined by Canadian courts and by the decisions of police complaints commissions. Structures of accountability are in place that are designed to provide oversight of the activities of the police.

Key Points Review

1. Before police services were developed, laws were enforced on an informal basis by community residents.
2. Policing in Canada is carried out at four levels: federal, provincial, municipal, and First Nations.
3. The RCMP is a federal police force but also provides provincial and municipal policing services under contract. Several features distinguish it from provincial and municipal police services.
4. Increasing the gender and cultural diversity of police services is a priority in recruiting.
5. There are a number of different models of police recruit training across Canada.
6. Private security officers outnumber police officers by a ratio of two to one.
7. The duties and activities of police officers generally fall under these categories: crime control, order maintenance, and crime prevention.

8. A number of factors influence the role and activities of police officers.
9. Police powers must be balanced with citizens' rights.
10. The Charter of Rights and Freedoms has had a significant impact on the powers and activities of the police.
11. Police officers are accountable for their actions to the law, to federal and provincial governments, to local police boards, and, ultimately, to the citizenry.

Key Term Questions

1. What does the **RCMP Act** do?
2. What is **contract policing**?
3. Describe the RCMP's **Division Staff Relations Representative (DivRep) Program.**
4. Identify the **basic** and **preferred qualifications for police candidates** required by police services in the recruitment process.
5. What is meant by the **working personality of the police**?
6. What is an **information** and what is its relationship to an **arrest warrant**?
7. Explain the differences between a **summary conviction offence** and an **indictable offence**, and describe how the police power to arrest relates to each.
8. What is a **search warrant**, and in what types of situations are such warrants required?
9. What is the **one-plus-one** use of force standard?

References

Bolton, P.M. (1991). *Criminal Procedure in Canada*, 10th ed. North Vancouver, BC: Self-Counsel Press.

Bridge, M. (2005, June 15). "Security Firm Starts Street Patrol." *Vancouver Sun*, B1, B6.

British Columbia Office of the Police Complaint Commissioner. (2005). *Taser Technology Review: Final Report*. Victoria.

Burbidge, S. (2005). "The Governance Deficit: Reflections on the Future of Public and Private Policing in Canada." *Canadian Journal of Criminology and Criminal Justice*, 47(1):63–87.

Calgary Police Service. (2004). *2004 Annual Report*. Calgary. Retrieved from http://www.calgarypolice.ca/facts/2004_annual_report.pdf

Canadian Association of Police Boards. (2004a, April). "Private Security Firms May Get OK." *CAPB National*. Ottawa.

———. (2004b, September). "Police Ranks More Diverse." *CAPB National*. Ottawa.

Canadian Police Research Centre. (2005). *Review of Conducted Energy Devices*. Ottawa: Canadian Association of Chiefs of Police.

Canadian Press. (1994, October 3). "Freeing of Guilty Man Lauded: Police Did Not Tell Him Why They Wanted a Blood Sample." *London Free Press*, A1.

Cardinal, M. (1998). *First Nations Police Services in Alberta: Review*. Edmonton: Minister of Justice and Attorney General.

Commission for Public Complaints Against the RCMP. (2005). *Annual Report, 2004–2005*. Ottawa.

Donais, T. (2004). "Peacekeeping's Poor Cousin: Canada and the Challenge of Post-Conflict Policing." *International Journal,* 59(4):943–63.

Erwin, S. (2005). "Ontario Police Training to Focus on Crystal Meth." Retrieved from http://cnews.canoe.ca

Goldsmith, A. (1990). "Taking Police Culture Seriously: Discretion and the Limits of the Law." *Policing and Society* 1(1):91–114.

Gordon, J. (2005, February 7). "RCMP 'Thwarting' Complaints: Watchdog." *Vancouver Sun,* A1, A3.

Griffiths, C.T., and S.N. Verdun-Jones. (1994). *Canadian Criminal Justice,* 2nd ed. Toronto: Harcourt Brace.

Griffiths, C.T., B. Whitelaw, and R. Parent. (1999). *Canadian Police Work.* Scarborough, ON: ITP Nelson.

Herbert, S. (1998). "Police Subculture Revisited." *Criminology* 36(2):343–69.

Kappeler, V.E., M. Blumberg, and G.W. Potter. (1996). *The Mythology of Crime and Criminal Justice.* Prospect Heights, IL: Waveland Press.

McLeod, R. (2002). *Parapolice: A Revolution in the Business of Law Enforcement.* Toronto: Boheme Press.

Middleton-Hope, J. (2004). "Misconduct among Previously Experienced Officers: Issues in the Recruitment and Hiring of 'Gypsy Cops.'" *Canadian Review of Policing Research,* 1:178–88.

Moore, D. (2003, December 22). "Training and Recruiting of Native Police Deficient, Audits Show." *Canadian Press.* Retrieved from http://www.canada.com.

Morris, R. (2004). "Ontario Police College Recruit Profile: September 1998 to September 2003." *Canadian Review of Policing Research* 1:205–19.

Murray, T. (2001). "Developing a Framework for Cooperative Police Management." Presentation to the Annual Conference of the Canadian Association of Police Boards. Online at http://www.capb.ca/services/conf_2001/conf11.shtml

Palango, P. (1998). *The Last Guardians: The Crisis in the RCMP … and in Canada.* Toronto: McClelland & Stewart.

Paoline, E.A. (2004). "Shedding Light on Police Culture: An Examination of Officers' Occupational Attitudes." *Police Quarterly* 7(2):205–37.

Rigakos, G.S. (2003). *The New Parapolice: Risk Markets and Commodified Social Control.* Toronto: University of Toronto Press.

Sauvé, J. and J. Reitano. (2005). *Police Resources in Canada, 2005.* Ottawa: Canadian Centre for Justice Statistics, Statistics Canada.

Skolnick, J.K. (1966). *Justice without Trial: Law Enforcement in a Democratic Society.* New York: John Wiley and Sons.

Statistics Canada. (2005). "Police Personnel and Expenditures." *The Daily.* Ottawa.

Taylor-Butts, A. (2004). "Private Security and Public Policing in Canada, 2001." Catalogue no. 85-002-XPE. *Juristat,* 24(7). Ottawa: Canadian Centre for Justice Statistics, Statistics Canada.

Williams, J.W. (2000). "Interrogating Justice: A Critical Analysis of the Police Interrogation and Its Role in the Criminal Justice Process." *Canadian Journal of Criminology,* 42(2):209–40.

WEBSITES

Blue Line
http://www.blueline.ca

The *Blue Line* magazine website focuses on Canadian policing issues and contains links to various police agencies throughout Canada's provinces and territories.

Canadian Association of Police Boards
http://www.capb.ca

The web page for police boards across the country that provide governance for municipal police services. The site provides the association's newsletters, bulletins on recent developments in police and criminal justice, and links to a variety of police and justice-related sites.

Canadian Police Link
http://www.policelink.ca

A compendium, organized by province/territory, of links to provincial and municipal police forces. Also includes a Canada page with RCMP and other police-related links as well as a list of upcoming events and conferences.

Canadian Professional Police Association
http://www.cppa-acpp.ca

The CPPA website includes sections on justice reform, publications, and media releases as well as links to various police-related sites. Back issues of the CPA's quarterly publication, *Express* magazine, can be downloaded from the site.

Canadian Security Intelligence Service
http://www.csis-scrs.gc.ca

The official site of Canada's security agency contains public reports, an Economic and Information Security page, extensive background information on CSIS, and a menu of articles (titled "Commentary") published by the Research, Analysis and Production branch of CSIS.

Police Association of Ontario
http://www.pao.on.ca

The PAO represents the 13,000 police and civilian members of Ontario municipal police services. Its website includes information on careers and upcoming events, public survey results, and a document library.

Royal Canadian Mounted Police
http://www.rcmp-grc.gc.ca

This regularly updated site provides a comprehensive introduction to the RCMP (structure, history, programs and services, and so on) as well as an extensive collection of reports and publications.

learning objectives

After reading this chapter, you should be able to

- Compare and contrast traditional policing with community policing.
- Identify and discuss the principles of community policing.
- Identify and discuss the core elements of community policing.
- Discuss the effectiveness of police strategies.
- Discuss the role of discretion in police decision making.
- Identify the factors that may influence the decision making of the police.
- Define what is meant by bias-free policing and racial profiling.
- Identify the issues arising from the relations between police and visible and cultural minorities and police and Aboriginal peoples.
- Discuss the role of DNA profiling, ViCLAS, and cold case squads in case investigations.

key terms

There is little doubt that Canadian police carry out their tasks in social, cultural, and political environments considerably more complex than those faced by their predecessors. Over the past two decades, police services have developed a wide range of strategies for preventing and responding to crime more effectively. Many of these are discussed in terms of community policing. Though often heralded as a radical departure, community policing is in fact a *return* to the community-focused roots of policing.

This chapter discusses the key elements of community policing and the ways in which this model has influenced how the police organize and deliver their services across the country. The specific strategies the police use to prevent and respond to crime are examined, with particular emphasis on the efforts of police services to build sustainable partnerships with communities and to utilize the latest technologies for detecting crimes and investigating cases. Also considered are police officers' use of discretion and the decision-making processes they follow in carrying out their tasks in the community. This includes police decision making in high-risk situations. The chapter concludes with a discussion of police relations with visible minorities and Aboriginal peoples, and of several key strategies for case investigation.

Since the terrorist attacks on the United States on September 11, 2001, police services have been facing increasing pressure to focus on public safety and security and to be more proactive in addressing specific threats; at the same time, they are expected to continue strengthening ties with other agencies and with the communities they serve. It may be that a new model of policing is emerging in the early twenty-first century—one that incorporates but also extends the principles of community policing.

THE EVOLUTION OF POLICE PRACTICE

The evolution of policing over the past century has been accompanied by changing philosophies as to the best way to respond to crime and social disorder. In Chapter 3 it was noted that during the early days of Canada, policing was closely tied to local communities. The founding of police forces and the expansion of the RCMP into provincial and federal policing contributed to the centralization of the police function; this is reflected in the arrangements for policing that exist in Canada today.

Even after the creation of formal police services in Canada, policing remained closely tied to communities; police officers patrolled communities on foot and were responsible for a variety of tasks. With the introduction of mobile patrol cars in the 1920s and 1930s, a **professional model of policing** emerged that was based on the three Rs: *random* patrol, *rapid* response, and *reactive* investigation. The central premise of random patrol, also known as the *watch system*, is that the mere presence and visibility of patrol cars serves as a deterrent to crime and at the same time makes citizens feel safer. During a typical shift, patrol officers respond to calls and spend the rest of their time patrolling randomly, waiting for the next call for service. In this model of policing any information that is gathered by the police is limited to specific situations and does not include an analysis of the problems that precipitate crime and social disorder. Little attention is given to proactive police interventions designed to

professional model of policing
a model of police work that is reactive, incident driven, and centred on random patrol

prevent crime and to address the underlying causes of crime in communities (Oppal, 1994).

Research studies have found, however, that levels of crime are generally unaffected by increases in the number of patrol cars, quicker response times by patrol officers, or the number of arrests made by patrol officers (Griffiths, Whitelaw, and Parent, 1999: 118). This lack of impact is due in part to the fact that many of the incidents to which the police respond are only symptoms of larger problems in the community. In fact, it is *how* police resources are allocated and deployed that makes a difference. If the police respond only when they are called and deal only with the incident at hand, the reasons *why* the incident occurred in the first place remain unaddressed, and this increases the likelihood that similar incidents will happen again. The emergence of modern-day community policing was precipitated in part by the recognition that the police cannot prevent and respond to crime on their own; they require the assistance of a variety of agencies and organizations as well as community residents.

COMMUNITY POLICING

The 1980s witnessed the reemergence of an approach to policing that focused on the community. What is now known as community policing has its roots in nineteenth-century England. In 1829 the British Home Secretary, Sir Robert Peel, established the Metropolitan Police, the first full-time police force in London. To counter public hostility to the concentration of power in this force, Peel set high standards for recruitment and training and for the manner in which police officers were to carry out their mandate. These standards came to be known as Peel's Principles of Law Enforcement, and they are presented here in Box 4.1 (Reith, 1956). Although these principles are nearly 200 years old, they continue to this day to provide the foundations for policing.

Defining Community Policing

community policing
a philosophy, management style, and organizational strategy centred on police–community partnerships and problem solving to address problems of crime and social disorder in communities

There is much confusion among both the public and the police as to what **community policing** actually means. Over the past decade the term has come to refer to both a philosophy of police work and the mechanics of operational practice. Any definition of community policing must acknowledge that this model of police work incorporates many elements of traditional police practice while expanding the role, activities, and objectives of police services and patrol officers.

As a concept, community policing has the following characteristics:

1. *It is an organizational strategy and philosophy.* Community policing is based on the idea that the police and the community must work together as equal partners to identify, prioritize, and solve problems such as crime, drugs, fear of crime, social and physical disorder, and general neighbourhood decay, with the goal of improving the overall quality of life in the area.
2. *It requires a department-wide commitment.* The community policing philosophy requires that all personnel in the police service—both civilians and sworn members—balance the need to maintain an effective police response to incidents of crime with the goal of exploring new, proactive initiatives aimed at solving problems before they arise or escalate.

focus

BOX 4.1

Peel's Nine Principles of Law Enforcement

Principle #1: The basic mission for which the police exist is to prevent crime and disorder.

Principle #2: The ability of the police to perform their duties is dependent upon the public approval of police actions.

Principle #3: Police must secure the willing co-operation of the public in voluntary observation of the law to be able to secure and maintain the respect of the public.

Principle #4: The degree of co-operation of the public that can be secured diminishes proportionately to the necessity of the use of physical force.

Principle #5: Police seek and preserve public favor not by catering to public opinion, but by constantly demonstrating absolute impartial service to the law.

Principle #6: Police use physical force to the extent necessary to secure observance of the law or to restore order only when the exercise of persuasion, advice, and warning is found to be insufficient.

Principle #7: Police, at all times, should maintain a relationship with the public that gives reality to the historic tradition that the police are the public and the public are the police; the police being only members of the public who are paid to give full-time attention to duties which are incumbent upon every citizen in the interests of community welfare and existence.

Principle #8: Police should always direct their action strictly towards their functions, and never appear to usurp the powers of the judiciary.

Principle #9: The test of police efficiency is the absence of crime and disorder, not the visible evidence of police action in dealing with it.

Source: http://en.wikipedia.org/wiki/Peelian_Principles

3. *It rests on decentralizing and personalizing police services.* Decentralization offers line officers the opportunity, freedom, and mandate to focus on community building and on community-based problem solving, so that each and every neighborhood can become a better place in which to live and work (Trojanowicz and Bucqueroux, 1998).

Community policing is about much more than the introduction of new programs to a community: it involves substantial changes in the organization and delivery of police services, as well as an expansion of the roles and responsibilities of line-level police officers.

Community policing can thus be defined as a philosophy, a management style, and an organizational strategy centred on police–community partnerships and problem solving to address problems of crime and social disorder in communities.

The Principles of Community Policing

Community policing is based on the three Ps: prevention, problem solving, and partnership (with the community). The basic idea is that the police and the community constitute a partnership that brings together the resources and talents of each to identify and solve problems. The key principles of community policing include the following:

- Citizens are responsible for actively involving themselves in identifying and responding to problems in their neighbourhoods and communities.
- The community is a source of operational information and crime control knowledge for the police.
- Police are more directly accountable to the community.
- Police have a proactive and preventive role in the community that goes beyond traditional law enforcement.
- The cultural and gender mix of a police agency should reflect the community it serves.
- The operational structure of the police agency should facilitate broad consultation on strategic and policing issues (U.S. Department of Justice 2005; online at http://www.cops.usdoj.gov/print.asp?Item=477).

The philosophy of community policing and the specific initiatives that police services have developed have changed over the past two decades. Perhaps the best, and most succinct, description of the current focus of community policing is provided by the Office of Community Oriented Policing Services (COPS) in the U.S. Department of Justice:

> Community policing focuses on crime and social disorder through the delivery of police services that include aspects of traditional law enforcement, as well as prevention, problem-solving, community engagement, and partnerships. The community policing model balances reactive responses to calls for service with proactive problem-solving centered on the causes of crime and disorder. Community policing requires police and citizens to join together as partners in the course of both identifying and effectively addressing these issues (U.S. Department of Justice, 2005).

This description is a good illustration of how community policing has morphed into an approach that goes far beyond the original focus on community crime-prevention programs.

The differences between community policing and the traditional model of police work with respect to relationships with the community are outlined in Figure 4.1.

Having defined what community policing *is*, it is equally important to state want it is *not*. Among other things, community policing is not:

- a panacea for solving all of a community's problems of crime and disorder;
- a replacement for many traditional police services and crime prevention strategies, including reactive police response and the investigation of serious crimes;
- a single police initiative (although specific programs can be developed within a community policing strategy);

FIGURE 4.1   Police–Community Relations: Community Partnership versus Justice Process

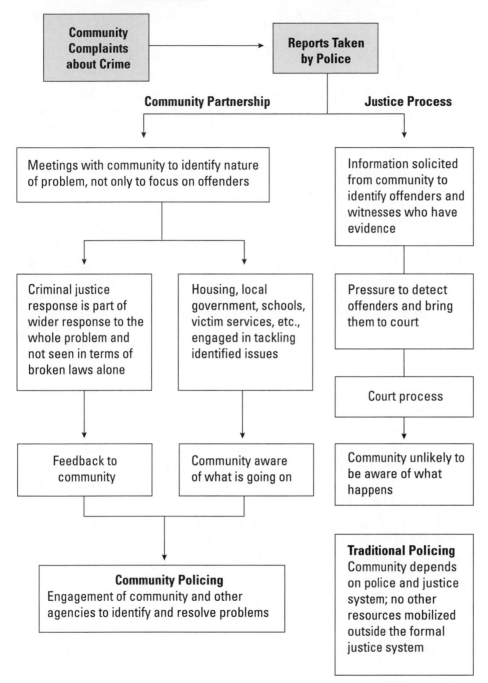

Source: C.G. Nicholl, *Community Policing, Community Justice, and Restorative Justice: Exploring the Links for the Delivery of a Balanced Approach to Public Safety* (Washington, DC: Office of Community Oriented Policing Services, U.S. Department of Justice, 1999), 50. Online at http://www.cops.usdoj.gov/Default.asp?Item=290. Reprinted by permission of the U.S. Department of Justice, Office of Community Oriented Policing Services.

- a generic, "one size fits all" policing model that can be applied without adaptation to the specific needs of individual communities;
- a program or series of initiatives that can be imposed on traditional police organizational structures;
- a policing strategy that is appropriate for addressing all types of criminal activity; *or*
- a substitute for having the capability in a police service to fight more sophisticated types of criminal activity; however, even these strategies may be more effective if there are police links with agencies and community residents.

THE CORE ELEMENTS OF COMMUNITY POLICING

> **core elements of community policing**
> the organizational and tactical strategies and external relationships of a police service that employs a community policing model

Perhaps the best way to capture the broad range of organizational, operational, and community partnership activities that are now included under the rubric of community policing is to identify the core elements that compose it. The core elements of community policing can be grouped into three areas: organizational, tactical, and external.

Organizational Elements

- *Community policing philosophy adopted throughout the organization.* This is reflected in the mission statement, policies, and procedures of the police service.
- *Decentralized decision making.* Individual line officers are given the discretion to solve problems and make operational decisions.
- *Fixed geographic accountability in tandem with generalist responsibilities.* Most decisions about staffing, command, deployment, and tactics are geographically based. Also, personnel are assigned to fixed geographic areas for extended periods of time in order to facilitate communication and partnerships between officers and the community.
- *Utilization of volunteer resources.* Citizens are actively encouraged to get involved in various police initiatives and activities.
- *Use of technology and analytical capacities.* The point of both is to facilitate information generation and the effective allocation of departmental resources.

Strategies: recruitment and deployment of volunteers in community police stations and storefronts; zone or team policing; intelligence-led policing

Tactical Elements

- *Enforcement of laws.* Specific strategies are applied to prevent and respond to crime.
- *A proactive, crime-prevention orientation.* Police collaborate with community partners in crime prevention initiatives.
- *Problem solving.* Police and the community form partnerships to address underlying issues that contribute to crime and social disorder.

Strategies: problem-oriented policing; zero tolerance policing; crime attack strategies; integrated service teams and neighbourhood service teams; crime prevention strategies

External Elements
- *Public involvement and community partnerships.*
- *Partnerships with government and other agencies.*

Strategies: police partnerships with key community stakeholders; private-sector initiatives (U.S. Department of Justice, 2005).

These core elements underscore that the philosophy of community policing affects (a) how police services that utilize a community policing model are structured, and how they deliver their services, and (b) the relationships between the police service and the community. The materials also remind us that community policing involves a broad range of proactive and reactive strategies as well as specific crime prevention and response initiatives.

The three areas outlined above are discussed in more depth below.

Organizational Elements

In traditional police services the management structure is a pyramid. In this top-down system, senior managers issue directives to be carried out by patrol officers. A one-way flow of information—from senior officers to patrol officers—ensures obedience to the goals and objectives established by senior officers.

Community policing turns the traditional pyramid on its head. The role of senior police managers is to provide leadership and support for the patrol officers who are implementing community policing on a daily basis. The focus shifts from achieving the goals set by senior managers to building communities and solving problems. In short, implementing community policing requires police services to "invert the pyramid" (see Figure 4.2).

FIGURE 4.2   Inverting the Police Pyramid

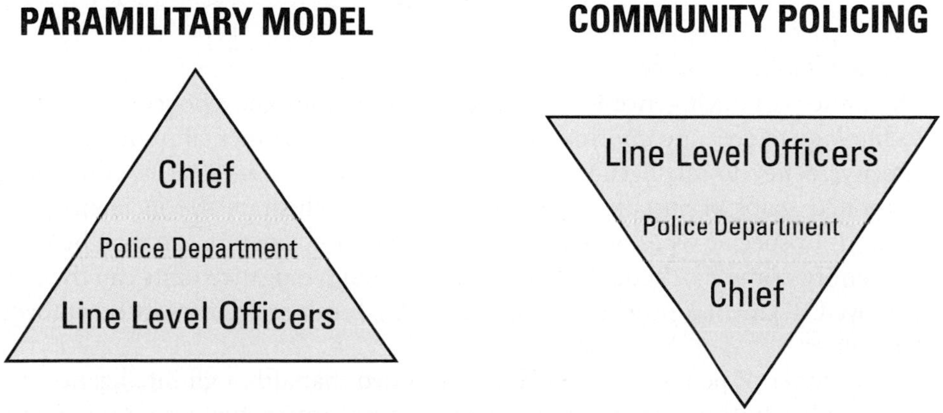

PARAMILITARY MODEL

Chief

Police Department

Line Level Officers

COMMUNITY POLICING

Line Level Officers

Police Department

Chief

Source: Reprinted from *Community Policing: How to Get Started*, 2nd ed., by R. Trojanowicz and B. Bucqueroux, with permission. Copyright 1998 Matthew Bender & Company, Inc., a member of the LexisNexis Group. All rights reserved.

Patrol officers are the cornerstones of community policing. They are given the support and authority to engage in problem-solving activities in the community and to work with community residents in identifying and responding to crime and social disorder. Officers are assigned to fixed geographic areas. This provides them with the opportunity to become familiar with the area, its issues, and its residents; to develop partnerships that draw on the community's resources; and to build community capacities to better address crime and disorder. The mission statements of North American police services generally incorporate the basic tenets of community policing; typically, these statements refer to police–community partnerships.

Many of the requirements of community policing are being met by implementing measures traditionally found in the private sector. Practices such as benchmarking, strategic planning, and community surveying are all components of police organizations. Terms such as *growth position, performance indicator,* and *performance outcome* are often heard in this context. Two key components of the corporate model of policing are *environmental scans* and *best practices.*

Environmental scans are studies designed to identify community, legislative, policy, and other forces in the community (here referred to as "the environment") that will result in demands on the police. A typical environmental scan involves gathering information on a number of factors external to the police service, including the following: demographic, social, and economic trends; crime trends; calls for police service; and the impact of legislative and policy changes. Many police services conduct scans annually to ensure a constant flow of information. On the basis of these data, changes in policies and operational practice can be made.

Best practices are organizational, administrative, and operational strategies that have proven to be successful in preventing and responding to crime. Many of these practices were first developed in the corporate sector and have since been adapted for use in police services. An example of best practices is the *community survey,* which provides police services with information about citizen satisfaction with the police, problems in the community that may require police attention, and the extent of criminal victimization (including the fear of crime).

intelligence-led policing
the application of criminal intelligence analysis to facilitate crime reduction and prevention

Intelligence-Led Policing

The strategy of **intelligence-led policing** is one example of how police services use technology to generate information and to deploy departmental resources more effectively. Key to intelligence-led policing are *crime maps*—that is, computer-generated maps of specific geographic areas that illustrate the incidence and patterns of specific types of criminal activity. This information can then be used to identify crime "hot spots," to which patrol and investigative units can then be deployed. Figure 4.3 provides a crime map for break and enters in downtown Vancouver.

A number of police observers have cautioned that although intelligence-led policing has potential, translating it from concept into actual practice involves a number of challenges, including resistance within the police organization itself and the need to develop working relationships between crime analysts and police operations personnel (see Cope, 2004; Ratcliffe, 2002).

FIGURE 4.3

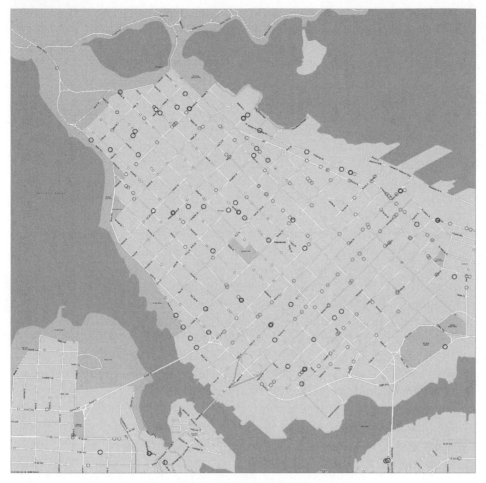

A crime map showing the location of break and enters in downtown Vancouver, British Columbia.

Source: Reprinted by permission of the Vancouver Police Department.

Tactical Elements

The tactical elements of community policing focus mainly on strategies for addressing crime and other community issues. The key to success here is proactive problem-solving with community partners. Possible strategies include the following.

Zero Tolerance Policing

A policing strategy that has gained popularity over the past decade or so is **zero tolerance policing**, also referred to as "order maintenance policing," "proactive policing," or "community policing with the gloves off." Central to this strategy is the idea that a strict order maintenance approach by the police in a specific area, coupled with high police visibility and presence as well as a strong focus on disorder and minor infractions, will result in a reduction of serious criminal activity. Often, the objective of these crackdowns is to disrupt open drug markets and the criminal activities (including theft of property) that are associated with them (Knox, 2001).

zero tolerance policing
an order maintenance approach that utilizes high police visibility and presence and that focuses on disorder and minor infractions with the goal of reducing more serious criminal activity

problem-oriented policing (POP)

a proactive strategy centred on developing strategies to address community problems

CAPRA model

a problem-solving approach used by the RCMP

Problem-Solving Policing

A key tactical strategy of community policing is **problem-oriented policing (POP).** This is based on the idea that policing should uncover the root causes of recurring crime and disorder and then address those causes directly. Central to POP is the "iceberg rule" (or "80/20 rule"), which posits that crime (the 20 percent of the iceberg) is only a visible symptom of invisible, much deeper problems (the 80 percent of the iceberg that lies below the surface of the water). The 80 percent represents the underlying conditions that allow the visible problems to exist. The SARA (scanning, analysis, response, and assessment) problem-solving model helps officers identify and respond effectively to problems, with the assistance of various agencies, organizations, and community groups.

Problem solving is central to the RCMP's **CAPRA model.** The letters stand for focusing on **C**lients, **A**cquiring and **A**nalyzing information, developing and maintaining **P**artnerships, generating an appropriate **R**esponse, and **A**ssessing the intervention. This model, which is taught to recruits at the Training Academy in Regina, incorporates the basic principles of community policing. It emphasizes identifying and responding to problems of crime and social disorder in the community by taking a problem-solving approach. It also highlights the importance of consultation and collaboration with community partners. The RCMP follows CAPRA in delivering policing to the communities it serves.

The OPP model for problem solving is called P.A.R.E. (**P**roblem Identification; **A**nalysis; Strategic **R**esponse, **E**valuation) and is depicted in Figure 4.4.

The P.A.R.E. model incorporates the same steps as problem-oriented policing (POP), and bears similarities to the CAPRA model.

FIGURE 4.4   The OPP P.A.R.E. Model of Problem Solving

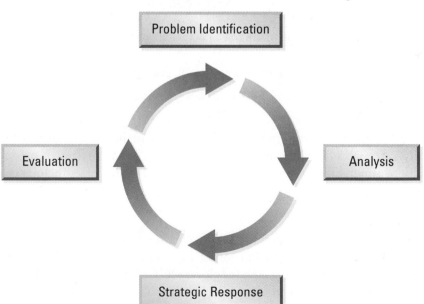

Source: OPP, online at http://www.opp.ca/cpdc/english/pare.htm. © Queen's Printer for Ontario. Reproduced with permission.

Crime Attack Strategies

Crime attack strategies are proactive operations carried out by the police to target and apprehend criminal offenders, especially those deemed likely to reoffend. One common strategy is the *tactical/directed patrol*, which involves saturating high-crime areas ("hot spots") with police officers, or targeting individuals who are engaged in a specific type of criminal activity. Hotspots are often identified through the process of intelligence-led policing. In addition, police services have developed a number of initiatives designed to target high-risk offenders. Examples are:

Serious Habitual/Significant Harm Offender Comprehensive Action Program (SHOCAP). SHOCAP is a multiagency program involving the Calgary Police Service, the probation authorities, the Crown, social services agencies, and corrections systems. Specifically, it is an information and case management program for youths and adults designated as serious habitual offenders. The Calgary Police Service's Serious Habitual Offenders Program (SHOP) monitors the activities of offenders both during custody and upon release in an attempt to reduce serious crime.

Durham Regional Police High Risk Offender Community Panel. This panel is composed of experts in the field of risk assessment and representatives from the private and public school systems. Its primary objective is to review the department's management strategies for high-risk offenders and to educate the public about the behaviours and risks these offenders present. Where appropriate, the panel may recommend that additional conditions be placed on high-risk offenders under supervision in the community. The purpose of this is to improve the monitoring abilities of the police and community corrections.

British Columbia Integrated Gang Task Force. This unit was established in 2005 and is composed of sixty municipal and RCMP officers. It is targeting Indo-Canadian gangs in the province, which have been responsible for a surge in violence and dozens of murders, most of them related to the drug trade.

Another proactive strategy that many Canadian police services use to manage high-risk offenders is *community notification*. This involves informing the media, crime victims, and the general public when certain offenders are released (usually from federal correctional facilities). The decision to issue a community notification is generally made by a committee and most often involves persons convicted of sex offences. This practice is examined in greater detail in Chapter 8.

Community Service Approaches

Community service approaches are designed to increase contact and cooperation between the police and the public in addressing problems of crime and disorder in communities. Foot and bike patrols, community police stations and storefronts, and team policing are some of the more common examples of these approaches. Team policing—also referred to as zone or "turf" policing—involves permanently assigning teams of police to small neighbourhoods in an effort to maximize interaction and communication with the community. Note that the assignment of patrol officers to a fixed geographic area is a component of the organizational component of community policing.

Foot patrols, which were first used in London, England, in the early 1800s, allow officers to leave their patrol (or scout) cars and acquaint themselves with

crime attack strategies
police patrol operations that are proactive and aimed at crime control

community service approaches
police strategies designed to increase police–public contact and cooperation in addressing problems of crime and disorder in communities

Chris Watte/Reuters/Landov

Ottawa Police Service bike patrol officers

the residents of the neighbourhoods they are policing. The increasing focus on foot patrols is an acknowledgment that blind faith in technology—which brought speedy scout cars loaded with sophisticated computers and communication systems—may have been misplaced. In fact, there is a widely shared view that overreliance on technology has isolated the police from the public. Technology has meant that instead of walking the beat and speaking with residents on a daily basis, officers now peer at computer screens in their cars to get information.

Foot patrols in urban areas across North America have produced some interesting results that suggest this strategy holds great potential. There is no clear evidence that foot patrols directly affect the levels of crime in neighbourhoods; they do, however, reduce people's fear of crime as well as calls for service. They also increase officers' familiarity with neighbourhoods and provide police and community residents with opportunities to collaborate in addressing crime.

Police services have applied a number of other strategies to connect with community organizations and residents. The Edmonton Police Service, for example, has Neighbourhood Empowerment Teams (NETs) that focus on developing community partnerships and community capacities. The goals in this are to prevent and respond to crime and social disorder and to improve social development in the community. A key component of the NETs is the setting up of storefront offices to provide a police presence in the community. An evaluation of the program found strong support among community residents and improved attitudes toward the police, particularly among minorities (Pauls, 2004). The impact on crime rates has been less clear, although the NETs have achieved significant progress toward the overall objective of improving community "wellness."

The Vancouver Police Department operates Neighbourhood Integrated Service Teams (NISTs). These are composed of police officers and representatives from other agencies (including provincial social services, the Vancouver School Board, the Vancouver Coastal Health Authority, and various community groups and organizations). These teams work to address specific crime, disorder, and quality-of-life issues that have been identified in the neighbourhoods. Through this collaboration, the teams are able to draw from a wide range of city and community resources and thus respond to the needs of neighbourhoods.

Police services and officers across the country involve themselves in a wide range of charitable events that not only raise money for important causes but also provide opportunities for officers to contribute to the community and to encounter community residents in a non–law enforcement capacity. One high-profile initiative is Cops for Cancer, which involves a wide range of fundraising activities—for example, officers have their heads shaved for donations (see photo on page 153). Another is the Law Enforcement Torch Run for the Special Olympics.

Crime Prevention Programs

Crime prevention programs are generally aimed at reducing crime, generating community involvement in addressing general and specific crime problems, and heightening citizens' perceptions of safety. There are three main approaches to crime prevention: primary, secondary, and tertiary. Police departments are most extensively involved in primary crime prevention programs, although they do participate in secondary and (to a lesser extent) tertiary crime prevention as well.

Primary Crime Prevention Programs These programs identify opportunities for criminal offences and alter those conditions to reduce the likelihood that a crime will be committed. They are most often aimed at property offences. In addition to problem-oriented policing (discussed earlier), primary crime prevention programs operated by police services include the following:

- *Operation Identification.* Citizens and businesses mark their property with ID numbers to make the disposal of stolen goods more difficult and to assist in the recovery and return of items by the police.
- *Neighbourhood Watch.* This program involves turning the "eyes and ears" of citizens in the community toward the crime prevention effort. Another goal here is to forge a sense of community among neighbourhood residents.
- *Citizen patrols.* Citizen patrols (by car and on foot) work in collaboration with police services and campus-based student patrols.
- *Crime Prevention Through Environmental Design (CPTED).* This program focuses on altering elements in the physical environment to discourage offenders. For example, access to areas is controlled to make victimization more difficult; architectural designs are employed that allow for surveillance; and environments are designed to foster a sense of community.
- *Media programs.* Media programs, such as the television show *America's Most Wanted,* are designed to educate the public about crime or to solicit the public's assistance in locating known criminals. Police services have developed a myriad of "hot tip" lines and televised reenactment programs (such as *Crime Stoppers*). These offer cash rewards for information leading to the arrest and conviction of offenders.
- *Closed-Circuit Television (CCTV) monitors.* Sophisticated systems utilizing moving cameras and camera operators who maintain radio contact with police officers have been installed in many public areas in Europe, the United States, and (to a lesser extent) Canada. These systems are designed to

crime prevention programs
initiatives designed to prevent or reduce crime and the fear of crime

focus

BOX 4.2

Point/Counterpoint: The Potential and Dangers of Closed Circuit Television

Just after 11 p.m. one Sunday evening, a CCTV operator noticed that someone was lying in the street. Checking the general area, he observed two people who appeared to be attacking people indiscriminately. One of the assailants knocked down someone who had been standing at the bus stop. As he fell, the victim struck his head on the curb. A bus arrived and the assailants got on, but the CCTV operator had already alerted police officers. Arriving on the scene just as the bus was leaving, the officers managed to stop the vehicle and arrest the assailants. The victim later died of his injuries.

This description of the use of CCTVs to apprehend criminal offenders in the English city of Newcastle upon Tyne encapsulates many of the issues at the centre of the debate over CCTVs (Brown, 1995: 24).

Point

Research evidence from England and the United States suggests that CCTVs may be effective in reducing property crime:

- A British survey found that only 6 percent of respondents felt that their privacy was threatened; also, they felt safer when CCTVs were installed.
- There were few public complaints when video cameras were installed in high-crime areas in Sudbury, Ontario, Hull, Quebec, and Kelowna, British Columbia.

- CCTVs do not appear to reduce crimes against persons; however, evidence gathered via video cameras is of great value to police who investigate these types of crimes.
- The only people who should worry about the intrusiveness of CCTVs are those who are engaged in illegal activity.

Counterpoint

- CCTVs pose a threat to the privacy of community residents—a threat that far outweighs any potential impact they might have on specific types of crime.
- There is no conclusive evidence that CCTVs installed in Canadian municipalities have reduced crime.
- Instead of preventing crime, CCTVs may simply disperse it to other areas.
- Although there is some evidence that CCTVs reduce property crime, they appear to have little or no impact on crimes against persons.
- Several communities (including Owen Sound, Ontario) have rejected CCTVs, citing concerns with invasion of privacy.

Sources: B. Brown, *CCTV in Town Centres: Three Case Studies,* London, UK: Police Research Group, Home Office (1995); M. Gill and A. Spriggs, *Assessing the Impact of CCTV,* Home Office Research Study, No. 292, London, UK: Research, Development and Statistics Directorate, Home Office (2005).

strengthen police response to incidents and to deter and prevent crime, particularly street mugging, car theft, and property damage. The tapes can also be used to identify perpetrators. CCTV holds considerable promise for crime prevention but also has potential perils (see Box 4.2). This strategy brings to the surface the tension between individual rights and efforts to ensure the safety and security of the community as a whole.

Secondary Crime Prevention Programs Secondary prevention initiatives focus on areas that produce crime and other problems; are often based on crime area analysis, including the targeting of high-crime areas; and attempt early identification and intervention with potential offenders. Examples of secondary prevention programs include:

- programs for youths, such as DARE to Stay Off Drugs and SADD (Students Against Drunk Driving);
- crime prevention initiatives in schools, some using conflict resolution and mediation;
- diversion programs for adult and young offenders; *and*
- wilderness experience programs for at-risk youth.

Tertiary Crime Prevention Programs Tertiary crime prevention initiatives use the criminal justice system to:

- direct responses toward actual youth and adult offenders;
- intervene to reduce the likelihood of offenders reoffending; *and*
- deter/incapacitate/rehabilitate offenders.

Such programs are typically operated by the justice system and include the intermediate sanctions (see Chapter 6) and correctional programs (see Chapter 7).

External Elements

A key feature of community policing is networking with community groups and organizations, with the private sector, and with various government agencies at the municipal, provincial, and federal levels. A review of police service websites reveals a myriad of partnerships designed to strengthen police–community relations, enhance efforts to address problems of crime and social disorder, and improve the overall quality of life in communities. Community resident associations and merchants' associations work with police services to address issues, sponsor forums, and publicize proactive initiatives. These may be supported by government grants, the private sector, and charitable organizations.

For example, the Ottawa Police Service (http://www.ottawapolice.ca) is involved in a variety of community partnerships throughout the city:

- The Somali Youth Basketball League (SYBL) is a volunteer, not-for-profit basketball league that provides a safe environment for Somali youth. It develops life and leadership skills among the participants and also provides positive role models.
- The Police Youth Centre provides counselling services, leadership programs, and a wide range of sports and recreation programs for children and youth aged six to nineteen. As well, the centre works with parents and parents' groups to address the challenges faced by youth in the community.

> **What do you think?**
>
> Which of the arguments in Box 4.2 do you find more persuasive? Explain your answer.
>
> Would you support the installation of CCTVs in your neighbourhood? In other neighbourhoods? In or near your workplace? If yes, under what conditions? If not, why not?

- In the Street Ambassador Program, volunteers—who wear shirts and name tags identifying themselves as ambassadors—assist merchants with problems such as panhandling and provide information and assistance to visitors to the area.

Volunteers provide a crucial link between police services and the community and are a vital component of many of the partnerships between the two. Volunteers assist with charitable events, staff police storefronts, and serve as victim/witness assistance workers.

THE FRAMEWORK FOR COMMUNITY POLICING IN ONTARIO

The commitment to community-based policing in Ontario can be traced back over thirty years to the Task Force on Policing in Ontario (1974). That task force found that centralized command and control, a top-down model of management, and reliance on radios and police patrol vehicles all tended to widen the gap between the police and the community. The questions the task force's report raised about the effectiveness of conventional policing strategies and tactics provided the impetus for the initial moves toward community policing in the early 1980s.

Ontario's police forces operate under one key piece of legislation, the Police Services Act, and an accompanying regulation, the Adequacy and Effectiveness of Police Services Regulation (commonly referred to as the "Adequacy Standards"). The Police Services Act contains provisions designed to ensure:

(a) that community policing is integrated into all segments of policing; *and*

(b) that government can monitor the performance of police services in providing adequate and effective policing, and, if necessary, intervene to correct any perceived deficiencies in service delivery.

The Adequacy Standards establish uniform service standards for police services across the province in a number of areas, including crime prevention, law enforcement, victim assistance, and emergency response services.

Four key components of the OPP delivery of policing services reflect the philosophy of community policing (see Box 4.3).

THE EFFECTIVENESS OF POLICE STRATEGIES

clearance rate
the proportion of the actual incidents known to the police that result in the identification of a suspect

Traditionally, the effectiveness of the police has been measured by the **clearance rate,** defined as the proportion of the actual incidents known to the police that result in the identification of a suspect, whether or not that suspect is ultimately charged and convicted. The shift toward community policing and the development of strategic plans, operational objectives, and performance measures, however, has resulted in an expansion in the ways in which the effectiveness of the police is measured. As well, many police services have developed a number of key performance indicators in an effort to assess many of the other types of activities in which the police are involved.

focus

BOX 4.3

Community Policing Capacities of the OPP

The OPP has undertaken a wide range of initiatives as part of its commitment to community policing, a number of which can be considered best practices to be emulated by other police services across the country. These include:

Community Policing Development Centre (CPDC). The CPDC is designed to support the development of police–community partnerships, enhance the problem-solving capacities of patrol officers, and identify and encourage the use of best practices in crime prevention and crime response.

CPDC News. This newsletter provides updates on the activities of the CPDC, examines specific community policing strategies, and highlights the achievements of OPP officers and detachments around the province.

Community Policing Network (CPNET). CPNET is a computer-based repository for information on past and current community-based policing projects. OPP officers enter a specific community problem, and CPNET provides suggestions for possible solutions based on previous problem-solving situations.

Policing for Results Survey. This survey provides community residents with the opportunity to express their views on policing services and to identify community concerns, needs, and priorities. The survey is carried out via telephone interviews with a sample of community residents; the responses are then entered into a computer database that is used by the police and municipal authorities to improve the delivery of policing services and enhance police–community partnerships.

Source: OPP, online at http://www.opp.ca/cpdc/english.
© Queen's Printer for Ontario. Reproduced with permission.

There are three main reasons why clearance rates should not be used as the only indicator of police effectiveness:

1. *Police officers do not spend most of their time pursuing criminals.* Most of their time is spent on non-enforcement activities such as maintaining order, providing services, and preventing crime.

2. *Not all police services and police officers work in the same types of communities.* Police officers carry out their tasks in a variety of task environments. Some officers are assigned to remote and rural areas, while others work in suburban and urban areas. Clearance rates do not reflect variations in the levels and types of crime in these different areas. As Pare and Ouimet (2004: 116) note: "Police departments with relatively difficult to solve crime mixes are strongly disadvantaged in comparison to police departments with easier to solve crime mixes."

3. *Police officers do not all engage in the same type of police work.* Some officers work out of community police stations; others work in specialty units such as gang squads and auto theft task forces. Using only one measure of police effectiveness does not allow for these differences to be taken into account.

Another measure that is often used to assess the effectiveness of the police—the *crime rate*—can lead to problems of interpretation. For example, does an increase in official crime rates mean the police are ineffective? Or does it mean they are catching more criminals? Another problem with using official crime rates to assess police effectiveness is that the focus is on "crime fighting" to the exclusion of other measures of police performance.

A number of other measures of police performance have been developed in recent years. These include:

- levels of community and victim satisfaction with the police and feelings of safety, as measured by surveys (see Chapter 2);
- the success of a police service in achieving its stated goals and objectives and fulfilling its mission statement;
- the success of the police in achieving specific performance objectives, such as a reduction in response times for 9-1-1 calls or effective target hardening and problem solving with respect to specific types of crime in identified problem areas in the community;
- the extent to which the police are involved in developing innovative programs to address issues related to community diversity (for example, issues relating to the gay and lesbian community, to visible minorities, and to Aboriginal people);
- the degree to which the police are involved in interagency partnerships with social service agencies, non-governmental organizations, and community groups; *and*
- the nature and extent of involvement of community volunteers in various police programs and services.

In contrast to the United States, there have been in Canada few evaluations of specific policing initiatives and activities. It is generally unknown, for example, whether municipal police departments (and the RCMP where it provides municipal and provincial policing services under contract) deploy their patrol officers in a manner that maximizes available resources, best meets the demands of the community, and effectively responds to specific patterns of criminal activity. As well, there is little research available on the work of special investigative units and on the most effective models for deploying investigative resources. Furthermore, most of the focus has been on street-level crime, with little attention paid to the effectiveness of enforcement strategies in the areas of financial crimes, organized crime, and drug enforcement.

The Canadian criminologist Thomas Gabor (2003: 6) has noted that there have been few Canadian evaluations of law enforcement efforts to target organized crime and white-collar criminals. And, he notes, the performance measures that are typically used—such as the number of persons convicted and the amount of assets seized—are insufficient. Gabor argues that it is important to gather information on other factors, including enforcement costs, the number of investigations that lead to convictions, and the degree to which specific policing initiatives are effective in disrupting organized crime. Needless to say, much more research remains to be conducted on Canadian policing before more definitive statements can be made about the effectiveness of the police.

A young cancer patient shaves the head of a police constable in Peterborough, Ontario, as part of the Cops for Cancer fundraising campaign.

Does Community Policing Work?

Keeping in mind that the term community policing is used to describe a wide range of policing strategies, research studies (conducted primarily in the United States and England) provide strong evidence that community policing approaches generally are beneficial both for the police organization and for the community, and that they improve police–community relationships. More specifically, studies have found that community policing can:

- create positive public perceptions of, and confidence in, the police;
- generate stronger feelings of community attachment among residents;
- improve the morale and job satisfaction of police officers;
- create positive police attitudes toward the community; *and*
- have some impact on the rates of crime (Adams, Rohe, and Arcury, 2002; Schneider, Rowell, and Bezdikian, 2003).

Do Police Interventions Prevent Crime?

It is difficult to determine whether crime prevention programs actually work. There have been very few controlled evaluations of crime prevention initiatives— a surprising fact, given the widespread use of crime prevention programs across the country. Traditionally, police effectiveness has been measured in terms of whether the crime rate has gone down for the targeted offence category. However, within a community-policing framework, other measures would be as appropriate, including whether the program increased positive police–community contact and whether citizens felt safer in their neighbourhoods.

Crime displacement is another issue to be considered in determining whether a crime prevention program has been effective. Crime displacement—also

> **crime displacement**
> the movement of criminal activity from one area to another that results from the implementation of an effective crime prevention program

known as "crime spillover"—is the movement of criminals and their activity from one potential target to another. The implementation of a crime prevention program in one neighbourhood may cause criminals to move to an area that does not have the program. Instead of reducing crime, the program has just moved it. One way to reduce crime displacement is to implement crime prevention programs on a communitywide basis rather than only in specific areas. Also, it may be necessary to target a wide range of criminal activity instead of focusing only on specific types of crime. Box 4.4 summarizes key research findings with regard to selected police-sponsored crime prevention programs. Note that this summary is based primarily on research conducted in the United States and England.

This brief review of the effectiveness of crime prevention programs suggests that these initiatives have been much less successful than is commonly assumed by both the police and the public. Remember, however, that the research findings are from studies that examine programs operating in a wide variety of policing environments, and that each program—as well as the community in which it operates—has its own unique attributes.

There are several possible reasons why crime prevention strategies have not been as successful as the police and public might have hoped:

- *Citizen participation.* Research findings suggest that only about 10 percent of households participate in crime prevention programs. Ironically, citizens who do participate tend to live in neighbourhoods with few problems; in other words, they are among those *least* at risk of victimization. For the full potential of crime prevention initiatives to be realized, it is essential that there be participation by residents in those neighbourhoods affected by high rates of crime and trouble.

- *Program continuity and stability.* Historically, it has been difficult for the police to sustain community interest and participation in crime prevention programs such as Neighbourhood Watch; programs often become dormant as a consequence.

- *Visible minority and Aboriginal communities.* Little attention has been paid to developing crime prevention programs that address the needs of specific ethnic and cultural groups. This is a particularly disturbing fact, given the multicultural character of Canadian society and the problems that often surround relations between police and visible minorities. These problems are exacerbated by the fact that visible minorities are dramatically underrepresented in police services in Canada (see Chapter 3).

- *Program transferability.* Most crime prevention programs have been designed for and implemented in urban or suburban settings, and some question exists as to whether these strategies can be effective in rural and remote areas of the country, particularly in Aboriginal communities. Most often, crime prevention programs have been developed first, and then applied in a wide range of settings, from urban areas to the remote North. The obvious differences among these environments suggest that programs should be tailored to suit the needs and requirements of individual communities.

focus

BOX 4.4

The Effectiveness of Selected Organizational and Tactical Strategies

Intelligence-Led Policing

- There is an absence of published research studies on the effectiveness of this strategy, although it has considerable potential to be a component of crime reduction efforts.

Zero Tolerance Policing

- American research suggests that an order maintenance approach, centred on preventing disorder in the community, can result in a reduction in the levels of fear among citizens and in a corresponding increase in their quality of life.

- Police crackdowns on open street drug markets have generally been unsuccessful over the long term.

- There is some evidence from police initiatives in Vancouver that police crackdowns can improve the quality of life and citizens' perceptions of safety in neighbourhoods.

Neighbourhood Watch

- There is no evidence that this program prevents crime.

- Many residents, particularly in high-crime areas, are reluctant to become involved with their neighbours.

- It may increase the fear of crime among community residents.

Problem-Oriented Policing (POP)

- There have been no formal evaluations of POP in Canada.

- American research has found that some POP initiatives have no apparent impact but others reduce certain types of crime and improve police–community relationships.

Preventing Crime in High-Crime Areas

- Property marking, and changes to building design and pedestrian routes, have been found to reduce residential break and enters, robberies, and assaults in some jurisdictions.

- Closing or selling properties that were venues for drug offences has reduced calls for service in some jurisdictions.

- There is some evidence that CCTV and employing multiple clerks in twenty-four-hour stores may reduce robberies and property theft.

Sources: Bennett, 2000; C.T. Griffiths, Y. Dandurand, V. Chin, and J. Chan, 2004; C.T. Griffiths, B. Whitelaw, and R. Parent, 1999; Scott, 2003).

Research findings on the effectiveness of selected organizational and tactical police strategies in preventing and responding to crime and social disorder are presented in Box 4.4. This includes a number of the more common crime prevention programs.

In addition to the previously noted issues that may affect specific crime prevention programs, there are a number of factors that may compromise the effectiveness of crime reduction strategies. These include poor planning, a failure to focus on the long term, and not developing sustainable partnerships with the community (Read and Tilley, 2000). The trend in recent years has been for police services to utilize best practice strategies to improve the effectiveness of their crime prevention and crime reduction strategies.

POLICE DISCRETION AND DECISION MAKING

Police officers make a variety of decisions in carrying out their tasks in the community, the majority of which are routine in nature. Some decisions, however, are high in risk as well as high in consequences. Notable among these are the decision to use deadly force (see Chapter 3) and the decision to engage in a hazardous pursuit with a suspect (discussed below).

The Exercise of Discretion

discretion

the ability of a police officer to choose among several possible courses of action in carrying out a mandated task

A patrol officer who is faced with the need to make a decision and who chooses between different options is exercising **discretion**. Discretion is an essential component of policing because no set of laws or regulations can prescribe what a police officer must do in each and every circumstance. Because it is impossible for officers to enforce all laws all of the time, they practise *selective* or *situational enforcement*. As the seriousness of the incident increases, however, the amount of discretion an officer can exercise decreases.

For police personnel, the authority to use discretion is set out in statutes such as the Criminal Code. For example, if an individual is found committing an offence, he or she *may be* arrested; arrest, then, is not a strict obligation on the part of the police. The decisions a police officer makes may ultimately be scrutinized by the courts or the public, particularly when it is alleged that the officer abused his or her discretionary powers and in doing so violated a person's legal rights.

In cases of domestic violence, some jurisdictions have "mandatory charge" or "zero tolerance" policies that curtail police discretion. These policies require police officers to arrest the suspect when it appears that an assault has occurred, even if the alleged victim does not want an arrest to be made. Until the early 1980s, a woman who had been assaulted by an intimate partner (usually with no witnesses) had to initiate the prosecution herself by laying an information with a justice of the peace. This requirement made the victim vulnerable to intimidation by the offender, and it was not uncommon for victim-complainants to refuse to cooperate with the police and the Crown attorney (in some cases, women were jailed for refusing to testify at trials). Even under today's zero tolerance policies, it is difficult for the Crown to proceed without the victim's testimony. Ironically, zero tolerance policies have resulted in cases where the victims of domestic violence have been charged with obstruction of justice for failing to cooperate with prosecutors. And there is concern that for a variety of reasons, female victims of domestic violence will not call the police if there is a zero tolerance policy in place; this may place them at a higher risk of revictimization.

Patrol officers bring to their work a set of cognitive lenses through which they make determinations about the people and events they encounter. They use a conceptual shorthand consisting of **typifications** and **recipes for action** to tailor their decision making to the particular area and population being policed. A visual cue such as a poorly dressed individual in an upscale neighbourhood would attract the attention of officers on patrol, as would a behaviour or activity considered out of place in a particular area. The risk is that racial or economic profiling may result. Officers who are assigned to a fixed geographical area for an extended period of time develop an intimate knowledge of its persons and places as well as extensive contacts with community groups, agencies, and organizations that are facilitative of police–community partnerships and the identification of and response to problems.

Factors Influencing Patrol Officers' Decision Making

Field research studies indicate that a number of factors may influence the decision making of police officers in encounter situations. These include, but are certainly not limited to, the following:

- *The setting in which the encounter occurs.* Across Canada, there is a wide variety of policing or task environments, ranging from small Inuit villages in the Arctic to large urban centres. All of these present different challenges. There may also be different environments within a particular urban area. Toronto, for example, has numerous distinct districts, including the old Chinatown, which is centred on the corner of Dundas Street and Spadina Avenue; Rosedale, a wealthy enclave just north of the central business district; and Jane–Finch in the northwest, which is home to mixed ethnic and cultural groups living in high-density public housing, has a young and transient population, and suffers from a high crime rate. Policing environments in Vancouver include Chinatown; the upscale Point Grey district near the University of British Columbia; the skid row area (more commonly referred to as the Downtown Eastside), which has a large population of drug addicts and sex trade workers; and the West End, a neighbourhood of high-density apartment blocks near English Bay.

    Police officers posted to small detachments in remote areas of the country must cope with isolation and cultural and language barriers. Policing in rural and remote communities is "high visibility–high consequence" policing: the activities of the patrol officers (including how they exercise discretion) are highly visible to the community, and the consequences of their decisions for the victim, the offender, and the community may be more pronounced than in larger communities, where there is a greater degree of anonymity and where the police enjoy more resources.

- *The policies of the police department.* In carrying out their tasks, patrol officers are influenced by the priorities of the department, the philosophy and management style of the chief and senior police administrators, and the resources available. For example, the management style of the police department's leaders will affect the development and implementation of policies and procedures within the department, how the competing demands on the agency are prioritized, and the extent to which patrol officers are encouraged to exercise discretion and seek solutions to problems they encounter.

typifications
constructs based on a patrol officer's experience that denote what is typical about people and events routinely encountered

recipes for action
the actions typically taken by patrol officers in various kinds of encounter situations

- *The policing style of officers.* The policing style of individual patrol officers may influence how they exercise discretion in encounter situations. Research studies have shown that the age, gender, length of service, ethnicity, and level of education of patrol officers can influence the decisions they make in any given situation. For example, as a group, female patrol officers are less likely than male officers to provoke violence in encounter situations; make fewer arrests than their male counterparts; are less likely to be involved in serious misconduct; and may be more effective in their interactions with the general public than male officers (LeBeuf, 1996; Martin, 1997).

- *Complainant preferences and suspect characteristics.* A consistent finding in the studies of police decision making is that the wishes of the complainant have a significant impact on the decision making of police officers in encounter situations. Similarly, the characteristics of the suspect may influence police decision making. Suspects who are non-white, male, and of lower socioeconomic status are more likely to be checked by patrol officers through CPIC (Canadian Police Information Centre); they are also more likely to be searched by officers. In addition, suspects who are uncooperative or disrespectful are more likely to be charged or arrested than those who behave in a deferential and civil manner. A 2000 study conducted in twenty-six communities across Canada by the RCMP, the Canadian Association of Chiefs of Police, and the Canadian Centre for Substance Abuse found that 53 percent of people arrested for criminal offences were under the influence of alcohol or drugs at the time the crimes were committed (Arnold, 2001). Needless to say, being impaired significantly increases the likelihood of being charged or arrested.

- *Seriousness of the alleged offence.* As would be expected, the seriousness of the alleged offence is strongly related to the action taken by police officers in encounter situations. Those suspects alleged to have committed serious crimes are more likely to be arrested than those alleged to have committed minor offences. As the seriousness of the alleged crime increases, the amount of discretion available to officers is diminished, as is the influence of complainant preference and other extralegal factors such as the suspect's ethnicity and demeanour. The selective enforcement of the law is illustrated by the wide variations in charge rates for possession of marijuana that exist among provinces and among police services.

High-Risk Decisions in the Community: Hazardous Pursuits

As Peace Officers, when you envision yourself in that position where you have to make a decision one day to perhaps take a human life, you don't often think about it being through a vehicle, but let's face it, you're driving around a big bullet, and it can kill. I never figured that I'd ever be involved in a situation as a police officer, contributing to someone's death in a chase. The hardest thing in this whole process was looking at his parents and thinking that they're still dealing with this everyday. To take a human life over a $40,000 vehicle? It's wrong for him to be there, it's wrong for him to be in the stolen vehicle, it was wrong for him not to stop when he was initially instructed to stop, but it cost him his life and it wasn't worth it.

We lost in the situation, everyone came out as losers. The members who were involved are all scarred for life, the family certainly has a significant loss in their life, the vehicle we were trying to save—that was a write-off, so what did we gain from it—nothing. (Quote from an RCMP officer involved in the pursuit of a stolen vehicle that resulted in the death of the suspect; cited in RCMP Public Complaints Commission, 1999.)

Many of the decisions police officers make, such as whether to issue a traffic citation, have no significant consequences (except for the fact the driver of the vehicle must pay a fine and/or receive demerit points on his or her driving record). Others, though, have significant and potentially fatal consequences for police officers, criminal suspects, and the members of the community. A prominent example is the decision to pursue a fleeing vehicle that refuses to stop when initially detected by a police scout car.

Hazardous pursuits are almost mandatory in police action movies and often result in spectacular scenes of stunt driving, colliding vehicles, and feats of daring. But the outcomes of these pursuits in real life are often much more sobering. These pursuits most often develop as a consequence of impaired driving, a stolen vehicle, or dangerous driving. There are thousands of hazardous pursuits in Canada every year, a high percentage of which involve youth driving stolen cars, who are either joyriding or intend to sell the car on the underground market. Approximately one-third of pursuits end in a collision in which the suspect(s), innocent bystanders, or other drivers (including the police) are injured or killed. A number of police services, including the Calgary Police Service, employ helicopters for pursuits; this reduces the risk to both officers and the general public.

The following two news releases from the Fredericton Police Force in 2005 (http://www.frederictonpolice.com; reprinted by permission of the Fredericton Police Force) refer to the all-too-common scenario of a hazardous pursuit of a stolen car being driven by teenagers. Fortunately, in both cases, there was only damage to vehicles.

On January 17, 2005 at approximately 2300 hrs officers responded to a theft of vehicle on Governor Lane. The stolen vehicle drove by police upon their arrival. Police pursued along Prospect St west. The pursuit ended in the area of Monteith Dr. with a police vehicle sustaining damage. Two 17-year-old Fredericton youths were arrested. Both young offenders will be appearing in court in the morning of January 18, 2005 on various charges.

On Thursday April 7, 2005 at approximately 11:35 PM, the Fredericton Police Force received a complaint of a stolen vehicle from the south side of the City. At approximately 1:45 AM, Friday April 8, 2005, the vehicle was located in the City and a short hazardous pursuit ensued, resulting in extensive damage to the stolen vehicle. No one was injured during the pursuit. A short time later, a 16 year old Fredericton young person was arrested and charged with possession of stolen property. He was released on a Promise to Appear for May 26, 2005 at 9:00 AM in Fredericton Youth Court.

These are the types of cases in which the question is asked whether it is worth the risk to pursue a stolen vehicle, which in most instances will result in a charge of possession of stolen property. The potential risks associated with hazardous pursuits are graphically illustrated in the following case, which resulted in fatalities:

> On a quiet Sunday night in Saskatoon, a Mercury Sable carrying Trevor Nordholm (a high school principal) and his wife slowly accelerated as the light at 2nd Avenue and 22nd Street West turned green. Seconds later, their car was broadsided by a stolen 1985 Dodge Aspen that was being chased by police. The impact crushed the front end of the stolen car, throwing the passenger in the front seat into the windshield; a partly smoked cigarette was later found wedged into broken glass. Nordholm, whose side of the car had taken the brunt of the impact, was pronounced dead on arrival at the hospital. His wife died a few hours later (Appleby, Ha, and Matas, 1999: A3).

There has been mounting pressure on provincial governments and police agencies in Canada to adopt guidelines governing hazardous pursuits. Critics of more restrictive pursuit policies argue that they provide criminals with a licence to violate the law without fear of apprehension. Proponents contend that the risks to life and property do not justify high-speed chases except in the most serious cases.

In recent years, more restrictive pursuit policies have been adopted in many provinces. In Ontario, for example, regulations of the Police Services Act require the police to meet a three-part standard before initiating or continuing a pursuit: (1) the officer must believe that a criminal offence has been or will be committed; (2) there are no alternatives to pursuit; and (3) the need to apprehend the suspect(s) is greater than the risk to public safety. During the chase, the risks to public safety must be constantly reassessed. Strategies for reducing the risks associated with hazardous pursuits include using helicopters to direct pursuits by patrol cars and placing spike belts and "stop sticks" in the path of the speeding vehicle. The absence of national statistics on hazardous pursuits is hindering the effort to develop a clearer picture of their nature and extent, the types of situations that result in them, and their outcomes.

THE POLICE AND VISIBLE/CULTURAL MINORITIES

It was noted in Chapter 1 that a key challenge facing the criminal justice system is the cultural diversity of Canadian society. Police services are under pressure to increase the numbers of visible and cultural minorities in their ranks, to provide effective policing services to visible minority communities, and to ensure that police officers are bias-free when exercising discretion and decision making.

In recent years, two flashpoints between the police and visible minority communities have been **bias-free policing** and racial profiling. Bias-free policing requires police officers to make decisions "based on reasonable suspicion or probable grounds rather than stereotypes about race, religion, ethnicity, gender or other prohibited grounds" (Canadian Association of Chiefs of Police, 2004: 7). Bias-free policing requires the equitable treatment of all people of diversity—gays and lesbians, for example. In contrast, racial profiling is most commonly associated with police encounters with visible minorities. Both concerns have been heightened in the years since the 9/11 terrorist attacks on the

bias-free policing
the requirement that police officers make decisions on reasonable suspicion and probable grounds rather than on the basis of stereotypes about race, religion, ethnicity, gender, or other prohibited grounds

United States; Muslims and other people who seem to be of Middle Eastern origin have been singled out for discriminatory treatment by security personnel at airports, by customs officers, and by the police (Turenne, 2005).

A number of definitions have been developed for **"racial profiling,"** and the specific definition that is used significantly affects the focus of research studies as well as the interpretation of research findings and the determination of whether a police department and/or its officers are engaged in racial profiling. In the United States, Fridell and colleagues (2001: 5) have written that racial profiling occurs "when law enforcement inappropriately considers race or ethnicity in deciding with whom and how to intervene in an enforcement capacity." This definition leaves open the possibility that there are times when police officers may "appropriately" consider race or ethnicity when making decisions regarding when, with whom, and how to intervene in carrying out their policing duties. The Ontario Court of Appeal, in *R. v. Brown* [2003] O.J. No. 1251 (discussed below in Box 4.5), defined racial profiling as involving "the targeting of individual members of a particular racial group, on the basis of the supposed criminal propensity of the entire group." These two examples illustrate the differences in how racial profiling is defined.

> **racial profiling**
> the targeting by police of individual members of a particular racial group on the basis of the supposed criminal propensity of the entire group

The debate over racial profiling in Canada has been hindered, and complicated, by vague definitions as well as by a lack of solid research into police decision making and police–minority relations (see Gabor, 2004; Melchers, 2003; Wortley and Tanner, 2003). This makes it difficult to determine the extent to which the police discriminate against visible minorities and whether it is the police as a whole, or individual officers, that are engaged in these discriminatory practices. There are, however, documented cases of individual police officers engaging in behaviour that has been found to be racist and discriminatory. Two of these cases are presented in Box 4.5.

A recent report by the Ontario Human Rights Commission (2003) found that cultural and visible minorities widely perceived that the police engage in racial profiling. A number of police services have acknowledged that racial profiling does occur and have introduced measures to address it, including upgrading training for officers, identifying those officers who are at risk of engaging in racial profiling, and improving interaction with the community. In Montreal, for example, a policy requires all police precincts to develop initiatives for improving contact with the community, particularly minority youth.

The Police and Aboriginal Peoples

A key feature of Canadian criminal justice is the overrepresentation of Aboriginal people at all stages of the justice system. The high rates of Aboriginal arrests in many regions of the country have raised the question as to whether police officers discriminate against Aboriginal people. Although there is no evidence that Aboriginal people are systemically discriminated against by the police, there have been serious incidents in a number of jurisdictions that have subsequently been found to be the result of discriminatory actions on the part of police officers.

In particular, a number of high-profile incidents have occurred in Saskatoon, Saskatchewan, where observers eventually coined the term "starlight tour" to describe the police practice of picking up impaired Aboriginal people in the

focus

BOX 4.5

Two Cases of Racial Profiling and Discrimination

The Dee Brown Case
(Ontario Court of Appeal)

In the early morning hours of November 1, 1999, a Metro Toronto police officer stopped a vehicle on the Don Valley Parkway that was exceeding the speed limit and that, according to the officer, had on two occasions wandered outside the lines of the lane in which it was travelling. The driver failed a roadside screening device test administered by the officer and was taken to a police station, where a breath test was administered. The driver failed this test, registering at 140 mg per 100 ml of blood—considerably higher than the legal limit of 80 mg of alcohol per 100 ml of blood.

The driver of the vehicle was Dee Brown, an African American and former member of the Toronto Raptors basketball team. At trial, Brown's defence lawyer argued that his client had been stopped as a result of racial profiling by the police officer: Brown was a young black man driving an expensive vehicle. The defence lawyer argued that because of this, the arrest of Brown was contrary to Section 9 of the Charter of Rights and Freedoms and the results of the breath tests should not be admitted into evidence. Despite these arguments, the trial judge found Brown guilty and fined him $2,000. In his remarks at sentencing, the judge called Brown's use of racial profiling by the police as a defence "distasteful" and suggested that he apologize to the police officer who had arrested him.

Brown appealed this decision to the Ontario Superior Court of Justice, which overturned the conviction and ordered a new trial, concluding that there was sufficient evidence to support the claim of racial profiling and that the trial judge had displayed a reasonable apprehension of bias in not considering the claim. A new trial was ordered. The Crown appealed this decision to the Ontario Court of Appeal. In 2003, the Court of Appeal released its decision, which reaffirmed the decision of the Superior Court that a new trial should be ordered. In its decision, the Court of Appeal set out a definition of racial profiling and a process for determining whether a police officer had engaged in this practice. The decision of the Ontario Court of Appeal was the first instance in which an appellate court stated its belief in the existence of racial profiling by police. Following this decision, the Crown decided to not retry Brown.

The Constable and the Boxer:
A Case of Police Discrimination
(Nova Scotia Human Rights Commission)

On April 12, 1998, Kirk Johnson, a Black Canadian former Olympic boxer, was stopped in his vehicle by a Halifax Regional Police officer. The car, which had Texas licence plates, was being driven by Johnson's cousin and had been passed by the constable prior to being stopped. After examining the registration and licence information provided by Johnson, the constable ordered the vehicle impounded and issued a ticket to Johnson for operating a vehicle without valid registration and with no proof of insurance.

Even though a supervising officer who had arrived on the scene determined that the ticket had been issued in error, Johnson's car was towed, leaving Johnson and his cousin on the roadside. The car was later returned, and the next day the ticket was invalidated.

Johnson filed a complaint with the Nova Scotia Human Rights Commission in December 1998, alleging that the police discriminated against him when they ticketed him and impounded his car. His lawyer argued during the hearing that the actions of the constable and his supervisor were the result of a "'deeply ingrained' culture of racism in the Halifax police department" (Auld, 2003: 3).

Based on an examination of the facts of the case, the hearing officer found that the conduct of the constable was sufficient to establish that discrimination had occurred: "One could infer that Constable Sanford, once having discerned their race, acted on an inadmissible stereotype of black criminality in deciding to stop the vehicle." Furthermore, the attending constable "did not display the reasonable tolerance and tact required of someone in his position and I infer that race was a major factor in this professional failing." He further concluded that the constable who stopped Johnson and his cousin had discriminated against Johnson in the process of ticketing and towing his vehicle (Nova Scotia Human Rights Commission, 2003: 1).

The hearing officer ruled that Johnson was entitled to damages in the amount of $10,000. He also ordered the Halifax Regional Police Service to hire two consultants to conduct an assessment to determine whether there was a need for race and diversity training in the department; to improve its antiracism policies; and, within sixty days of the judgment, to develop the capacity to begin gathering racial data on traffic stops. In 2004, the chief of the Halifax Regional Police apologized to Johnson for the actions of the police constable and admitted that he was the victim of racial discrimination (the complete transcript of the decision is available online at http://www.gov.ns.ca/humanrights).

Sources: T.K. Pain, "Evidence Ruled Capable of Supporting Racial Profiling Allegation" (2003). Retrieved from http://www.torontocriminaldefence.com/articles. Reprinted by permission of Tusher K. Pain, Criminal Defence Lawyer; A. Auld, "N.S. Human Rights Agency Rules Police Discriminated Against Boxer," Canadian Press (2003). Retrieved from http://www.canada.com; Nova Scotia Human Rights Commission, "Johnson Victim of Racial Discrimination, Board Finds" (2003). Retrieved from http://www.gov.ns.ca/news/details.asp?id=20031223001

city, transporting them to outlying areas, and dumping them. In at least one case, these actions were directly responsible for the unlawful confinement of an Aboriginal person (see Box 4.6).

Police Initiatives to Improve Police–Aboriginal Relations

A number of police services have developed specific initiatives to improve the delivery of policing services to Aboriginal communities and to reduce the conflict that has often surrounded contacts between the police and Aboriginal peoples.

The RCMP has for many years operated provincewide advisory groups; the national Aboriginal Advisory Group; the Aboriginal Policing Program, which is designed to increase the number of Aboriginal Constables; and the First Nations Summer Student program, which provides Aboriginal youths with the opportunity to be employed as supernumerary special constables.

The Toronto Police Service has established a peacekeeping unit staffed by Aboriginal officers. These officers, who often do not wear uniforms, are part of an initiative to improve relations between the police and the city's estimated 65,000 Aboriginal people. The Vancouver Police Department operates a Native Liaison Unit that has received high marks from the urban Aboriginal community for facilitating positive relationships between the police and Aboriginal people in the city.

focus

BOX 4.6

"Starlight Tours" in Saskatchewan

In January 2000, two Saskatoon police officers picked up an Aboriginal man, Darrell Night, drove him to an industrial park on the outskirts of the city, and abandoned him in extreme winter weather. Luckily, Night was assisted by a security guard. He made his way back to the city, where he subsequently filed a complaint with the police. On the basis of his testimony, two city police officers were convicted at trial of unlawful confinement, fired from their positions, and sentenced to eight months in jail. The court rejected a request by the two officers that they be sentenced by an Aboriginal sentencing circle. In 2003, the Saskatchewan Court of Appeal upheld the convictions and the officers began serving their sentences.

The *Night* case raised suspicions that the Saskatoon police had transported and dumped other Aboriginal people outside the city, some of whom had frozen to death. Similar incidents included the discovery of the frozen bodies of Rodney Naistus on January 29, 2000, (a day after Night had been dumped) in the same industrial area, and of Lawrence Wegner, found frozen to death on February 3, 2000, in a field outside the city. Naistus was naked from the waist up; Wegner was not wearing shoes and had no jacket, even though it was winter. Subsequent investigations by the RCMP were not able to determine the circumstances surrounding the deaths of the two men.

The cases, however, focused attention on the death of an Aboriginal teenager, Neil Stonechild, whose frozen body had been found in a field on the outskirts of Saskatoon ten years earlier, on November 29, 1990. Stonechild was last seen alive by his friend Jason Roy; at the time, Stonechild was struggling with two Saskatoon police officers, who forced him into the back of a police cruiser. The temperature on the night Stonechild disappeared was −28°C. In February 2003, the province's justice minister announced a commission of inquiry into Stonechild's death. In its final report (Wright, 2004; available online at http://www.stonechildinquiry.ca/finalreport/default.shtml), the commissioner, the Hon. Mr. Justice D.H. Wright, found that Stonechild was in the custody of the police on the night he disappeared and that the injuries that were on his body were caused by handcuffs. However, there was no evidence presented that the two police constables actually dropped Stonechild off outside the city, and therefore, the circumstances surrounding his death remain undetermined. Wright, however, was severely critical of the initial investigation conducted by the Saskatoon police, and rejected the version of events offered by the police. Despite this, the absence of evidence precluded criminal charges being laid against the officers who were last seen with Stonechild. For an account of the Stonechild case, see *Starlight Tour: The Last, Lonely Night of Neil Stonechild* (Reber and Renaud, 2005).

These cases heightened tensions between Aboriginal people (particularly Aboriginal youths) and the police and seriously undermined earlier efforts by the Saskatoon police to improve police–Aboriginal relations.

Sources: CBC News Online, "Cold Case: The Lawrence Wegner Story," (2003); G. Smith, "The Death of Neil Stonechild," *The Globe and Mail* (2004), A1, A13; Hon. Mr. Justice D.H. Wright, (commissioner), *Commission of Inquiry into Matters Relating to the Death of Neil Stonechild*, Regina: Department of Justice, Province of Saskatchewan (2004). Available online at http://www.stonechildinquiry.ca

CASE INVESTIGATIONS

The increasing complexity of criminal cases and the globalization of crime have challenged police services to develop specialized capacities for case investigation, including the use of scientific techniques. The field of forensic investigation, in particular, has generated considerable public interest, due in large measure to the success of television series such as *CSI* and *Cold Case Squad*. Needless to say, most case investigations take longer than an hour to solve.

Advances in science and technology have resulted in the development of new strategies and tools, which the police are using in criminal investigations. Three of the more important are DNA typing, the Violent Crime Linkage Analysis System (ViCLAS), and cold case squads.

DNA Typing

DNA (deoxyribonucleic acid) is found in the cells of all living things. With the exception of identical twins, no two people have the same DNA type. The perpetrators of violent crimes can often be identified through **DNA typing**, which involves comparing biological samples taken from suspects with biological specimens left at crime scenes. Blood, saliva, semen, single hair roots, used facial tissue, cigarette butts, and licked envelope flaps can all be used for DNA typing. In one case, police in Vancouver used DNA evidence from the deployed air bag of a car to make an arrest in a fatal hit-and-run collision. Genetic links were found between the suspect (who had limped away from the crash scene) and facial skin cells found on the air bag. The suspect was subsequently convicted (Bailey, 2000: A1).

> **DNA typing**
> the use of genetic information in case investigations

Under Section 487.04 of the Criminal Code, the police may obtain DNA warrants—which allow the collection of DNA samples from crime suspects—only for certain specified offences, including (but not limited to) murder, manslaughter, assault, sexual assault, and sexual exploitation. DNA typing has helped secure convictions in hundreds of violent crimes, eliminate potential crime suspects, and establish the innocence of persons who were wrongly convicted and sentenced to prison.

To facilitate the collection and use of DNA evidence, the National DNA Data Bank was created in 1998 with the enactment of the DNA Identification Act. Managed by the RCMP, this facility contains the following:

- *Crime Scene Index (CSI)*. This index contains DNA evidence gathered at the scenes of unsolved crimes, making it possible to link those crimes by matching the DNA found at different crime scenes.

- *Convicted Offender Index (COI)*. This index contains DNA profiles derived from bodily substances taken from adult and young offenders convicted of specified offences, making it possible to compare the DNA profiles of individuals with the profiles in the crime scene index.

In recent years there have been efforts to expand the facilities for conducting DNA typing and to improve training for technicians (http://www.nddb-bndg.org).

"First, they do an on-line search."

Violent Crime Linkage Analysis System (ViCLAS)

ViCLAS is a Canadian-developed, computer-based system that contains information on all sexual or predatory-type homicides and on all sexual assaults or attempts of a predatory nature, including stranger-to-stranger assaults, date rapes, and child sex crimes. The database also contains information on missing persons where foul play is suspected and on all non-parental abductions or attempted abductions. ViCLAS is based on the premise that repeat offenders follow similar patterns and that persons committing homicides and sexual offences have identifiable and often predictable characteristics and motivations.

Case investigators enter information into ViCLAS by completing a lengthy questionnaire that is designed to allow subsequent analysis of established criminal patterns, traits, and pathologies. Each case that is entered into ViCLAS is examined for possible linkages with other crimes committed anywhere in Canada. If a potential linkage is identified, the originating police agency is notified of other similar cases and is able to follow up.

There are ViCLAS centres in every province except Prince Edward Island, which is served by Nova Scotia. Seven of these sites are maintained by the RCMP; the OPP, the Sûreté du Québec, and the Service de police de la Communauté urbaine de Montréal (Montreal Urban Community Police) maintain one each.

Cold Case Squads

The development of sophisticated techniques to gather and analyze DNA specimens and the creation of computer-based systems such as ViCLAS have allowed police investigators to revisit—and in many cases solve—crimes that

ViCLAS

a data system based on the analysis of victim information, suspect description, modi operandi, and forensic and behavioural data are years or even decades old. Many urban police services have created **cold case squads**, which focus exclusively on unsolved serious crimes and which coordinate the collection of new evidence that may help identify suspects. An increasing number of unsolved cases are being concluded with the arrest and conviction of the perpetrators:

> **cold case squads**
> specialized police units that focus on unsolved crimes

- In 1999, thirty years after two young mothers were raped and shot to death in their homes in Caledon, Ontario, a man was arrested for the crimes.
- Police in British Columbia arrested a man in 1999 and charged him with the murder of a twelve-year old girl thirty-four years earlier.
- In 2002 an Ontario man was arrested for the murder, twenty-five years earlier, of two Vancouver teenagers.

A cold case that was solved through the use of DNA is presented in Box 4.7.

focus   BOX 4.7

The Convenience Store Murder Case

On March 22, 1992, in Sydney, Nova Scotia, a convenience store clerk named Marie Lorraine Dupe was stabbed to death during a robbery. The assailant left the store with cash and cartons of cigarettes, but disappeared into a snowstorm, which prevented police dogs from picking up his trail. Left behind at the crime scene were several pieces of evidence, including a coffee cup and several cigarette butts. Owing to the state of forensic science at the time, however, these items were of little use to the investigation, although they were kept by police investigators. Despite a massive investigation by local police, which included interviewing 200 people, the crime remained unsolved for a decade.

In 2001, advances in forensic technology allowed the creation of a DNA profile from the items that had been discarded at the crime scene, and the profile was added to the National DNA Data Bank. Nine years after the crime date, hundreds of kilometres from the original crime scene, Ernest Gordon Strowbridge was convicted in an Ontario court of assault causing bodily harm. The presiding judge ordered Strowbridge to provide a biological sample for DNA analysis under the provisions of the DNA Identification Act. His DNA profile was entered into the National DNA Data Bank and resulted in a "hit" with a profile from the DNA on one of the cigarette butts found at the scene of the convenience store murder in Sydney. After an extensive investigation by the OPP and the Cape Breton Regional Police, Strowbridge confessed to murdering Marie Dupe. At trial, DNA evidence was introduced which indicated that the odds of his DNA matching that of another male were one in 1.5 trillion. Confronted with this and other evidence from the investigation, Strowbridge pleaded guilty to second-degree murder and was sentenced to life imprisonment.

Source: National DNA Data Bank of Canada, *Annual Report, 2003–2004* (2004), 32–33. Online at http://www.nddb-bndg.org. Reprinted by permission of National DNA Data Bank of Canada.

Summary

The discussion in this chapter has centred on the approaches and strategies used by police in their efforts to prevent and respond to crime and social disorder. At the outset of the chapter it was noted that community policing is the general framework within which police services in Canada carry out their tasks. The core elements of community policing were identified and discussed. This was followed by an examination of the effectiveness of the various strategies used by the police. A review of the research evidence indicated that these strategies are, perhaps, less successful than commonly assumed and that there are numerous challenges in measuring and improving police performance.

The discussion then turned to an examination of police discretion and decision making and the factors that can influence the decisions made by police officers. Hazardous pursuits were considered to illustrate the fact that although most police decisions are routine, police officers must sometimes make decisions that have profound consequences. The discussion also explored police relations with visible minorities and with Aboriginal people and the issues of bias-free policing and racial profiling. The chapter concluded with a brief examination of case investigation techniques, including DNA typing, the ViCLAS system, and cold case squads.

Key Points Review

1. The traditional (or professional) model of police work is based on random patrol, rapid response, and reactive investigation.
2. Community policing has evolved into a general approach to delivering policing services—one that includes both proactive and reactive policing strategies.
3. The core elements of community policing are *organizational*, *tactical*, and *external*.
4. The implementation of community policing has implications for how policing services are organized and delivered, as well as for the role of the community.
5. Ontario has developed a legislative framework within which police services can develop and implement the principles of community policing.
6. In moving toward a community policing model, many police services have adopted strategies and measures traditionally found in the private sector.
7. Problem-oriented policing—a strategy that puts the philosophy of community policing into practice—involves four steps represented by the acronym SARA (scanning, analysis, response, assessment).
8. Although it is difficult to determine whether crime prevention programs actually work, many programs are less successful than is widely believed.
9. A key feature of policing is the exercise of discretion.
10. Factors that may influence the decision making of the police include the setting in which the encounter occurs, the policies of the police service, the participants in the encounter, and the gravity of the alleged offence.

11. Bias-free policing and racial profiling are two issues that surround police relations with visible/cultural minorities, including Aboriginal peoples.
12. Hazardous pursuits are one example of high-risk decisions that police officers make in the community.
13. Case investigations are assisted by DNA typing, the Violent Crime Linkage Analysis System (ViCLAS), and cold case squads.

Key Term Questions

1. Describe the key features of the **traditional (or professional) model of policing.**
2. Identify and discuss the philosophy and key principles of **community policing.**
3. Identify and discuss the **core elements of community policing.**
4. What is **intelligence-led policing?**
5. Discuss the police strategies involved in **zero tolerance policing.**
6. Define and discuss the key features of **problem-oriented policing (POP).**
7. What is the RCMP **CAPRA model** and how is it related to problem-oriented policing?
8. Compare and contrast **crime attack strategies, community service approaches,** and **crime prevention programs.**
9. Discuss the use of **clearance rates** as a measure of police effectiveness.
10. What is **crime displacement** and what is its significance to efforts to determine the effectiveness of crime prevention programs?
11. What is **discretion,** and why is this concept important in any study of police decision making?
12. What are **typifications** and **recipes for action?** How do these concepts contribute to our understanding of the decision making of the police?
13. Define **bias-free policing** and **racial profiling,** and then discuss why these concepts are important in the study of police relations with visible/cultural minorities and Aboriginal people.
14. Define **DNA typing,** the Violent Crime Linkage Analysis System (**ViCLAS**), and **cold case squads**, and discuss the contributions these are making to police case investigations.

References

Adams, R.E., W.M. Rohe, and T.A. Arcury. (2002). "Implementing Community-Oriented Policing: Organizational Change and Street Officer Attitudes." *Crime & Delinquency,* 48(3):399–430.

Appleby, T., T.T. Ha, and R. Matas. (1999, March 24). "Police-Chase Deaths Prompt Calls for Action." *Globe and Mail,* A3.

Arnold, T. (2001, February 26). "53% of People Arrested under the Influence: Study." *National Post,* A7.

Auld, A. (2003, December 23). "N.S. Human Rights Agency Rules Police Discriminated Against Boxer." *Canadian Press.* Retrieved from http://www.canada.com

Bailey, I. (2000, January 5). "DNA Collected from Air Bag Links Man to Fatal Crash." *National Post*, A1.

Bennett, T. (2000). "Drugs and Crime: The Results of the Second Developmental Stage of the NEW-ADAM Programme." Home Office Research Study no. 205. London: Research Development and Statistics Directorate.

Brown, B. (1995). *CCTV in Town Centres: Three Case Studies*. London, UK: Police Research Group, Home Office.

Canadian Association of Chiefs of Police. (2004). "Bias-Free Policing." In *Resolutions Adopted at the 99th Annual Conference*. August. Vancouver, BC. Retrieved from http://www.cacp.ca

CBC News Online. (2003). "Cold Case: The Lawrence Wegner Story." Retrieved from http://www.recomnetwork.org/articles/03/04/16/0043239.shtml

Cope, N. (2004). "Intelligence Led Policing or Policing Led Intelligence? Integrating Volume Crime Analysis into Policing." *British Journal of Criminology,* 44(2):188–203.

Dandurand, Y., C.T. Griffiths, V. Chin, and J. Chan. (2004). *Confident Policing in a Troubled Community. Evaluation of the Vancouver Police Department's City-wide Enforcement Team Initiative.* Vancouver: Vancouver Agreement Coordination Unit and the City of Vancouver.

Fridell, L., R. Lunney, D. Diamond, and B. Kubu. (2001). *Racially Biased Policing: A Principled Response.* Washington, DC: Police Executive Research Forum.

Gabor, T.I. (2004). "Inflammatory Rhetoric on Racial Profiling Can Undermine Police Services." *Canadian Journal of Criminology and Criminal Justice,* 46(4):457–66.

———. (2003). *Assessing the Effectiveness of Organized Crime Strategies: A Review of the Literature.* Ottawa: Department of Justice Canada. Retrieved from http://www.justice.gc.ca/en/ps/rs/rep/2005/rr05-5/p9.html

Gill, M., and A. Spriggs. (2005). *Assessing the Impact of CCTV.* Home Office Research Study no. 292. London, UK: Research, Development and Statistics Directorate, Home Office.

Griffiths, C.T., B. Whitelaw, and R. Parent. (1999). *Canadian Police Work.* Scarborough, ON: ITP Nelson.

Knox, F. (2001). "Clarifying Police Tolerance." *Police Journal,* 74(4):292–302.

LeBeuf, M-E. (1996). *Three Decades of Women in Policing: A Literature Review.* Ottawa: Canadian Police College.

Martin, S.F. (1997). "Women Officers on the Move: An Update on Women in Policing," In R.G. Dunham and G.P. Alpert (eds.), *Critical Issues in Policing: Contemporary Readings,* 3rd ed. Prospect Heights, IL: Waveland Press.

Melchers, R. (2003). "Do Toronto Police Engage in Racial Profiling?" *Canadian Journal of Criminology and Criminal Justice,* 45(3):347–66.

National DNA Data Bank. (2004). *Annual Report, 2003–2004.* Ottawa. Retrieved from http://www.nddb-bndg.org

Nova Scotia Human Rights Commission. (2003, December 23). "Johnson Victim of Racial Discrimination, Board Finds." Retrieved from http://www.gov.ns.ca/news/details.asp?id=20031223001

Ontario Human Rights Commission. (2003). "Paying the Price: The Human Cost of Racial Profiling." Toronto. Retrieved from http://www.ohrc.on.ca

Oppal, The Honourable Mr. Justice W.T. (Commissioner). (1994). *Closing the Gap: Policing and the Community—The Report (Volume 1).* Victoria: Attorney General of British Columbia.

Pain, T.K. (2003). "Evidence Ruled Capable of Supporting Racial Profiling Allegation." Retrieved from http://www.torontocriminaldefence.com/articles/EpZukpAkpAltwoIdNx.php

Pare, P.-P., and M. Ouimet. (2004). "A Measure of Police Performance: Analyzing Police Clearance And Charge Statistics." *Canadian Review of Policing Research,* 1:111–17.

Pauls, M. (2004). "An Evaluation of the Neighbourhood Empowerment Team (NET): Edmonton Police Service." Ottawa: Public Safety and Emergency Preparedness Canada. Online at http://ww2.psepc-sppcc.gc.ca/policing/publications

Ratcliffe, J.H. (2002). "Intelligence-Led Policing and the Problems of Turning Rhetoric into Practice." *Policing and Society,* 12(1):53–66.

RCMP Public Complaints Commission. (1999). *Police Pursuits and Public Safety.* Ottawa. Retrieved from http://www.cpc-cpp.gc.ca/DefaultSite/Reppub/index_e.aspx?ArticleID=94

Read, T., and N. Tilley. (2000). "Not Rocket Science? Problem Solving and Crime Reduction." *Crime Reduction Research Series,* Paper no. 6. London, UK: Research, Development and Statistics Directorate. Home Office, Policing and Reducing Crime Unit.

Reber, S., and R. Renaud. (2005). *Starlight Tour: The Last, Lonely Night of Neil Stonechild.* Toronto: Random House Canada.

Rcith, C. (1956). *A New Study of Police History.* London: Oliver and Boyd.

Schneider, M.C., T. Rowell, and V. Bezdikian. (2003). "The Impact of Citizen Perceptions of Community Policing on Fear of Crime: Findings from Twelve Cities." *Police Quarterly,* 6(4):363–86.

Scott, M. (2003). *The Benefits and Consequences of Police Crackdowns.* Washington, DC: Office of Community Oriented Policing Services, U.S. Department of Justice.

Smith, G. (2004, October 27). "The Death of Neil Stonechild." *Globe and Mail,* A1, A13.

Task Force on Policing in Ontario. (1974). *The Task Force on Policing in Ontario: Report to the Solicitor General.* Toronto: Ministry of Attorney General.

Trojanowicz, R., and B. Bucqueroux. (1998). *Community Policing: How to Get Started,* 2nd ed. Cincinnati, OH: Anderson Publishing.

Turenne, M. (2005). *Racial Profiling: Context and Definition.* Quebec: Commission des droits de la personne et des droits de la Jeunesse.

U.S. Department of Justice. (2005). *What Is Community Policing?* Washington, DC: Office of Community Oriented Policing Services (COPS). Retrieved from http://www.cops.usdoj.gov/print.asp?Item=477

Wortley, S., and J. Tanner. (2003). "Data, Denials, and Confusion: The Racial Profiling Debate in Toronto." *Canadian Journal of Criminology and Criminal Justice,* 46(4):367–89.

Wright, Hon. Mr. Justice D.H. (Commissioner). (2004). *Commission of Inquiry into Matters Relating to the Death of Neil Stonechild.* Regina: Department of Justice, Province of Saskatchewan. Available online at http://www.stonechildinquiry.ca/finalreport/default.shtml

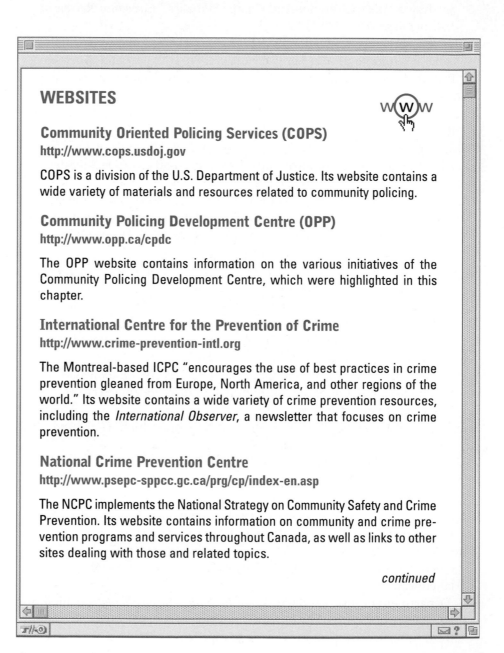

WEBSITES

Community Oriented Policing Services (COPS)
http://www.cops.usdoj.gov

COPS is a division of the U.S. Department of Justice. Its website contains a wide variety of materials and resources related to community policing.

Community Policing Development Centre (OPP)
http://www.opp.ca/cpdc

The OPP website contains information on the various initiatives of the Community Policing Development Centre, which were highlighted in this chapter.

International Centre for the Prevention of Crime
http://www.crime-prevention-intl.org

The Montreal-based ICPC "encourages the use of best practices in crime prevention gleaned from Europe, North America, and other regions of the world." Its website contains a wide variety of crime prevention resources, including the *International Observer,* a newsletter that focuses on crime prevention.

National Crime Prevention Centre
http://www.psepc-sppcc.gc.ca/prg/cp/index-en.asp

The NCPC implements the National Strategy on Community Safety and Crime Prevention. Its website contains information on community and crime prevention programs and services throughout Canada, as well as links to other sites dealing with those and related topics.

continued

continued

National DNA Data Bank
http://www.nddb-bndg.org/main_e

This website contains a number of links to legislation, training, news releases, real cases, and the annual report. An excellent resource for keeping abreast of developments in the use of DNA in case investigations.

Research, Development, and Statistics Directorate, United Kingdom Home Office
http://www.homeoffice.gov.uk/rds/index.html

An excellent website that contains links to research publications on criminal justice, including policing. The majority of the publications are available online.

learning objectives

After reading this chapter, you should be able to

- Discuss the structure of the Canadian courts, including specialized courts and circuit courts.
- Describe the work of the Supreme Court of Canada and the debate over judicial activism.
- Identify the principles of Canadian criminal law.
- Describe the prosecution process.
- Discuss the issues surrounding security certificates.
- Identify and discuss three models for the delivery of legal aid services.
- Describe plea bargaining and note the issues surrounding its use in case processing.
- Identify the various modes of trial.
- Describe the trial process.
- Identify and discuss the general strategies used by defence counsel.
- Discuss the needs and rights of victims in the context of the criminal court process.
- Discuss the causes of case delay and the impact of such delays on the justice system.
- Discuss the issues surrounding wrongfully convicted persons.
- Describe the impact of megatrials on the courts and the justice system.
- Discuss the provisions for accountability and review of judicial conduct.

key terms

The criminal courts play an important, multifaceted role in Canada's criminal justice system. Although the process for disposing of cases has changed little over the past two centuries, the cases coming into the courts are more complex than they once were, the legal issues are more challenging, and workloads are heavier. Many observers attribute these increased workloads and the resulting strains to the impact of the Charter of Rights and Freedoms.

The courts are responsible for determining the guilt or innocence of accused persons and for imposing an appropriate sentence on those who are convicted. They are also responsible for ensuring that the rights of accused persons are protected; this often involves monitoring the activities of the various agents of the criminal justice system (including the police and systems of corrections). The decisions of the courts reflect ongoing efforts to balance the rights of the accused with the need to protect society. In addition, Canadians believe that the principle of *judicial independence* is essential to the proper functioning of the courts. This principle holds that citizens have the right to have their cases tried by tribunals that are fair, impartial, and immune from political interference. The principle of independence also applies to juries in criminal cases.

Figure 5.1 provides a breakdown of the cases in adult criminal courts during 2003–04.

Impaired driving and common assault are the most frequent cases heard (each represents 11 percent of all offences).

The courts are important to the criminal justice process and to Canadian society as a whole, yet they are something of a mystery to many Canadians. This is due in some measure to the fact that the deliberations of judges and the activities of Crown counsel and defence lawyers are much less visible than the activities of the police. To understand how cases are prosecuted, let us first examine the structure of the court system.

FIGURE 5.1   Cases Heard in Adult Criminal Court, 2003–04

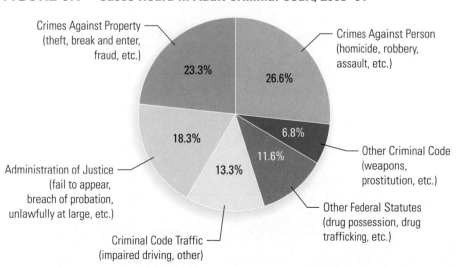

Source: Adapted from the Statistics Canada publication, *Juristat*, Adult Criminal Court Statistics, 2003/04, vol. 24, no. 12, Catalogue 85-002, December 10, 2004, p. 13.

Canada does not have a uniform court system. This often leads to considerable confusion when the various provincial and federal courts are discussed. This chapter will attempt to describe the system of courts in a clear and concise manner; that said, students would be well advised to familiarize themselves with the structure and names of the various courts in their own jurisdiction. Each province and territory maintains a website that provides detailed information on its court system (see, for example, http://www.ontariocourts.on.ca; http://www.manitobacourts.mb.ca).

THE PROVINCIAL COURT SYSTEM

There is some variation in the specific names given to the provincial courts across the country; even so, the system is much the same in all jurisdictions. In every province and territory the court system has two levels: provincial and superior. Nunavut is the sole exception to this: the Nunavut Court of Justice is a "unified" or single-level court. This means that the powers of the Territorial Court and the Northwest Territories Supreme Court—which formerly had jurisdiction in Nunavut—have been combined into one superior court that can hear any type of case (http://www.nunavutcourtofjustice.ca/unifiedcourt.htm).

With the exception of Nunavut, there are four levels of courts that deal with criminal cases: provincial/territorial courts, provincial superior courts, provincial appellate courts, and the Supreme Court of Canada (see Figure 5.2). Each of these is described next.

Provincial Courts

The provincial courts are the lowest level of courts; nearly all criminal cases are begun and disposed of in them. Their judges are appointed by the provinces, which also fund these courts and have jurisdiction over them. Provincial court

FIGURE 5.2   Four Levels of Courts Dealing with Criminal Cases

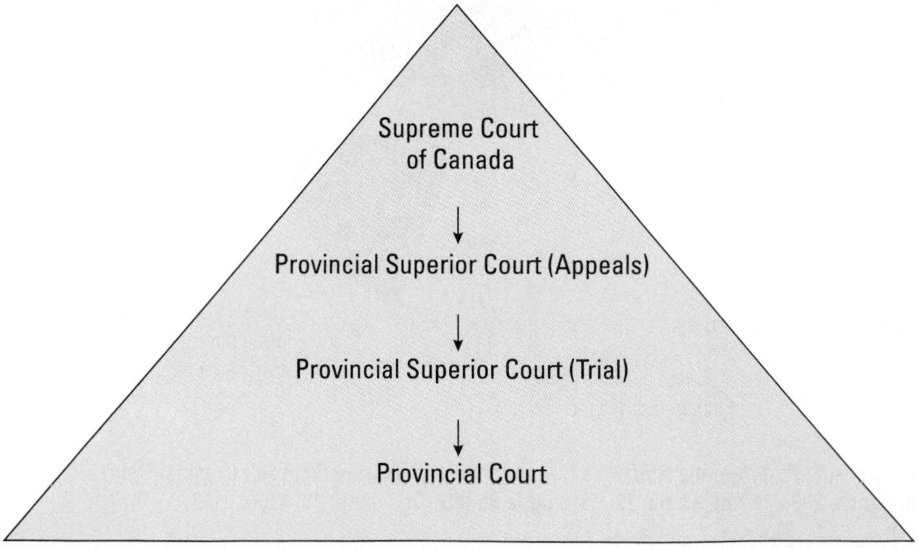

Supreme Court
of Canada

↓

Provincial Superior Court (Appeals)

↓

Provincial Superior Court (Trial)

↓

Provincial Court

judges sit without juries. These courts also hear cases under the Youth Criminal Justice Act, as well as cases involving alleged offences against provincial statutes. Provincial courts may also include family courts and small claims courts. Provincial court judges (along with justices of the peace) may preside over preliminary inquiries, which are held to determine whether there is sufficient evidence to warrant a trial.

Historically, the provincial courts dealt with less serious cases. This has changed in recent years, however; the judges in these courts now hear increasingly serious offences. As well, provincial court judges are confronted with specialized populations that may strain court resources and challenge judges to apply more appropriate sentences. Some observers now argue that the traditional distinction between the provincial courts and the superior courts has blurred somewhat in recent years. For example, research has found that although the superior courts hear proportionately more serious offences and more cases involving multiple offences, provincial courts hear more of these cases in terms of absolute numbers (Webster and Doob, 2003).

Specialized Provincial Courts

A recent trend in Canada is that the provincial courts are becoming more specialized. A number of provinces have created domestic violence courts, drug courts, and Aboriginal courts.

Domestic Violence Courts

These courts have been established in a number of urban centres, including Toronto, London, Winnipeg, and Calgary. They are designed specifically to hear criminal law charges relating to domestic violence. In Ontario the domestic violence courts take two main approaches to handling cases: early intervention and counselling; and coordinated prosecution. When no significant injuries have occurred, no weapons were used, and the accused has pleaded guilty, a counselling/ early intervention approach is taken. As a condition of bail prior to sentencing, the accused is required to complete the Partner Assault Response program, a specialized intervention program designed to teach non-abusive ways to resolve conflict. If offenders do not successfully complete the program, or if they reoffend, new charges are laid.

In cases where the accused is a repeat offender or inflicts serious injury, a coordinated-prosecution approach is taken. This involves specialized investigations by the police to obtain evidence, and the prosecution of offenders by Crown attorneys who specialize in domestic violence cases. In all cases that are handled by domestic violence courts, support for victims is provided by the Victim/Witness Assistance Program (http://www.attorneygeneral.jus.gov.on.ca/english/about/vw/vwap.asp).

In Yukon, the Domestic Violence Treatment Option Court involves a team of professionals, including RCMP officers, counsellors, Crown and defence lawyers, probation officers, and judges. The court provides the person who has been convicted of domestic violence with the opportunity to participate in various treatment programs prior to sentencing. Victims of domestic violence are provided with support, assistance, and referral information (http://www.justice.gov.yk.ca/pdf/pubs/dvtocourt.pdf).

Aboriginal Courts

Section 718.2(e) of the Criminal Code (see Chapter 6) requires judges to consider sentencing options other than incarceration, particularly for Aboriginal offenders. The principle that the judiciary should make efforts to explore alternative sentencing options—including the use of restorative justice—was affirmed by the Supreme Court of Canada in *R. v. Gladue* ([1999] 1 S.C.R. 688). To address the needs of Aboriginal offenders more effectively, several provinces have created courts specifically for Aboriginal people.

The Gladue Court (Toronto). The "Gladue Court," an Ontario Court of Justice, meets twice a week and deals with the cases of Aboriginal people who have been charged in downtown Toronto. It handles bail hearings, remands, trials, and sentencing. The judge, the Crown, and the defence lawyers, court clerks, and court workers are all Aboriginal persons. When the cases are processed, every attempt is made to explore all possible sentencing options and alternatives to imprisonment.

Tsuu T'ina Nation Peacemaker Court (Alberta). This initiative on the Tsuu T'ina Nation near Calgary involves a Peacemaker working with the Crown counsel to identify cases that could appropriately be diverted to the community's peacemaking program. The cases so identified are adjourned by the court so that a Peacemaker can open a restorative justice process involving the victim, the accused, elders, and community residents. A plan of action is designed, the purpose of which is to facilitate healing and address the issues associated with the harm done. Interventions include apologies, restitution payments, alcohol or drug treatment programs, and the requirement that the offender hold a traditional feast. Once this process is completed, the case returns to court, where the presiding judge considers outcomes of the peacemaking process in passing sentence (Ehman, 2005).

Drug Treatment Courts

The first drug treatment court was established in Toronto in 1998; Vancouver followed suit in 2001. Drug treatment courts provide an alternative forum for responding to offenders who have been convicted of drug-related offences. Offenders avoid incarceration by agreeing to abide by specified conditions, including participation in a drug abuse treatment program and regular drug testing. Cases are diverted to the drug court by Crown counsel, who determine whether the accused is eligible. Persons charged with violent offences are not eligible.

Total abstinence from drugs is not mandatory; however, in order to remain in the program, offenders must report relapses to program staff and demonstrate a reduction in their level of drug dependency. On completing the program, those who have been charged with less serious offences may have the charges stayed or withdrawn, while those charged with more serious offences may receive a period of probation. Those offenders who fail to abide by the conditions of the program are processed through the regular courts. Note that although offenders who are sent to a drug treatment court are being diverted from the traditional court system, the drug treatment court is still a court of law (see Chiodo, 2001).

An ongoing evaluation of the Toronto Drug Court indicates that among participants who successfully complete the program, there are reduced rates of criminal offending and drug use (see http://www.torontodrugtreatmentcourt.ca). In 2005 the federal government announced that additional drug courts would be established in Edmonton, Regina, Winnipeg, and Ottawa.

Provincial/Territorial Circuit Courts

In many northern and remote areas, judicial services are often provided via circuit courts. Circuit court parties, composed of a judge, a court clerk, a defence lawyer, a Crown counsel, and perhaps a translator, travel to communities (generally by air) to hold court. Many communities are served every month; others are visited quarterly or even less often if there are no cases to be heard or if the weather or mechanical problems with the court plane prevent a scheduled visit. The most extensive provincial/territorial circuit court systems are in the Northwest Territories, northwestern Ontario, northern Quebec, and Nunavut.

Concerns about the circuit court system have been voiced with respect to the following: the lengthy court dockets resulting from the backlog of cases; time constraints on the court party, which often preclude effective Crown and defence preparation and result in marathon court sessions, frequently lasting up to twelve hours; the shortage of interpreters, in that Aboriginal accused may understand little English or French and even less of the legal terminology spoken in court; and the general difficulties arising from the cultural differences between Canadian law and its practitioners and Aboriginal offenders, victims, and communities.

Circuit court judges often face a difficult decision: Should they remove the convicted person from the community and place him or her in confinement

A circuit court sitting in the community of Paulatuk, Northwest Territories

hundreds or even thousands of kilometres away? To address this concern, the circuit courts are encouraging community elders to participate in the court process and are supporting the development of community forums for dispute resolution as well as alternatives to incarceration. Restorative justice strategies are often applied in this environment. However, circuit court judges must often balance the need to develop culturally and community-relevant approaches to conflict resolution and case processing with the need to ensure that the rights and safety of crime victims are protected. This is especially so in cases involving women and young girls who have been the victims of spousal or sexual assault.

In an attempt to make the justice system more relevant to Aboriginal people, Saskatchewan has established a Cree Court in Prince Albert. This court travels to remote communities to hear cases. Its judges and lawyers are Cree speakers, and it is often attended by a Cree-speaking probation officer. Translators are provided when necessary. This court makes it possible for crime victims, witnesses, and defendants to speak in their own language.

The Cree Court and similar initiatives are designed to address the serious issues that surround the delivery of justice services in many rural and remote Aboriginal communities (see Box 5.1).

focus

BOX 5.1

Circuit Court Day, Northern Saskatchewan

August 15, 2002: It's a nice day, so people amble about outside, waiting for the judge to arrive. A big fellow wears a black shirt which taunts: "I DID NOT ESCAPE. THEY GAVE ME A DAY PASS." On one side of the building is located the community hall, which serves as both bingo parlour and courtoom—when a trial runs too long, the bingo players bang on the door to be let in—and on the other side is the village office. All the windows have bars or are covered with wire mesh, which is why it seems such a dismal place. The railing on the stairs has mostly fallen away, and the floor of the entryway has a gaping hole in it. How someone hasn't broken their leg is a wonder. On the outside wall of the village office, in bright blue paint, is scribbled "FUCK" in huge letters; on the community-hall side, there's a smaller

"Fuck" painted in the same painful blue. Piles of garbage and rubble are scattered around.

The interior is not much better. The walls are streaked and need a paint job, the grey-white floor has tiles missing. An old, faded red Christmas decoration hangs from the Exit sign, which has not lit up in years. The smell of cigarettes, smoked during frantic rounds of bingo, hangs in the air. A steel door right near the judge's chair opens onto the "executive washroom," a small space with toilet, sink, and one chair. This is the defence lawyer's consultation room, where he or she discusses a client's case—often for the first time. The lawyer sits on the toilet, the client on the chair, or the other way around.

The first group of accused file in. Since they are being held in custody, all are handcuffed and shackled, looking haggard from lack of sleep . . . The captives sit in chairs directly behind the prosecutor, which, he admits, makes him very nervous. Quietly, a toddler escapes from his stroller and runs toward his father. Despite his fetters, he lifts the child to his knee and kisses him. Another young prisoner, wearing a red Indian Posse bandana, sits and smooches with his girl, who is about seven months pregnant. She is oblivious to his chains. Another shackled captive explains that he is trying to get back into school: the judge listens as he munches on an apple. An attractive young woman is called to the witness stand, which consists of a rickety chair. She wobbles, obviously inexperienced at walking with her legs chained. The court is told that, under the influence of alcohol, she stabbed her husband twice. The wounds were not life-threatening. She has a history of depression—twice she has tried seriously to commit suicide—and no previous record. She is given a suspended sentence, and ordered to attend an alcohol treatment centre. The RCMP officer undoes her handcuffs and shackles and she joins the crowd in the back of the room.

Source: M. Siggins, *Bitter Embrace: White Society's Assault on the Woodland Cree* (Toronto: McClelland & Stewart, 2005), 291–92. Used by permission, McClelland & Stewart Ltd.

Superior Courts

The superior courts are the highest level of courts in a province and are administered by the provincial government; however, superior court judges are appointed and paid by the federal government. The name of the superior court generally identifies its location (for example, the Manitoba Court of Queen's Bench). About 10 percent of criminal cases are heard in the superior courts.

Superior courts generally have two levels: trial and appeal. These two levels may be included in the same court, with two divisions (trial and appeal), or they may involve two separate courts. In Ontario, however, the Court of Appeal is independent and separate from the Superior Court of Justice and the Ontario Court of Justice, which are the main two trial courts in the province.

The trial-level superior court hears cases involving serious criminal offences; the appeal-level superior court hears criminal appeals (and civil appeals as well) from the superior trial court. The trial court may be known as the Supreme Court or the Court of Queen's Bench; the appeal court is usually called the Court of Appeal. These courts hear cases involving the most serious offences, such as murder. Trials at this level may involve juries. After a case has been decided at the trial level, the accused has the right to appeal the verdict or the sentence, or both, to a higher court. Appeals of provincial court decisions may have to be heard first in a superior court. Appeals from the trial divisions of the superior courts go directly to the provincial court of appeal. There is one court of appeal in each province and territory, except in Quebec and Alberta, where there are two. In nine provinces, these courts are called the Court of Appeal (for example, the British Columbia Court of Appeal or the Quebec Court of Appeal). In Prince Edward Island, appeals are heard in the Appeal Division of the Supreme Court of that province. While many preliminary matters are dealt with by a single judge, certain final hearings require at least three judges to hear the appeal, and the final decision rests with the majority.

FEDERAL COURTS

If at least one appellate judge dissents (that is, does not agree with the majority), the unsuccessful party may pursue another appeal at the federal level. The "court of last resort"—the Supreme Court of Canada—is located in Ottawa but hears cases from all provinces and territories. The Supreme Court was established under the Constitution Act (1867), which authorized Parliament to establish a general court of appeal for Canada, although the bill creating the Court was not passed until 1875. The governor in council appoints the nine judges of the Supreme Court; those chosen must be superior court judges or lawyers with at least ten years' standing at the bar in a province or territory. The appointees are selected from the major regions of the country; however, three of the judges on the court must be from Quebec (http://www.scc-csc.gc.ca). The decisions of the Supreme Court are final and cannot be appealed. However, in some instances Parliament has passed legislation in response to a decision of the Supreme Court that has effectively changed the result of the decision. This occurred in the case of *R. v. Feeney* ([1997] 3 S.C.R. 1008) (see Box 5.3).

Two other federal courts are the Federal Court and the Tax Court. The Federal Court has a Trial Court and a Court of Appeal, and hears all cases that concern matters of federal law, including copyright law, maritime law, the Human Rights Act, the Immigration Act, and appeals from the National Parole Board.

Recent Supreme Court Decisions

The cases that are decided by the Supreme Court of Canada often involve interpretations of the Charter of Rights and Freedoms or complicated issues in private and public law. In many cases that come before the Supreme Court, either the defendant or the Crown asks for permission, or "leave," to appeal the decision of a lower court. In some instances the federal government asks the Supreme Court for a legal opinion on an important legal question, a process that is referred to as a "reference." In 1998 the federal government asked the Court to decide whether Quebec could secede unilaterally from Canada under the Constitution and whether international law gives the Province of Quebec the right to secede unilaterally from Canada (*Reference re Secession of Quebec* ([1998] 2 S.C.R. 217)). More recently the federal government asked the Supreme Court for a non-binding opinion as to whether the government could redefine marriage to allow for same-sex marriages. The court ruled (*Reference re Same-Sex Marriage* [2004] 3 SCC 698) that the federal government could do so; this resulted in legislation giving gays and lesbians the right to marry.

As noted in Chapter 1, there is an inherent tension between individual rights as set out in the Charter of Rights and Freedoms and the need to protect the general public. This tension is illustrated in *R. v. Sharpe* ([2001] 2 S.C.R. 45). In *Sharpe*, the Supreme Court upheld the law relating to the possession of child pornography (with certain exceptions; see Box 5.2). In other cases, laws have been struck down. In *Morgentaler v. R.* ([1988] 1 S.C.R. 30), for example, the Court held that the procedures for obtaining a therapeutic abortion as defined in

focus

BOX 5.2

R. v. Sharpe: A Case of Competing Rights

In *R. v. Sharpe* ([2001] 2 S.C.R. 45), the accused was charged with two counts of possession of child pornography under section 163.1(4) of the Criminal Code and two counts of possession of child pornography for the purposes of distribution or sale under section 163.1(3). Among other materials, Sharpe had in his possession pictures of young boys engaged in sexual activities and a collection of child pornography stories (titled "Kiddie Kink Classics") that he had written.

At trial, the B.C. Supreme Court acquitted Sharpe of the charge of possession of child pornography. The acquittal was later upheld by the B.C. Court of Appeal, which stated that the Criminal Code section on possession of child pornography was "one step removed from criminalizing simply having objectionable thoughts."

The case was appealed to the Supreme Court of Canada by the Province of British Columbia. The federal government, most provincial governments and police associations, and a variety of child advocate and child protection organizations argued that the need to protect children from sexual exploitation outweighed any protections that might be offered to Sharpe under the Canadian Charter of Rights and Freedoms. In a unanimous ruling, the Supreme Court upheld the law that makes it a crime to possess child pornography. Sharpe was convicted of possessing more than 400 photographs that met the legal definition of child pornography. "Freedom of expression," the Chief Justice stated, "is not absolute," given the constitutional limitations provided under s. 1 of the Charter, which expressly permit the Court to consider "reasonable limits in a free and democratic society."

The controversial part of the Supreme Court's decision was its creation of two exceptions. The first of these asserted the right to protect private works of the imagination or photographic depictions of one's own body; the second permitted the possession of child pornography by those who create sexually explicit depictions of children for their own personal pleasure. The Supreme Court of Canada directed that Sharpe be retried on the charge of possessing child pornography and be required to prove that his case met the requirements of one of the two exceptions. Some critics asserted that the Court's decision was tantamount to a legalization of child pornography. In March 2002, a B.C. Supreme Court justice ruled that Sharpe's written work, which contained descriptions of child sex and violence, had "artistic value," and Sharpe was acquitted (*R. v. Sharpe,* [2002] B.C.S.C. 423).

Source: CBC News Online. (June 22, 2004)"The Supreme Court and Child Porn: Saving Children or Thought Control?" Retrieved from http://www.cbc.ca/news/background/childporn

section 287 of the Criminal Code infringed on the right to security of the person because of the uneven availability of services across the country. And in *R. v. Zundel* ([1992] 2. S.C.R. 731), the Court held that the offence of spreading false news (s. 181) and even hate literature is constitutionally invalid because it infringes the fundamental freedoms of thought, belief, opinion, and expression and is not a reasonable limit in a democratic society. Although the laws referred to in these cases are still part of the Criminal Code, they cannot be used to prosecute anyone.

The decisions of the Supreme Court in Charter-related cases can also affect legal procedures. In *R. v. Stinchcombe* ([1991] 3 S.C.R. 326), for example, the Court held that the prosecution must give all relevant evidence gathered by the police to the defence to permit a defendant to make a full answer and defence to the charges.

The Court has been criticized on the one hand for engaging in social activism in its decision making, and on the other hand for being too deferential to law enforcement and to the laws enacted by Parliament. A review of Supreme Court decisions handed down in 2001 found a split in the Court over the powers granted to police officers and prosecutors (Chwialkowska, 2002a). A study of Supreme Court decisions in cases involving Charter challenges found that governments won their cases just over 62 percent of the time (Choudhry and Hunter, 2003). However, another study that examined a broader range of cases heard by the Court since 2000, in which persons argued that their constitutional rights had been violated, found that the appellant won 52 percent of the time, compared to an average of 33 percent in the 1990s (Monahan, 2004). These conflicting findings suggest that the extent to which the Supreme Court can be considered an activist court depends to a large extent on which cases are examined and on which criteria are being applied to assess "activism."

See Box 5.3 for a brief summary of several important recent Supreme Court decisions.

focus
BOX 5.3

Selected Decisions of the Supreme Court of Canada, 1997–2005

R. v. Feeney ([1997] 3 S.C.R. 1008). As part of a murder investigation, the police knocked on the door of the accused's home. Receiving no answer, they entered the home, woke the accused, and arrested him after seeing blood on his shirt. The officers informed the accused of his right to counsel, but not of the immediate right to counsel, and seized evidence from his home that was subsequently used to obtain a search warrant to retrieve additional evidence. At trial,

the accused was convicted of second degree murder. His appeal was unanimously dismissed by the B.C. Court of Appeal. The issue for the Supreme Court was whether the police had violated the accused's right under the Charter to be free from unreasonable search or seizure and his right, upon being arrested, to have access to legal counsel without delay and—either being so—what evidence, if any, should be excluded under section 24(2) of the Criminal Code. From an examination of the facts in the case, the Court concluded that the legal requirements for a warrantless arrest following a forced entry into a person's private premises had not been met. More specifically, the arresting officer had not believed he had reasonable grounds to arrest prior to the forcible entry.

The Court held that in order to protect the privacy rights of Canadians under the Charter of Rights and Freedoms, the police must obtain a search warrant before entering a dwelling to arrest or apprehend a suspect. In response to this decision, Parliament amended the Criminal Code to require that as a general rule, peace officers must obtain a warrant prior to entering a dwelling to apprehend or arrest someone, but police may enter dwellings and arrest or apprehend without a warrant in those circumstances where entry is required to prevent bodily harm or death or to prevent the loss or destruction of evidence in the case.

R. v. Gladue ([1999] 1 S.C.R. 688). The accused, an Aboriginal woman, pled guilty to manslaughter for the killing of her common law husband. The sentencing judge considered a number of mitigating and aggravating circumstances in the case in determining that a three-year period of custody was the appropriate sanction. The fact that she and her victim were Aboriginal was not taken into account by the sentencing judge. The Court of Appeal dismissed the convicted woman's appeal of her sentence. The Supreme Court upheld the sentence on appeal. However, in its reasons for judgment, it reiterated that section 718.2(e) requires that in passing sentence in cases where a term of incarceration would normally be imposed, judges must consider the unique circumstances of Aboriginal people and are under an obligation to consider all alternatives to incarceration.

R. v. Tessling ([2004] 3 S.C.R. 424). In this case, the RCMP had used an airplane equipped with a forward-looking infrared (FLIR) camera to overfly properties owned by the accused. This camera can record images of heat radiating from a building, although it is not able to identify the source of the heat. With data from the FLIR and information from two informants, the police secured a search warrant, entered the house, and found a large quantity of marijuana and a number of guns. On appeal, the Court held that the overflight did not violate the accused's constitutional right to be free from unreasonable search and seizure.

R. v. Kerr ([2004] 2 S.C.R. 371). This case involved an inmate in a maximum security prison who had received death threats from another inmate (the victim) who was a member of a criminal gang inside the institution. The victim was attacked by the accused with a homemade knife, and a physical altercation occurred in which the victim was fatally stabbed by the accused. In setting aside a conviction for "possession of a weapon for a dangerous purpose" imposed by the Court of Appeal, the Supreme Court concluded that the accused, in using the weapon for self-defence, had not endangered the public: "On the day of the altercation, the accused possessed his weapon for the purpose of defending himself against an imminent attack by specific individuals. His purpose was not, in all of the circumstances, dangerous to the public peace, since the attack was clearly unavoidable."

R. v. Clark ([2005] 1 S.C.R. 6). The accused was observed by neighbours masturbating near an uncovered window in the living room of his home. The police were called, and he was charged under section 173(1) of the Criminal Code, subsections (a) "in a public place in the presence of one or more persons" and, (b) "in any place, with intent thereby to insult or offend any person." At trial, the accused was convicted under subsection (a), having been found to have converted his living room into a public place. Upon appeal from the Court of Appeal in British Columbia, the Supreme Court overturned the decision and entered an acquittal for the accused, holding that the accused's acts were not committed in a "public place."

What do you think?

Check out these cases in Box 5.3 in further detail at http://www. lexum. umontreal.ca/csc-scc/en. Do you agree with the decision of the Court in each of these cases? Explain.

stare decisis

the principle by which the higher courts set precedents that the lower courts must follow

Recall from Chapter 1 that interest groups often play a role in the formulation and application of the criminal law. Research conducted on Supreme Court cases decided during the years 1996 to 1999 found that in 54 percent of the cases, the court permitted intervenors (persons or parties not directly involved in the case) to file written materials and in some instances to make oral arguments (York University, 2000).

THE SYSTEM OF CRIMINAL LAW

The legal system operates under precise albeit not always logical rules. Canada inherited the British system of common law; as a consequence, our law—both civil (except in Quebec) and criminal—is found both in statutes and in judicial precedents (the latter referred to as *case law*). In other words, many laws—such as those in the Criminal Code—are written down or codified. But through their decisions in cases, judges can interpret, modify, extend, restrict, or strike down statutory laws.

As Figure 5.2 illustrated, the courts are organized in a hierarchy, with the Supreme Court of Canada at the top. The principle whereby higher courts set precedents that lower courts must follow is known as **stare decisis** (Latin for "to stand by what was decided"). Underlying this principle is the idea that like cases should be treated alike. Especially when the law is not precise, judicial interpretation can add clarification so that all courts are playing by the same rule book, so to speak. Once the Supreme Court of Canada rules on a thorny legal issue, all courts below it are bound to apply that ruling in subsequent cases.

Our adversarial system has many rules of procedure and evidence governing criminal prosecutions. Some of these common-law rules have been enshrined in the Charter of Rights and Freedoms. The more important principles include the following:

- *Presumption of innocence.* A defendant is deemed innocent of the charge(s) until either convicted or acquitted.

- *The Crown bears the burden of proof.* It is the task of the Crown to prove guilt, not the responsibility of the accused to prove his or her innocence.

- *Doli incapax (Latin for "too young for evil").* A child under twelve cannot be held criminally responsible or prosecuted for criminal acts.

- *Insanity.* No one is criminally responsible and liable to punishment if incapable of knowing the act was wrong owing to a mental disorder.

- *Attempts are crimes.* Those who attempt crimes (going beyond merely the planning stage) commit an offence and are generally subject to half the penalty that the completed act would draw.

Remember also the rights afforded to criminal defendants in the Charter of Rights and Freedoms (see Box 3.4). Additional principles of Canadian criminal law are presented in Box 5.4.

focus

BOX 5.4

Principles of Canadian Criminal Law

- *actus non facit reum nisi mens sit rea.* An act does not make a person guilty unless he or she has a guilty mind.

  Each crime has two components. The first is *actus reus,* or the act of doing something. The second is *mens rea,* or the guilty intent. To be convicted of most crimes (but not all), a person must have done something criminal, and usually (but not always) must have intended to do it. A few offences impose strict or absolute liability, in that the Crown need not prove *mens rea.* Possession of burglary instruments (s. 351) is one example. Children under twelve and the insane are deemed unable to form *mens rea,* and therefore will not be held criminally responsible for their actions. There are provisions that apply to the criminal conduct of children but these provisions are found in child protection statutes rather than in the Criminal Code.

- *nullum crimen sine lege, nulla poena sine lege.* No crime without a law, no punishment without a law.

  This principle means that the rules cannot be changed in the middle of the game. Laws cannot be applied retroactively.

- *ignorantia juris non excusat.* Ignorance of the law is no excuse.

  There is a formal expectation that every citizen be familiar with all the laws and therefore able to distinguish between legal and illegal behaviour. This expectation is a fiction because the law is constantly changing and, at any given point in time, is subject to debate and differing interpretations. However, the legal system would grind to a halt if defendants were able to claim that they had no idea their alleged offences were illegal.

- *nemo tenetur seipsum accusare.* No one is compelled to incriminate himself.

  Criminal suspects and defendants have the right to remain silent during the police investigation. If they are forced or threatened to make a confession, that statement will be inadmissible in court. In addition, a criminal defendant may choose not to testify in his defence. This principle is enshrined in the Charter.

- *nemo debet bis vexari pro eadem causa.* No one should be twice troubled by the same cause.

  This principle is more commonly known as "double jeopardy." An alleged offender cannot, under most circumstances, be tried twice for the same offence. In contrast to the American criminal justice system, however, an alleged offender in Canada can be retried after being acquitted if the Crown successfully appeals the decision by claiming problems with the correct application of the law at the trial.

The Classification of Offences

At this point, it is important to note that there are three categories of criminal offences and to understand the differences between them. In the Criminal Code there are summary conviction offences, indictable offences, and hybrid (or elective) offences. *Summary conviction offences* are generally less serious offences that are triable before a magistrate or judge. Examples include impersonating a police officer, keeping or being found in a common bawdy house (brothel), water skiing in the dark, and injuring animals other than cattle. In olden times, these crimes were called petty offences or misdemeanours—terms that better convey the less serious nature of these crimes relative to indictable offences. *Indictable offences* are serious criminal offences that may carry maximum prison sentences of fourteen years or more. Examples include murder, robbery, aggravated sexual assault, perjury, passing counterfeit money, breaking out of prison, conducting a pyramid scheme, and hostage taking.

There are a number of important differences between summary and indictable offences. Summary conviction proceedings must begin within six months after the offence occurs; there is no such time limit for indictable offences except as determined by the common law in interpretations of the right to a speedy trial under the Charter. Summary conviction offences can only be tried by a provincial court judge sitting alone; indictable offences can be tried in a number of courts, depending on such factors as the seriousness of the alleged offence and the court chosen by the accused. A person who is charged with an indictable offence generally has the choice of being tried by a provincial court judge or by a superior court judge, with or without a jury. In many indictable cases, a **preliminary hearing** is held to determine whether there is sufficient evidence to proceed with the trial.

Most summary offences, upon conviction, carry a maximum penalty of a fine (not to exceed $2,000), or six months in a provincial correctional facility, or both. There are a few summary conviction offences that have as a maximum penalty eighteen months' imprisonment. Much more severe penalties—including a life sentence—can be imposed on conviction of an indictable offence.

Hybrid (or **elective**) **offences** generally fall between summary offences and indictable offences in terms of their seriousness. There are upwards of seventy offences in the Criminal Code that can be classified as hybrid. For example, under Section 86 of the Criminal Code, pointing a firearm is a hybrid offence:

86. (1) Every one who, without lawful excuse, points a firearm at another person, whether the firearm is loaded or unloaded,
    (a) is guilty of an indictable offence and liable to imprisonment for a term not exceeding five years; *or*
    (b) is guilty of an offence punishable on summary conviction.

Other hybrid offences are sexual assault, driving while disqualified, uttering death threats, and selling used goods as new. Once such a charge is laid, the Crown attorney decides whether to proceed summarily or by indictment. The power of the prosecutor to select the mode of trial in relation to hybrid offences has implications for the procedural rules that will be followed as well as the severity of the sanctions to which the accused may be subjected.

Many factors influence a prosecutor's decision regarding how to proceed on a hybrid offence. The seriousness of the alleged offence is considered. Another factor is the limitation period. Summary conviction prosecutions must be commenced within six months of the alleged offence. If more than six months have elapsed, a Crown attorney who wants to pursue the matter must proceed by indictment. Another consideration by the Crown is that summary matters are resolved more quickly by the courts, although the maximum penalties available for summary conviction offences are lower.

preliminary hearing
a hearing to determine whether there is sufficient evidence to warrant a criminal trial

hybrid (or elective) offences
offences that can be proceeded summarily or by indictment—a decision that is always made by the Crown

THE PROSECUTION PROCESS

There are a number of steps involved in bringing a case to criminal court. In considering the prosecution process, it is important to note that there is considerable case attrition and that many cases do not progress very far into the system. The police or the Crown send many offenders to alternative measures programs—a process known as **diversion.**

diversion

programs designed to keep offenders from being prosecuted and convicted in the criminal justice system

Offenders can be sent to diversion programs at various stages in the criminal justice process. There are pre-charge diversion programs, post-charge diversion programs (which are generally operated by probation officers or community agencies), and post-sentencing diversion programs (which allow offenders to avoid incarceration). There are diversion programs for young offenders in all provinces and territories, and many parts of the country have diversion programs for adult offenders as well.

The overall objective of diversion programs is to keep offenders from being processed further into the formal criminal justice system. But they have more specific objectives as well:

- to avoid negative labelling and stigmatization;
- to reduce unnecessary social control and coercion;
- to reduce recidivism;
- to provide services (assistance); *and*
- to reduce the costs of the justice system (Palmer, 1979).

Most diversion programs require offenders to acknowledge responsibility for their behaviour and to agree to fulfill certain conditions; for example, they may be required to pay restitution to the victim or to complete a specified period of community service. Some diversion programs are directed toward offender groups charged with more serious offences, but most of them target first-time, low-risk offenders. This raises the issue of net widening (that is, involving offenders who would have otherwise been released outright by the police or not charged by Crown counsel). The concerns are that diversion programs may be bringing people into the justice system who would otherwise have been dealt with on an informal basis, and that diversion programs may increase the workload and costs of the justice system (Bonta, 1998).

B.C.'s Prostitution Offender Program is an example of a diversion program. A joint initiative of the City of Vancouver, the Vancouver Police Department, and the John Howard Society, this program is offered to people who have been arrested for the offence of communication for the purposes of prostitution and is for sex trade consumers ("johns"). The program is designed to educate offenders about health issues, the role of pimps, and the impact of prostitution on sex trade workers, their families, and the community (http://www.jhslmbc.ca/popbc.html).

Diversion programs are widespread, yet there have been few studies of their effectiveness.

Participants in the Process

Crown attorneys are lawyers who represent the Crown (or government) in court and who are responsible for prosecuting criminal cases. The responsibility for prosecuting cases is shared between the provinces and the federal government, with provincially appointed Crown attorneys prosecuting Criminal Code offences and federally appointed Crown attorneys prosecuting persons charged with violating other federal statutes such as the Controlled Drugs and Substances Act. In Yukon, the Northwest Territories, and Nunavut, federally appointed Crown attorneys are responsible for prosecuting all cases.

Crown attorneys are involved in a range of activities. They provide advice to police officers at the pre-charge stage; they prepare for trial (for example, they collect evidence from the police and other sources, research case precedents, and interview victims, witnesses, and experts who may be called to testify); and they prepare for post-trial appeals. Crown counsel are also involved in negotiating pleas, developing trial strategies, managing witnesses, arguing conditions of bail, recommending sentences to the court, and appealing sentences deemed too lenient. Crown attorneys must also remain up-to-date on changes in the law and in judicial precedent, including decisions in Charter cases.

In Chapter 1 it was noted that one trend in Canadian criminal justice is increasing workloads. This is evident in the work of Crown counsel, many of whom process up to fifty cases a day and work ninety hours a week. And while caseloads are increasing, resources are dwindling. In the words of one Crown counsel: "You know I've winged many cases. I've seen me do trials when the first time I ever read the file was when I was calling my first witness 'cause I never had time. Just didn't have time to prepare for it. I'll call my witness, and while he's walking up to the stand I'll read his statement, and then I'll find out what he's got to say and then I'll question him. I did that many times" (in Gomme and Hall, 1995: 194).

Increasingly, prosecutors must deal with sensitive cases involving sexual offences, family violence, and the victimization of children. New technologies, such as DNA evidence, require prosecutors to have specialized knowledge (or access to it). In some regions of the country, prosecutors travel with circuit courts or to satellite court locations and thus often have little time for case preparation. Other challenges are the cultural and language barriers that are encountered in northern and remote Aboriginal communities as well as in some urban centres.

Other Court Personnel

Besides lawyers and judges, other court personnel play important roles in the processing and disposition of cases. Court administrators—also known as court registrars or court clerks—perform a variety of administrative tasks. For example, they appoint staff, manage court finances, sign orders and judgments, receive and record documents filed in the court, and certify copies of court proceedings. On request, the court reporter can make a verbatim (word for word) transcript of everything that is said during the trial. This is possible because the proceedings are tape recorded (Department of Justice Canada, 2005).

Sheriffs support the court by assisting in jury management, escorting accused and convicted persons, and providing security in the courtroom. In some provinces they serve legal documents, seize goods, and collect fines.

Laying an Information and Laying a Charge

The police are usually responsible for laying informations, which are then ratified or rejected by the Crown. An information is a document that briefly outlines an allegation that a person has contravened a criminal law in a certain location during a specified period. Multiple offences are divided into separate counts.

Not all cases must be brought before a JP. For certain offences, police officers are authorized to issue summons, traffic offence notices, appearance notices, and promise-to-appear notices. In such cases, accused persons are released *on their own recognizance*, which means they are responsible for ensuring that they appear in court on the designated date.

The information may be laid either after the suspect has been informed (as in the case of an arrest without a warrant or the use of an appearance notice) or before (see Figure 5.3). Remember from Chapter 3 that there are a limited number of circumstances in which the police can arrest without a warrant; there is a presumption that an appearance notice will be used for most cases. On receiving the information, the JP may not agree that the informant has made out a case; in practice, however, this rarely happens. If the JP determines there is sufficient reason to believe that a crime has been committed, he or she will issue either a warrant for the arrest of the person named in the information, or a summons that directs the named person to appear in provincial court on a specified date.

FIGURE 5.3   Compelling the Appearance of the Accused

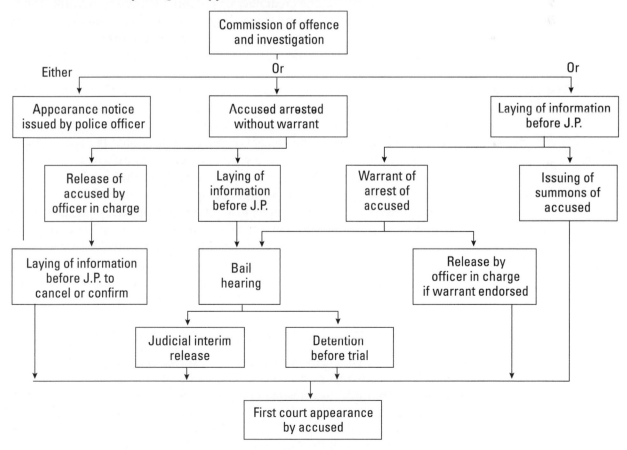

Source: A.W. Mewett and Shaun Nakatsuru, *An Introduction to the Criminal Process in Canada*, Fourth ed. (Toronto: Carswell, 2001), p.72. Reprinted by permission of Carswell—a division of Thomson Canada Limited.

The police and the Crown exercise a considerable amount of discretion in deciding whether to lay a charge (Layton, 2002). Charges are not laid in one-third of all violent crimes and property crimes that are cleared by the police. Reasons for not charging include the following: the victim or complainant is reluctant to cooperate; the suspect or an essential witness dies; or the suspect was committed to a psychiatric facility or was under twelve. The judiciary, including the Supreme Court of Canada, have been reluctant to review prosecutorial decision making; however, the Supreme Court has held that provincial law societies are permitted to review such decisions to ensure adherence to professional standards (*Krieger v. Law Society of Alberta,* [2002] 3 S.C.R. 372).

Figure 5.4 illustrates the disparity that exists across the country in terms of the charging decisions of prosecutors—in this case for possession of marijuana.

Continuing with the marijuana theme: Figures compiled by criminologist Darryl Plecas during a study of marijuana-related enforcement in British Columbia found that in 2003, a significant number of busts of growing operations (grow-ops) did not result in criminal charges being laid. This, even in situations involving a large number of marijuana plants (see Figure 5.5).

Legal, administrative, and political factors may also influence the decision to lay a charge. Legal considerations include the reliability and likely admissibility of available evidence and the credibility of potential witnesses. Administrative factors include the workload and case volume of the Crown counsel's office, as well as the time and cost of prosecution relative to the seriousness of the crime. Political considerations include the need to maintain the public's confidence in the justice system.

In New Brunswick, Quebec, and British Columbia, the Crown must give approval before the police can lay a charge. Once the decision has been made to lay a charge, a police officer can initiate the process by laying an information before a JP. When doing so, the officer is called an "informant." In practice, most informants are police officers, but any person can lay an information if he or she "on reasonable grounds" believes that a person has committed an offence.

Not all crime victims support charges being filed. Victims of domestic violence or spousal assault, for example, may refuse to cooperate with the Crown for a variety of reasons, including fear of retaliation, economic insecurity, and family pressures. Also, victims who are also involved in criminal activities (such as gang members) may be understandably reluctant to appear in court and to provide testimony against accused persons. In these circumstances, crime victims may not make use of the specialized services that are available.

Compelling the Appearance of the Accused in Court

After a prosecution has been initiated, the next step is to ensure that the accused appears in court to answer the charge. This can be accomplished in a number of ways—for example, by arresting and placing the accused person in remand custody until the court appearance, or by allowing the accused person to remain at liberty in the community with a promise to appear on the court date. If the accused person does not appear, the judge can issue an arrest warrant. Figure 5.3 illustrates the various ways for compelling an accused person to appear in court.

What do you think?

What factors may explain the differences in the charging patterns across the country? Does this variation in charging practices cause you concern?

What do you think?

What might be some possible explanations for the lack of charges in cases involving marijuana grow-ops in British Columbia? What do these possible reasons tell us about (a) the enforcement practices of the police, and (b) the charging practices of police and prosecutors?

FIGURE 5.4   Possession of Marijuana Charges, 1999

Share of possession of cannabis incidents that led to charges, 1999
(Those not charged include cases diverted out of the courts and seizures without charges.)

Province	Incidents	Charges	Percentage
B.C.	10,094	1,739	17.2%
Alberta	2,837	1,909	67.3%
Saskatchewan	1,348	1,026	76.1%
Manitoba	995	664	66.7%
Ontario	14,214	10,057	70.8%
Quebec	6,817	3,762	55.2%
N.B.	1,091	680	62.3%
N.S.	1,152	720	62.5%
P.E.I.	121	94	77.7%
Nfld.	570	324	56.8%

Source: Statistics Canada

Possession charges vs. marijuana users

Province	Pct. users	Estimated no. of users	No. of charges	Charges per 100,000 pop.	Charges per 1000 marijuana users
B.C.	12%	352,680	1,739	43.23	4.93
Alberta	8%	165,840	1,909	64.39	11.51
Saskatchewan	7%	53,690	1,026	99.83	19.11
Manitoba	9%	78,660	664	58.07	8.44
Ontario	5%	433,650	10,057	87.35	23.19
Quebec	9%	521,640	3,762	51.22	7.21
N.B.	6%	36,180	680	90.07	18.79
N.S.	8%	59,440	720	76.61	12.11
Nfld.	4%	18,320	324	59.89	17.69

(The sample size for P.E.I. was too small to include in the Alcohol and Drug survey.)
Source: % of users: 1994 Statistics Canada Alcohol and Drug Survey; Estimated no. of users: Based on Statistics Canada's 15+ population estimates for 1994; No. of charges: Canadian Centre for Justice Statistics, 1999 figures

Possession of marijuana charges per capita for B.C. municipalities, 1999

City	Charges	Pop.	Charges per 100,000 residents
Victoria	86	76,121	112.98
Prince George	39	80,943	48.18
North Vancouver*	62	130,768	47.41
Kelowna	41	97,372	42.11
Surrey	89	332,836	26.70
Burnaby	48	189,513	25.33
Delta	23	101,202	22.73
Richmond	34	162,245	20.96
Coquitlam	16	110,633	14.41
Vancouver	74	554,900	13.34

*City and district

Source: C. Skelton, "Vancouver: The Next Amsterdam," *Vancouver Sun* (May 26, 2001), A1, A8. Reprinted by permission of the *Vancouver Sun.*

FIGURE 5.5   Percentage of Growing Operation Seizures That Did Not Result in Charges, by Scale of Operation, 2003

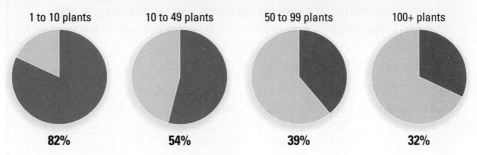

1 to 10 plants   10 to 49 plants   50 to 99 plants   100+ plants

82%   54%   39%   32%

Raids, but no charges

There has been a dramatic increase in the number of "no-case seizures" in B.C., where police raid a marijuana growing operation and seize the plants, but do not investigate further and do not recommend charges.

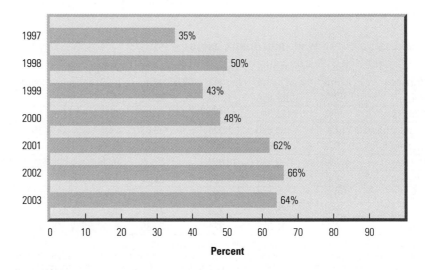

Year	Percent
1997	35%
1998	50%
1999	43%
2000	48%
2001	62%
2002	66%
2003	64%

Source: C. Skelton, "Out of Control: Criminal Justice System 'On the Brink of Imploding,'" *Vancouver Sun*, (March 11, 2005), B3. Reprinted by permission of the *Vancouver Sun*.

Appearance Notice

If the alleged offence is not serious and the police have no reason to believe that the accused will fail to appear in court, an appearance notice can be issued followed by the laying of an information. The appearance notice sets out the details of the allegation against the accused person, provides the court date, and warns the accused that failure to appear in court is a criminal offence. If the charge is an indictable or elective offence, the appearance notice directs the person to appear at a specific location to be fingerprinted, pursuant to the Identification of Criminals Act. If the suspect is a young person, the appearance notice emphasizes the right of accused youths to legal representation.

Summons

Another option is for the police to lay the information first, in which case the JP will likely issue a summons, which briefly states the allegation and directs the person to appear in court on a certain day. The fingerprint demand is also made, where applicable, as is the statement about youths' right to counsel under the Youth Criminal Justice Act. The summons is then served on the accused, usually by a police officer. If the accused does not appear in court, and if there is proof that he received the summons, the judge may issue a bench warrant for his arrest; in addition, the accused may be charged criminally with failing to appear in court.

Arrest

If the situation dictates, the police can arrest without a warrant (see Chapter 3) and *then* lay the information. Or, if there is time, an officer may seek an arrest warrant from a JP. Following the arrest, the next decision to be made is whether to release the accused from police custody or keep him in custody. Remember that the Charter protects people from arbitrary detention. The presumption is that everyone will be released from police custody after arrest.

There are only three circumstances in which immediate release might *not* occur:

- The charge pertains to a serious indictable offence carrying a maximum sentence of more than five years in prison.
- The police have reasonable grounds to believe the person will not appear in court.
- The police have reasonable grounds to believe it is necessary, in the public interest, to detain the accused.

The public interest is defined as the need to establish the person's identity, to secure or preserve evidence, or to prevent the continuation or repetition of the offence or the commission of another offence.

A notable exception involves situations where a person is arrested pursuant to a security certificate (see below).

Release by the Police

When issuing an arrest warrant, the JP usually gives the police some direction as to whether the accused person should be detained or released. When an arrest is made without a warrant, the police have the authority to release some accused persons from police custody; however, in some circumstances a bail hearing before a JP or a judge is required. When the offence is summary or elective (or one of a specified list of less serious indictable offences, including theft under $5,000), the arresting officer can simply issue an appearance notice or explain that a summons will be sought. In these circumstances, even those persons who have been arrested need not be placed in police custody. Research conducted in Ontario indicates that there is considerable variation across the province in the numbers of persons detained by the police prior to trial (Attorney General of Ontario, 1999).

For indictable offences carrying a maximum prison sentence of five years or less, the officer in charge of the police lockup has the authority to release the person from police custody. Several means are available to the officer in charge to compel the accused's later appearance in court. Beginning with the least consequential, they are a promise to appear, an undertaking to appear, and a recognizance not exceeding $500 (with or without deposit).

Judicial Interim Release

judicial interim release (or **bail**) is overseen by a judicial functionary—usually a JP—but by a superior court judge if the offence is a serious one, such as murder. A person in police custody who is not released by the officer in charge must be brought to court within twenty-four hours or as soon as is reasonably possible. In addition, if the arrest warrant was issued by a JP, the police must bring the arrested person before a JP unless release was authorized when the warrant was issued.

The JP or judge must determine whether the accused will be released or will remain in custody until the case is disposed of. Section 11(e) of the Charter stipulates that any person charged with an offence has the right not to be denied bail without reasonable cause. If the Crown chooses to oppose the release of the accused, the Crown must demonstrate, at a *show cause hearing*, that detention of the accused until the trial date is necessary. In support of the recommendation that the accused be held in custody, the Crown can produce evidence of prior criminal convictions, other charges currently before the courts, or previous instances of failing to appear in court. In some cases, *reverse onus* applies; in other words, the accused must "show cause" why a release is justified. These situations include when the alleged indictable offence occurred while the person was on bail for another charge. In 1998, one-third of all detained persons seeking pretrial release in Ontario had to appear in court three or more times before a ruling was made; overcrowded court dockets and a lack of personnel to participate in bail hearings are among the probable reasons for the delays (Attorney General of Ontario, 1999: 2).

If the JP or judge decides to release the accused, the conditions under which that release will take place must be determined. Again, the Crown must show cause why conditions should be attached to the release. Accused persons may be required to relinquish their passport, report periodically to the police, avoid contact with the complainant or other witnesses, or not possess firearms. Young offenders may be required to live with a responsible person who agrees to guarantee that they will appear in court.

Accused persons may also be asked to enter into a recognizance in which they agree to forfeit a set amount of money if they fail to appear in court. Generally, there is no requirement that the money be produced before the accused is released. However, a monetary deposit may be required if the accused is not normally a resident of the province or lives more than 200 kilometres away.

Another option is to release the accused on a recognizance in which a *surety* promises to forfeit a set amount of money if the accused fails to appear in court. A surety is a friend or relative who agrees to ensure the accused person's appearance for trial. In most cases involving a surety, a deposit is not required. However, if a large sum of money is involved, the existence of collateral to guarantee the payment may have to be demonstrated. If a surety withdraws support, the accused will be placed in custody unless another surety is immediately available.

judicial interim release (or bail)
the release by a judge or JP of a person who has been charged with a criminal offence pending a court appearance

In some parts of Canada, accused persons who are released on bail may be subject to bail supervision by probation officers and/or electronic monitoring (see Chapter 6). Accused persons who violate the conditions of release or who fail to appear in court at the designated time may have new charges filed against them.

Note that bail in Canada is different from the bail often seen on American television. In the United States, a deposit of money is required in order to guarantee a person's appearance in court; this practice is followed only in exceptional cases in Canada. This explains the proliferation of bounty hunters in the United States, who pursue persons who "skipped" bail and did not appear in court. Canadian courts are generally sensitive to the possibility that cash bail requirements could leave accused persons of modest means to languish in custody while their more affluent counterparts remain free while awaiting trial.

Pretrial Remand

A small number of accused persons are denied release and remanded into custody. This is done through the issuance of a warrant of committal by a JP or judge. Other accused are placed in custody after violating their release conditions. Remand admissions to provincial/territorial institutions, then, include accused persons who have been charged with an offence and ordered by the court to be held in custody until their court appearance. These individuals have not been sentenced to custody or to a community sanction; they are being held because it is feared they will not appear for court or may reoffend before their court date, or because they have not arranged bail. Accused persons on remand constitute 35 percent of all custodial admissions to provincial/territorial correctional facilities (Johnson, 2004).

Accused persons who are remanded in custody still have the option of reapplying to the court for release at some point before trial. In fact, all detention orders must be reviewed by a judge (after ninety days in the case of indictable offences) to determine whether continued detention is necessary. One of the options available to the judge at that point is to order that the case be expedited (heard as soon as possible). Generally, those accused persons who are remanded in custody should expect to have their cases disposed of on a priority basis.

Security Certificates

Under the Immigration and Refugee Protection Act, **security certificates** can be issued against non-citizens (visitors, refugees, or permanent residents) in Canada who are deemed to pose a threat to national security. These persons can then be held in detention, without charge, for an indefinite period of time. The certificates must be signed by both the Minister of Citizenship and Immigration and the Minister of Public Safety and Emergency Preparedness. Foreign nationals who have a security certificate issued against them are automatically detained; permanent residents may be also detained if it is determined that they are a danger to society or are likely not to appear for court proceedings. Otherwise, permanent residents can be released under strict bail conditions. The detention period is indefinite. The reasonableness of the security certificate is reviewed by a judge of the Federal Court. If the court upholds the security certificate, it becomes a removal order from Canada and the person

security certificates
a process whereby non-Canadian citizens who are deemed to be a threat to the security of the country can be held without charge for an indefinite period of time

is deported to his or her home country. The Federal Court's decision in cases involving security certificates is final and cannot be reviewed.

Security certificates have been part of the Immigration and Refugee Protection Act since 1978, but it is only since the 9/11 attacks on the United States in September 2001 that this process has generated controversy. Since that date, security certificates have been issued for five individuals, all of them Muslim men who were deemed to pose a security threat to Canada. Two of these men—Mohamed Harkat and Adil Charkaoui—were permanent residents of Canada.

Mohamed Harkat, originally from Algeria, was detained under a security certificate in 2002 after the Canadian Security Intelligence Service (CSIS) had watched him for five years and gathered evidence that, the federal government argued, indicated he had ties with terrorist movements in Afghanistan, Pakistan, and Chechnya. Harkat has been identified by CSIS as a member of al-Qaeda and as a "sleeper agent." In 2005 the Federal Court of Appeal upheld the use of a security certificate against Harkat (2005: FCA 393), thus increasing the likelihood that he will be deported to his home country. In January 2006, the Supreme Court of Canada is to hear Harkat's appeal that is based on the assertion that the security certificates are unconstitutional.

Adil Charkaoui, from Morocco, was also identified as an al-Qaeda sleeper agent by CSIS and determined to be a threat to the security of Canada. He was arrested in 2003 and held in detention until 2005, when he was released under bail conditions that included electronic monitoring. The Federal Court of Appeal dismissed his challenges to the security certificate process; however, in 2005 the Supreme Court of Canada agreed to hear his appeal, which is centred on the constitutionality of security certificates. The court is expected to hear the case in early 2006 in conjunction with that of Mohamed Harkat.

Amnesty International and other human rights groups have argued that the security certificate process violates fundamental human rights, including the right to a fair trial and the right to protection against arbitrary detention. Of concern is that much of the evidence in security certificate cases is heard *in camera* (behind closed doors), with only the Federal Court judge and government lawyers and witnesses present. Although persons who have been detained receive a summary of a portion of the evidence, the specific allegations against them and the sources of the allegations are not disclosed to the detainee. As well, evidence against the detainee may be presented in court without the detainee and his or her lawyer being present; this precludes a cross-examination of witnesses (http://www.amnesty.ca/take_action/actCertificates_300305.php).

The Supreme Court of Canada has ruled that the certificate process does not violate the principles of fundamental justice (*Chiarelli v. Canada*, [1992] 1 S.C.R. 711). And in 2004, in the case of Adil Charkaoui (2004: FCA 421), the Federal Court of Appeal held that the security certificate process is constitutional and does not violate the Charter of Rights and Freedoms. The controversy surrounding security certificates highlights the ongoing challenge of balancing individual rights with those of society. Increasingly, though, concerns are being raised that people are being detained under security certificates for long periods of time in provincial correctional facilities that are not designed for indefinite custody (Horsey, 2005).

What do you think?

What is your opinion of security certificates? Do you agree with the concerns expressed by Amnesty International? Explain.

Access to Legal Representation

All adults accused of crimes have the right to retain legal counsel. As noted in Chapter 3, the Charter of Rights and Freedoms stipulates that persons who are arrested and detained must be informed of this fact; and they must be permitted to contact a lawyer before giving a statement, if they so choose. The right to retain legal counsel levels the playing field, so to speak, between the accused and the police and Crown attorney. Most Canadians are unaware of their rights or the intricacies of this country's complex legal system. In our adversarial system, the police and prosecution enjoy the home field advantage, and the lawyer is on the defendant's team.

Most accused persons require legal representation, yet not all of them can afford a lawyer. There is no blanket right to state-paid legal representation. At arrest, the police officer recites this Charter warning: "It is my duty to inform you that you have the right to retain and instruct counsel without delay" (see Box 3.5). The right to *retain* counsel does not impose an absolute duty on provincial governments to provide all accused persons with *free* counsel. However, all persons who are arrested or detained must have the opportunity to access preliminary advice from duty counsel through a toll-free telephone line, where such services exist. According to the Supreme Court of Canada (*R. v. Prosper,* [1994] 3 S.C.R. 236), detainees must be told they may qualify for free counsel if they meet the financial criteria of the local legal aid plan.

Legal Aid

Clearly, it would be unacceptable for wealthy criminal defendants to have lawyers while poor defendants go unrepresented. At the same time, the universal provision of free legal representation would be expensive. There are also concerns about whether free representation can be as good as representation paid for by the accused.

Although every province/territory has a legal aid plan, Canada's jurisdictions vary greatly with respect to which types of cases qualify for assistance and which income levels are sufficiently low that an applicant is entitled to full or partial coverage. In recent years, several provinces have lowered the qualifying income levels as one means of stemming the dramatic rise in legal aid costs. Also, some types of cases no longer qualify for legal aid. It is not uncommon, for example, for applicants to be required to demonstrate that they face the very real prospect of being incarcerated for the offence. Across the country, legal aid services are delivered by lawyers in private practice, who are paid by a legal aid plan; by legal aid staff lawyers; and by lawyers working in legal aid clinics (Canadian Bar Association, 2005). This includes duty counsel services in provincial courthouses.

There are three models for the delivery of legal aid services by the provinces/territories:

1. *Judicare.* Historically the most common model. Defendants who pass a "means test" qualify for a legal aid certificate and can retain a private lawyer, who bills the legal aid plan for services rendered. The hourly rate for legal aid can be half what the lawyer would charge a paying client, and there is a

ceiling on how much the lawyer can bill the plan. Lawyers are not required to take legal aid cases.

2. *Staff.* In this model, salaried lawyers provide legal aid services through legal aid clinics. Generally less expensive than the judicare model, the staff model is becoming more common as provincial/territorial governments struggle to contain the rising costs of legal aid.

3. *Mixed.* In the mixed model, both private and legal aid staff lawyers provide legal services. In Ontario, services are delivered through a community clinic program. Some community clinics are geographically based, such as Westend Legal Services in Ottawa; others are based on cultural services, such as the Aboriginal Peoples Clinic in Toronto; and still others are based on one type of legal problem encountered, such as family law.

The delivery models include a certificate program based on the judicare model, a community clinic program, and a duty counsel program. The latter program includes staff duty counsel, twenty-four-hour hot line advice for persons in custody, and private lawyers who offer duty counsel services on a per diem basis (Legal Aid Ontario, 2004).

Overall spending on legal aid has been declining, owing to stricter eligibility criteria and restrictions in both the types of cases covered by legal aid and the fees paid to lawyers. Concerns have been expressed that these cutbacks could severely affect access to legal assistance for lower income individuals.

Legal Representation of Young Offenders

Besides the rights available to adults (the right to retain counsel, and the right to be so advised when arrested or detained), the Youth Criminal Justice Act states that youths have the right to counsel during bail hearings, transfer hearings (held to determine whether a youth should be raised to adult court), trials, sentencing hearings, and disposition reviews, and also when alternative measures are being considered. Moreover, the judge is required to ensure that a young person has counsel independent of his parents if it appears that the interests of the youth and those of the parents are in conflict. Youths who apply for legal aid may be turned down, just like adults. However, youths who fail to qualify can ask the judge to appoint counsel to act on their behalf.

Fitness to Stand Trial

A fundamental principle of the common law is that the accused person must be fit to stand trial. During the early stages of the court process, a lawyer may suspect that his or her client is suffering from some degree of mental illness. The existence of a mental disorder at the time of the offence may be integral to the defence strategy. However, mental disorder is a concern for another reason. Accused persons who cannot understand the object and consequences of the proceedings because of mental disorder are unfit to stand trial. In other words, they are unable to instruct their counsel or even fully appreciate that they are on trial.

At the request of the defence counsel or on its own initiative, the court may order that the accused person be assessed to determine fitness. That order is normally in force for no more than five working days, but a longer period can be ordered in "compelling circumstances."

Almost always, the fitness of an accused person to stand trial is assessed by a psychiatrist while the accused is either remanded in custody or at a hospital or psychiatric facility. Those found unfit to stand trial may be detained in a mental health facility until deemed fit to stand trial by a body such as the Ontario Review Board. Once the accused is found fit, the trial can resume. If a person never achieves a state of fitness, the Crown may conclude that it is no longer prudent to continue the criminal prosecution. In cases where the alleged offence is not serious, an accused who is found to be unfit to stand trial may simply be diverted into the provincial/territorial mental health system. There are other defences that speak to the capacity to "commit" the offence; these are discussed later in the chapter.

Arraignment and Plea

The arraignment of the accused takes place early in the process, if not at first appearance. The charges are read in open court, and the accused can enter a plea. The two most common pleas are "guilty" and "not guilty." If a plea of guilty is entered, the case goes directly to sentencing (see Chapter 6); a plea of not guilty results in the case being bound over for trial. Technically, every accused person—even those who are "guilty as sin"—can plead not guilty. Remember that in our adversarial system of justice, all accused persons are presumed innocent and the onus is on the Crown to prove guilt. Pleading not guilty, therefore, is not the same as claiming innocence.

Accused persons may plead not guilty because they are, in fact, innocent; because they have a plausible defence and want to exercise their right to a trial; and/or because their lawyer has advised them to do so. On the other hand, accused may plead guilty because:

- they are, in fact, guilty;
- they want to get the case over with as quickly as possible;
- their lawyers advised them to do so;
- they are satisfied with the sentence promised in plea negotiations; *and/or*
- they are taking the rap for someone else.

A not guilty plea can prolong the process for months, even years. Accused persons who are remanded in custody can be especially anxious to end the proceedings, because time served on remand is "dead time" in that they will not be given credit for it at sentencing.

Although most cases end with a guilty plea, they do not always begin that way. Accused persons often plead not guilty at the outset of the process, in part to strengthen their position in any plea bargaining that may take place. Accused who plead not guilty can change their plea to guilty at any point before the verdict.

Plea Bargaining

"In the halls of justice, the only justice is in the halls."—Lenny Bruce, comedian

Logically, it would seem to be to a defendant's advantage to take a chance and go to trial, in the hope that incriminating evidence will be declared inadmissible

plea bargaining
an agreement whereby an accused pleads guilty in exchange for the promise of a benefit

or a key witness will become unavailable. However, the court system would be overwhelmed if the majority of cases went to trial. This is one justification for plea bargaining—a controversial but pervasive practice that is not written in law or policy.

What is plea bargaining? Simply put, it is an agreement in which an accused gives up the right to make the Crown prove the case at trial in exchange for the promise of a benefit. For example, the Crown can promise the possibility of a lower sentence by withdrawing some charges; by reducing a charge to a *lesser but included offence* (that is, an offence that is similar but not as serious); by proceeding summarily rather than with an indictment; by asking the judge that multiple prison sentences run concurrently rather than consecutively; or by agreeing to a joint submission to the judge about sentencing. In addition (see Chapter 6), there is a pervasive belief—which may well be true—in the existence of a guilty plea "discount": that is, a defendant who pleads guilty can expect a lower sentence than if convicted after trial. Following is a description of the "Cave"—a cramped, windowless room in the Scarborough, Ontario, courthouse and the site of plea negotiations:

> The Cave exists for one purpose—for prosecutors to cut deals with defence lawyers and generally hasten cases. A circus-like atmosphere prevails in the main corridor directly outside the Cave. Police, lawyers, defendants and their families compete for standing space and wait for these cases to be called. The Cave door is almost always open, revealing two prosecutors who sit like expectant shopkeepers. Periodically, a defence lawyer or duty counsel ventures in to sound out a prosecutor about a particular case. The prosecutor may or may not end up having carriage of the case, but he or she can offer an informal view of what a fair plea bargain might involve (in Makin, 2003: F6).

As Box 5.5 illustrates, plea bargaining has both benefits and drawbacks.

Crime Victims and Plea Negotiation Although there has been increasing involvement of crime victims in sentencing hearings and parole hearings, less attention has been given to the potential role of crime victims in the plea negotiation process. Canadian criminologists Verdun-Jones and Tijerino (2005) argue that there are a number of benefits associated with giving crime victims the right to participate in plea negotiations. For example, victims can provide the Crown with information about the incident; they may emerge more satisfied with the criminal justice process; and the opportunity to participate in the plea negotiation process may help them heal.

Plea negotiations do have a number of potential benefits for crime victims. A guilty plea by the accused spares the victim the trauma of testifying in court, ensures that the case does not drag on for months or years, and eliminates the uncertainty over the final verdict of the judge or jury. Existing guidelines, policies, and laws encourage prosecutors to ascertain the views of victims and to inform them of the outcome of plea negotiations. The final responsibility for assessing the appropriateness of a plea agreement rests with Crown counsel, although ultimate jurisdiction for the outcome of the case still resides in the court. In Manitoba, crime victim legislation requires Crown counsel to consult with crime victims regarding any plea negotiation.

What do you think?

What is your position on plea bargaining? More specifically, are policies or guidelines required to provide a framework for plea bargaining, or is the status quo acceptable?

focus

BOX 5.5

Point/Counterpoint: Plea Bargaining

Point

Supporters of plea bargaining argue that it:

- saves time and taxpayers' money by encouraging guilty pleas;
- reduces the backlog of cases;
- spares complainants the difficult task of testifying;
- helps offenders take responsibility for their crimes by admitting guilt;
- does not compromise the administration of justice; *and*
- provides an opportunity to get evidence against co- or other defendants that might not otherwise be available to the police or Crown (see Box 5.6).

Counterpoint

Detractors counter that plea bargaining:

- brings the administration of justice into disrepute;
- does not follow any policy or guidelines and is therefore subject to abuse;
- places pressure on innocent defendants to "cop a plea" in order to avoid being found guilty at trial and receiving a more severe sentence;
- places pressure on persons who committed the offence(s) to plead guilty; *and*
- is a closed process that is not subject to public scrutiny and threatens the rights of accused persons.

Judges and Plea Bargaining Judges may be frustrated with the plea bargain process. One Ontario provincial court judge stated: "Pre-trial negotiations are fine when they are properly conducted. But my problem is that I question how many of them are properly conducted. And even if they are properly conducted, half the time I'm not told on the record why the lawyers came to the agreement they did. Justice is becoming less and less visible" (in Makin, 2003: F6).

In an interesting case from Nova Scotia, the provincial Supreme Court overturned the decision of a provincial court trial judge not to accept a joint recommendation (plea bargain) for the sentencing of a sex offender submitted by Crown and defence counsel (*R. v. Hamm* [2005] 230 N.S.R. (2nd) 41). The trial judge declined to accept the proposed sentence of house arrest (conditional sentence) for nine months, followed by three years' probation, noting the premeditated nature of the offence and the fact that the assaults on the thirteen-year-old victim occurred over the course of several months. The trial judge instead sentenced the offender to seven months in jail, to be followed by three years' probation. The Supreme Court judge, in reversing the trial judge's decision and imposing the sentence recommended by the Crown and defence, stated that the judge had erred in not giving counsel notice that he was not going to accept the proposed sentence and that he had not provided sufficient reasons for not accepting the joint recommendation.

Perhaps no single criminal case placed plea bargaining under the spotlight as emphatically as that of the convicted murderer, Karla Homolka (see Box 5.6).

focus

BOX 5.6

Karla Homolka's Deal: Canada's Most Controversial Plea Bargain

In July 1993 in Ontario, Karla Homolka pleaded guilty to two counts of manslaughter and received a twelve-year jail sentence. She was released from prison in 2005, having been denied parole and statutory release. Homolka's sentence was the result of a plea bargain between her lawyer and Crown counsel. Her husband, and co-offender in the kidnapping, torture, sexual assault, and murders of two fourteen-year-old girls, was Paul Bernardo, who would subsequently be convicted of two counts of first degree murder and additional sexual offence-related charges and, furthermore, be declared a dangerous offender. In 1995, Bernardo was sentenced to life in prison without the possibility of parole for twenty-five years.

In February 1993 the police executed a search warrant on the Bernardo home but failed to find videotapes that contained recordings of the sexual assaults on the two teenage girls, on Karla's younger sister (also fourteen years old), and on at least one other young woman. In May of that year, Bernardo's lawyer secured access to the home and located the videotapes hidden in the ceiling. He did not turn these tapes over to the police, but kept them and subsequently turned them over to Bernardo's new lawyer. In the meantime, in the same month in 1993 that the tapes were found, Homolka signed off on a plea agreement reached between her lawyers and Crown counsel. The agreement included the proviso that she would testify against Bernardo at trial. The videotapes were finally turned over to the police in 1994 and shown at Bernardo's trial. The recordings revealed that Homolka had played an active role in the sexual torture and subsequent deaths of the two

teenagers. Evidence presented at the trial also indicated that Homolka had drugged and helped Bernardo rape her sister, who died from an overdose of drugs administered by Homolka and Bernardo. There were calls for the plea bargain to be rescinded and for Homolka to be sent to trial, but the agreement of twelve years' imprisonment stood.

Throughout the twelve years of her imprisonment and after her release in July 2005, the debate continued over the plea bargain with

Francois Hudon/CP Photo Archive

An artist's sketch of Karla Homolka sitting in a courtroom in June 2005 prior to her release from prison after serving a 12-year sentence.

Homolka (called "a pact with the devil" by Bernardo's lawyer) and whether it was necessary to secure the conviction of Bernardo for the deaths and for a string of violent sexual assaults. Debate also continues over whether Homolka had instigated the murders or, as several psychiatrists testified, had been a "compliant victim of a sexual sadist." There is the widespread view that, given the actions of Homolka on the videotape recordings, she was equally responsible for the deaths of the two girls and should have received the same life sentence as Bernardo. And, that once her role in the deaths of the girls had been revealed on the videotapes, the plea bargain should have been nullified and her case sent to trial.

The case has been documented in several books, including *Karla: A Pact with the Devil* (Williams, 2003) and *Lethal Marriage: The Unspeakable Crimes of Paul Bernardo and Karla Homolka* (Pron, 1995).

Disclosure of Evidence

Understanding the strength of the Crown's case helps an accused person and his or her lawyer decide on a plea. Early in the process, the Crown must give the defence lawyer access to all evidence that might be presented by the prosecution in a trial, especially any potentially *exculpatory evidence* (evidence that might indicate the accused did not commit the crime). This process is called "disclosure of evidence" or "discovery." The failure to disclose evidence can trigger a Charter remedy because it impairs an accused person's right to make full answer and defence to the charges. However, the disclosure requirement does not work in reverse: the defence is not obliged to disclose material to the prosecution.

Mode of Trial

The "trier of fact" in a criminal case—usually a judge—decides whether the guilt of the accused person has been proved beyond a reasonable doubt. In a small number of cases, a jury of citizens makes this decision. Jury trials are virtually mandatory in some types of cases, an available option in many, and prohibited in others. Jury trials are not available for summary conviction offences; nor, with a handful of exceptions, are they available in youth court. In fact, there are key differences in the prosecution of summary conviction offences, indictable offences, and charges heard.

The key roles in criminal courts are played by the judge, the prosecutor or Crown counsel, the defence counsel, the witnesses, and the jury. The clerk assists in the administration of the trial, and the stenographer records all proceedings. The accused person is generally present throughout the proceedings, and may testify but is not required to do so. To avoid media scrutiny, a "famous person" being charged may be represented by counsel, negotiate a guilty plea through a plea bargain, and not appear at all. Figure 5.6 illustrates the court settings in traditional common law courts and in Quebec's criminal courts. To take a virtual tour of several provincial courts in Alberta, visit http://www.albertacourts.ab.ca.

The flow of cases through the court system is depicted in Figure 5.7.

FIGURE 5.6   Court Settings: Common Law and Quebec Courts

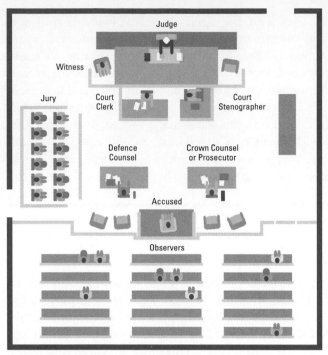

Traditional common law court setting

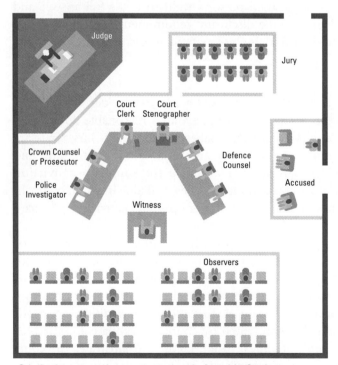

Criminal court setting most commonly found in Quebec

Source: Court Settings: Common Law and Quebec Courts. *Canada's System of Criminal Justice*, 1993, pp. 28–29. Department of Justice. Reproduced with the permission of the Minister of Public Works and Government Services Canada, 2006.

FIGURE 5.7   Flow of Cases through the Canadian Court System

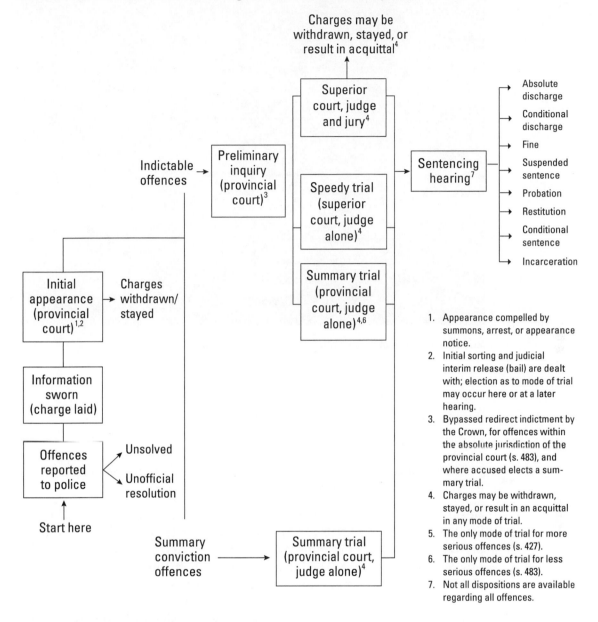

Source: Adapted from C.T. Griffiths and S.N. Verdun-Jones, *Canadian Criminal Justice*, 2nd ed. (Toronto: Harcourt Brace, 1994), 4. © 1994. Reprinted with permission of Nelson, a division of Thomson Learning: www.thomsonrights.com, FAX 800 730-2215.

Summary Trial

When the case involves a summary conviction offence, or when the Crown proceeds summarily, it is resolved in a provincial court. Summary trials do not involve juries, and the sentences are usually less severe.

Indictable Offences

When the accused is charged with an indictable offence, or the Crown proceeds by indictment on a hybrid (elective) offence, a different sequence of events unfolds. The Criminal Code defines three categories of indictable offences:

Felicity Don/CP Photo Archive

An artist's sketch of a jury with a sheriff sitting nearby

- offences under the absolute jurisdiction of provincial courts;
- offences under the absolute jurisdiction of superior courts; *and*
- electable offences.

The key difference is *election*—that is, the right of the accused to choose to be tried by a judge instead of a jury.

Absolute Jurisdiction of Provincial Courts Section 553 of the Criminal Code lists the less serious indictable and hybrid (elective) offences wherein the accused person has no choice but to be tried in a provincial court, even if the Crown proceeds by indictment. The offences include theft (other than cattle theft), obtaining money on false pretences, fraud, and mischief (where the subject matter of the offence is not a testamentary instrument and its value does not exceed $5,000). The list also includes keeping a gaming or betting house and driving while disqualified. There are no jury trials in provincial court.

Absolute Jurisdiction of Superior Courts Section 469 of the Criminal Code is a list of serious offences that are also non-electable offences. The list includes murder, treason, and piracy. These cases must be tried in a superior court before a jury unless both the accused and the provincial attorney general agree to waive this right.

The processing of non-electable offences begins with a preliminary hearing, sometimes called a preliminary inquiry. This (usually) short hearing is held to determine whether there is a *prima facie* case—that is, sufficient evidence to justify the time and expense of a criminal trial. A magistrate or provincial court judge listens to some (or all) of the Crown witnesses. The court may order a

The courtroom built for the Air India trial in Vancouver, British Columbia. Estimated cost: $7.2 million.

publication ban to protect the identity of any victim or witness and is required to order a publication ban to protect the identity of all victims of sexual offences and witnesses of sexual offenders who are less than eighteen years old (http://canada.justice.gc.ca/en/news/nr/1999/doc_24280.html).

The judge does not rule on the guilt of the accused at the preliminary hearing, but must decide if the Crown has evidence that could be used to prove guilt. If the judge does so decide, there is a *prima facie* case. If there is not a *prima facie* case, the judge will dismiss the case or at least dismiss the problematic charges against the accused. Usually, the matter is committed to trial, and a trial date is set. The accused person can waive the right to a preliminary hearing and go directly to trial. In rare cases, which generally involve more serious allegations, the provincial attorney general can skip the preliminary hearing and go straight to trial. This course of action is called "preferring the indictment."

Electable Offences Most indictable offences fall into neither of the two categories just described. These are the electable offences, and the accused person has three modes of trial from which to choose:

1. trial by a provincial/territorial court judge;

2. trial by a superior court judge sitting alone; *or*

3. trial by a superior court judge and a jury.

The Charter guarantees the right to a jury trial if the alleged offence carries a maximum sentence of more than five years' imprisonment. However, not every accused person wants a jury trial.

Once an accused person has elected, he or she can reelect another option or enter a guilty plea, in which case there will not be a trial. It is also possible (although this happens rarely) that the provincial attorney general may inter-vene and require a jury trial if the offence is punishable by more than five years' imprisonment and if the accused has chosen one of the first two options. Accused persons who choose option 1 do not have a preliminary hearing and waive their right to trial by jury. Accused persons who choose option 2 or 3 are entitled to a preliminary hearing unless they waive that right. Accused persons who abscond and who fail to appear for trial by jury on the appointed court date may lose their right to a jury trial.

Jury decisions must be unanimous. Unlike their American counterparts, Canadian jurors are prohibited from discussing their deliberations with the media. Each province/territory has legislation that sets out the qualifications for jurors and that provides other directives for selecting juries and guiding their activities. Jury duty is still regarded as a civic duty; with a few exceptions, a person called for jury duty will be required to serve.

There are three important differences between trial by jury and trial by judge alone:

- In jury trials, the jury decides on the true facts and determines the person's guilt; in trials with a judge alone, the judge determines the law and the facts.
- In a jury trial, the judge makes a "charge to the jury," during which he or she instructs the jurors about the law that applies to the case.
- Judges give reasons for their decisions. Lawyers use these reasons to help predict outcomes in future cases with similar facts. Jurors don't give reasons with their verdict (School Net, 2000).

Youth Court

Procedures for the prosecution of young persons are found in both the Criminal Code and the Youth Criminal Justice Act. Cases are heard by youth court judges, who are—depending on the jurisdiction—either provincial/ territorial court or family court judges. The prosecution of cases in youth court follows a process similar to the one already described for summary conviction offences, and the possible case outcomes are the same. The names of young offenders can be published if they are convicted of a crime and receive an adult sentence; are fourteen to seventeen years of age and have been sentenced for murder, attempted murder, manslaughter, or aggravated sexual assault; have a record of violent offences; or are at large and considered by a judge to be dan-gerous. Youth court cases do not involve preliminary hearings; also, there is no opportunity for jury trial, except for youths charged with murder (in such cases, a preliminary hearing is held).

---

stay of proceedings
an act by the Crown to terminate or suspend court proceedings after they have commenced

THE TRIAL

A trial takes place if the accused person, who pleads not guilty, does not change that plea and the Crown does not withdraw the charges or terminate the matter with a **stay of proceedings.** Especially in provinces/territories where the police have sole responsibility for laying charges, a Crown attorney may review cases early in the process and screen out those which might not succeed as well as those for which there is insufficient evidence to secure a conviction. Because of this practice of case screening, and guilty pleas on the part of accused persons, most cases do not go to trial.

Pretrial Conferences

In cases involving trial by jury, the presiding judge must order a pre-hearing conference, which is attended by the Crown, the defence counsel, and the judge. They can discuss any "matters to promote a fair and expeditious trial." In non-jury cases, pretrial conferences are optional. Informal "pretrials" are becoming increasingly routine. They take place in a judge's chambers and involve an off-the-record discussion of issues surrounding the case. These discussions provide an opportunity for plea bargaining, since the presence of a judge can promote a fair resolution between the two parties (which is preferable to eleventh-hour bargaining on the courthouse steps).

The Case for the Crown

The trial begins with the prosecution calling witnesses and presenting evidence in support of the position that the accused is guilty. For interpersonal offences, the testimony of the complainant may well be the Crown's key evidence. At the very least, the Crown attorney must produce evidence covering all the major elements of the offence. For example, in a murder case the Crown must show that someone died and that the death was culpable homicide (that is, not an accident or death by natural causes). There should be evidence linking the accused to that death (for example, eyewitnesses, fingerprints, DNA evidence, or circumstantial evidence such as a strong motive on the part of the accused). Expert witnesses may be called to interpret evidence or to present findings from the police investigation.

It is the task of the Crown to prove the guilt of an accused person beyond a reasonable doubt; if the Crown fails to do this, there can be no conviction.

The Case for the Defence

The defence attorney can cross-examine Crown witnesses and challenge the admissibility of Crown evidence. At the close of the Crown's case, the defence may enter either an insufficient-evidence motion or a no-evidence motion, suggesting to the judge that the state has not made its case and that there is no point to continuing the trial. If the judge agrees, the case is dismissed. If not, the defence presents its case.

As part of the case for the defence, the accused person may testify (give evidence) on his or her own behalf, but is not obliged to do so. For accused persons who testify in court, there are advantages and disadvantages. On the one hand, testifying gives defendants an opportunity to present their side of the story and establish credibility. On the other hand, a defendant who testifies opens the door to cross-examination by the Crown prosecutor, who will attempt to point out weaknesses and inconsistencies in the testimony (University of Alberta Law Students, 2000). In addition, if the defendant presents good character or reputation as a reason why he or she could not have committed the offence, the prosecution is free to enter into evidence any previous convictions. Otherwise, the jury or judge cannot learn if the accused has a prior criminal record (at least until the sentencing phase, if the defendant is found guilty).

There are three basic defences: (1) "you've got the wrong person" defences ("I did not commit the offence"); (2) excuse-based defences ("I committed the offence but had good reason to do so"); or (3) procedural defences ("a grievous police or prosecution error makes it impossible for me to get a fair trial").

The "You've Got the Wrong Person" Defence The first type of defence strategy centres on one of two possibilities: that the police arrested the wrong person, or that the complainant fabricated the allegation, thus no crime was committed. To support a claim of false accusation, the defence may present evidence verifying the defendant's alibi. One example of a verified alibi is establishing that the defendant was in jail when the offence was committed.

Excuse-Based Defences The second set of strategies involves the use of traditional defences such as self-defence, duress, or necessity. For example, those charged with murder can claim provocation to justify a reduction to the charge of manslaughter (here, provocation is a partial defence). Common excuse-based defences include consent, automatism, and not criminally responsible on account of mental disorder (NCRMD).

Consent Consent can be used as an excuse-based defence if the complainant voluntarily agreed to engage in the activity in question. A common example: two individuals can consent to a fistfight if both parties appreciate the risks and neither is seriously injured. However, lack of resistance to an assault or sexual assault does not constitute consent if the submission of the complainant was achieved by force, threats, fraud, or the exercise of authority. An example of this last is where there is a clear power imbalance, as in the case of teacher–student, doctor–patient, or parent–child relationships.

Consent can be real, or it can be apprehended if the accused mistakenly believed that a non-consenting complainant consented. In the past, some accused persons were able to argue successfully that they honestly believed the complainant was consenting to sexual activity—that "no meant yes." Parliament has responded by restricting the use of consent as a defence for sexual offences. Since 1988, for example, consent has not been available as a defence if the complainant was under fourteen at the time of the offence.

In 1992 the "no means no" law was enacted. If a sexual assault complainant expresses "by words or conduct, a lack of agreement to engage in the activity," consent to the activity is deemed not to have been obtained. Neither can consent be used as a defence if "the complainant, having consented to the sexual activity, expresses, by words or conduct, a lack of agreement to continue to engage in the activity." Also, consent cannot be voluntarily given by someone who is induced to engage in the activity with a person who is abusing a position of trust, power, or authority. Nor is apprehended consent a defence to a sexual assault if the accused person's belief in consent arose from self-induced intoxication or reckless or willful disregard, or if the accused did not take "reasonable steps" to ascertain whether the complainant was in fact consenting.

Some of the most controversial defences centre on the argument that the accused is not criminally liable because he or she could not have formed *mens rea* (see Box 5.3). This mental state could have been temporary and situational or the result of a long-term mental disorder. In order to convict in most cases, the judge or jury must believe that the action under scrutiny—the *actus reus*— was a voluntary exercise of the person's will. In a 1994 decision, the Supreme Court of Canada found a man not guilty of raping a woman because he had been so intoxicated that his actions were not voluntary (*R. v. Daviault*, [1994] 3 S.C.R. 63). This decision triggered a public outcry, and the federal government responded by amending the Criminal Code to specify that self-induced intoxication cannot be used to excuse certain types of interpersonal offences, including assault and sexual assault, even if *mens rea* is absent.

Automatism Automatism is another controversial excuse-based defence. In 1987 a Toronto-area man drove across town, fatally stabbed his mother-in-law, and promptly turned himself in to the police, confessing repeatedly to the crime. Despite overwhelming evidence that he had killed the woman, he was acquitted at trial. The jury accepted the defence evidence that the man had been sleepwalking and therefore could not have formed the requisite *mens rea*.

The defence of automatism does not always result in an outright acquittal. In 1996, Calgary socialite Dorothy Joudrie was tried for the attempted murder of her estranged husband to whom she had been married for thirty-nine years. The evidence was clear that she had shot him six times (he survived to testify as a Crown witness). Her defence was non-insane automatism. She claimed that she had no memory of the offence and that she had been in a dissociative or automatonic state brought about by years of mental and physical abuse at the hands of her husband, as well as by his leaving her for another woman. Had Joudrie's defence succeeded (like that of the sleepwalker), she would have walked out of the court a free woman. However, the jury—possibly believing that she was suffering from so-called insane automatism—handed down a finding of "not criminally responsible on account of mental disorder." See *R. v. Fontaine* ([2004] 1 S.C.R. 702) for a more recent case involving mental disorder automatism.

Not Criminally Responsible on Account of Mental Disorder (NCRMD) A classic case of NCRMD arose in 1995 when a forty-year-old Ontario man suddenly stabbed his three-year-old son, proclaiming "for the good of the world, I got to do this." He later told of an apocalyptic dream warning that everyone on earth would be "slaughtered by extraterrestrial super-powered devices." The accused stated: "If I didn't kill him he would be taken by the absolute power of Satan and the universe would take a de-evolutionary tailspin lasting for 1000 years." The man, a schizophrenic, believed the boy would come back to life afterward. Fortunately, the boy recovered. The man was charged with attempted murder.

There was no doubt he had stabbed the boy, but did he really understand what he was doing? The court found the man guilty of the offence, but not criminally responsible on account of mental disorder. In other words, a "disease of the mind" had rendered him incapable of appreciating the nature and quality of his actions or knowing they were wrong. The sleepwalking killer discussed earlier was not suffering from a mental disorder. Sleepwalking, or somnambulism, is not a disease of the mind, so the sleepwalker was not legally insane when he killed his mother-in-law.

Insanity is no longer a true defence. Before 1992, it could result in a finding of not guilty by reason of insanity. A person so declared is now usually detained in a mental health facility for an indeterminate period, with the date of release to be determined by a review board.

Battered Woman Syndrome Experienced by women who have suffered chronic and severe abuse, battered woman syndrome (BWS) is a condition characterized by feelings of social isolation, worthlessness, anxiety, depression, and low self-esteem. In the landmark case *R. v. Lavallee* ([1990] 1 S.C.R. 852), the Supreme Court of Canada accepted BWS as a defence, and it has since been used successfully in subsequent cases (see Tang, 2003).

In *R. v. Malott* ([1998] 1 S.C.R. 123), the Supreme Court of Canada dismissed the appeal of a woman who had been convicted of killing her abusive husband and who had used BWS as a defence at her trial. In its judgment, the Court stated that the trial judge had properly informed the jury with respect to the evidence on BWS and how such evidence related to the law of self-defence. The Court further noted: "'Battered woman syndrome' is not a legal defence in itself, but rather is a psychiatric explanation of the mental state of an abused woman which can be relevant to understanding a battered woman's state of mind."

In the mid-1990s, the federal government conducted a review of the cases of women convicted and incarcerated for killing abusive spouses, partners, or guardians. Based on the review's findings, several women were granted conditional pardons and early parole (Ratushny, 1997; http://www.canada.justice.gc.ca/en/dept/pub/sdr/rtush.html).

Procedural Defences The third category of defence strategies focuses not on the guilt or innocence of the accused, but rather on the conduct of the police or prosecution, or perhaps the validity of the law itself. In common parlance, this is known as "getting off on a technicality." The judge can rule on most of these issues before the trial even starts, but Charter arguments can sidetrack a trial until the issue is resolved.

Procedural defences fall roughly into four categories:

- *Challenging the validity of the applicable law.* Some successful procedural defences have attacked the constitutionality of the law used to charge the accused.

- *Challenging the validity of the prosecution.* Another strategy is to claim that the police or prosecutors acted unfairly in the investigation or charging of the accused. Entrapment and abuse of process (discussed in Chapter 3) are two examples of unfair conduct.

- *Contesting the admissibility of evidence gathered by the police.* If key evidence is excluded, not enough evidence may remain to prove guilt beyond a reasonable doubt. As noted in Chapter 3, a confession gained after an unlawful arrest may be ruled inadmissible if its use would bring the system of justice into disrepute.

- *Seeking a remedy for violation of a Charter right.* In extreme circumstances, the violation of an accused person's Charter rights can be remedied by the termination of the prosecution. There have been cases in which the Charter right to trial within a reasonable time was violated and a stay of proceedings was ordered by the presiding judge.

Appeal

There is the possibility of appeal once a case has been concluded in court. Not every case can be appealed; in fact, in the majority of cases an appeal is *not* filed. The right to appeal exists only in certain situations; in others, the Court of Appeal can grant leave (permission) to appeal. Either the Crown prosecutor or the defence lawyer can file an appeal. A distinction is made between grounds for appeal which involve questions of law, and those which involve questions of fact, and those which involve both. Note also that there are different appeal procedures for summary conviction and indictable offences.

Once an appeal has been launched, the incarcerated appellant may be released on bail until the appeal is heard. The judge who hears this request considers, among other things, the *prima facie* merits of the appeal itself; this is to ensure that frivolous appeals cannot routinely be used to defer the serving of a prison sentence.

An appeal may be directed at the verdict, or the sentence, or both. However, most appeals are directed at the sentence: the incarcerated appellant thinks it is too severe, or the prosecutor thinks it is too lenient. The appellate court assesses the sentence against the prevailing norms found in reported case law. In deciding the case, the court may raise the sentence, or lower it, or refuse to interfere with what the trial judge ordered.

An appeal of the verdict usually requires some demonstration that a legal error was made at the trial or that new, exculpatory evidence has been discovered. There are five possible outcomes of verdict appeals. The court can:

- decide not to hear the appeal;
- hear the appeal and dismiss it;
- substitute a conviction on a lesser but included offence (and probably reduce the sentence);
- direct that the offender be acquitted; *or*
- order a new trial.

Most appeals originate from the defence side. However, in Canada it is also possible for a Crown attorney to appeal the acquittal of an accused.

CRIME VICTIMS AND THE COURT PROCESS

In Chapter 2, it was noted that there has been a growing emphasis on the rights and needs of crime victims. This section examines those needs and rights in the context of the criminal court process.

Ensuring the Safety of Victims

Between arrest and sentencing, the crime victim(s) may be concerned about retaliation and intimidation by the offender. The legal mechanisms for protecting victims from pretrial intimidation include the following:

- *Obstruction of justice charges.* Attempts to "dissuade" a person from giving evidence (by threats or bribery, for example) can justify a charge of obstruction of justice (s. 139). This can include instances where a defendant makes harassing telephone calls before the case is resolved in the courts.
- *Non-association conditions of pretrial release.* It is routinely required that defendants agree not to communicate with any witnesses in the case, including victims. When the defendant is charged with a violent offence or with criminal harassment ("stalking"), the judge is required to consider imposing a non-communication condition and a condition restricting possession of firearms by the accused.
- *Peace bonds.* A person who has reasonable grounds to fear victimization by a specific party (injury or damage to property) can seek one of three measures—a civil restraining order, a common law peace bond, or a judicial recognizance order—under section 810 or 810.1 of the Criminal Code. Collectively, these remedies are referred to as peace bonds. They may be used even when the potentially dangerous party is not currently before the courts. Most often, they are used in cases involving family violence where an ex-partner continues to pose a risk to a woman and/or her children. They are also used in disputes between neighbours or between family members. Breaching a peace bond is a criminal offence.

Accused persons who violate these measures may have their bail revoked or be remanded to custody pending trial. Note that many women's groups are dissatisfied with the ability of these measures to protect women who have been abused, and encourage these women to make additional arrangements for their safety.

Victims as Witnesses

At trial, victims may be called upon to testify. They are summoned to court (by subpoena) and are paid a small fee just like any other witness. Some courthouses have on-site victim services to help witnesses on the day of court, but most do not.

Testifying in a public courtroom, in the presence of the alleged perpetrator, is an emotionally arduous task for victims. The Criminal Code contains concessions to crime victims who testify in court; most of these, however, are offered at the discretion of the trial judge. These provisions (see below) were developed in response to concerns that the victims of sexual offences were often victimized a second time by the experience of testimony and cross-examination.

Publication Bans

In trials involving certain offences, the judge can order that the "identity of the complainant or of a witness and any information that could disclose the identity of the complainant or of a witness shall not be published in any document or broadcast in any way." This applies when the case involves incest, extortion, a sexual assault, a sexual offence involving children, or what is commonly called loan sharking. The name of the accused may be included in the ban if release of the name would identify the victim. If requested by the prosecutor, the ban is mandatory. If the prosecutor has forgotten to ask, the judge must inform the victim, or any witness under eighteen, of the right to make such a request.

It is a summary conviction offence to contravene a publication ban. A few newspapers have been convicted and fined for violating publication bans.

"Rape Shield" Provisions

The term "rape shield" (an ill-chosen label) refers to the admissibility of evidence about a sexual assault victim's sexual history (with people other than the accused). Since their enactment in 1976, these contentious provisions have received intense judicial scrutiny and have been modified several times. Generally, the provisions make the sexual history of the victim inadmissible as evidence in court when defendants want to show that the complainant was "more likely to have consented" to the alleged offence or "is less worthy of belief." The restriction is not absolute, and defendants can argue that the information is necessary for their defence.

Courtrooms Closed to the Public

It is a fundamental principle of justice that trials must be public. Only rarely will a judge agree to "exclude all or any members of the public from the courtroom for all or part of the proceedings." To eliminate or reduce the number of observers, the judge must be convinced that "it is in the interests of public morals, the maintenance of order, or the proper administration of justice."

There are no limitations as to the types of cases from which the public may be excluded. However, the issue usually arises when a crime victim or witness is apprehensive about testifying in front of people—a common occurrence with sexual crimes. Potential embarrassment on the part of the victim is insufficient reason to clear a courtroom. It must be demonstrated that the witness could not relate the full details of the offence, perhaps because of extreme fear or the stress of a crowded courtroom. The primary concern is not with protecting the emotional well-being of the witness, but rather with ensuring that justice is done. If fear or stress on the part of the victim were to result in incomplete testimony, the "proper administration of justice" would be compromised. The applicant—usually the prosecutor—must prove that public exclusion is necessary.

Section 486(1.1) of the Criminal Code speaks directly to the need to protect child witnesses under fourteen in cases where the accused is charged with a sexual offence. There are other concessions besides this for child witnesses.

Concessions for Children Who Testify

To facilitate the testimony of children and of "vulnerable adults" such as the mentally disabled, a number of arrangements have been made.

Testimonial Aids Two testimonial aids are available for child victims who testify in court:

- A one-way screen may be placed on the witness box to block the child's view of the defendant while allowing all people in the court to see the witness.
- A closed-circuit television (CCTV) system may be used so that the child can be outside the courtroom but be able to hear and respond to questions from the courtroom.

Neither of these aids is used very often, probably because the threshold test to prove their need is quite high. The judge must believe that the child absolutely could not testify with an unobstructed view of the defendant. In addition, the availability of CCTV equipment is limited in many areas. As with public exclusion orders, the purpose of these mechanisms is not to minimize trauma for the child witness, but rather to increase the probability that the child will provide "a full and candid account" of the crime.

Support Person The judge may order that a "support person of the witness's choice be permitted to be present and to be close to the witness" when he or she is testifying. Persons who are expected to testify in the proceedings may be disqualified from acting in this capacity. Typically, the judge will order the support person and the witness not to communicate with each other while the testimony is given.

Videotaped Evidence A videotape of a child victim's statement, made within a reasonable time after the alleged offence, can be introduced as evidence at trial. The judge views the tape beforehand to identify any portions that should be edited out. The child, when testifying, must "adopt the contents" of the video by agreeing that the information relayed in the tape is truthful. The rationale is that it is easier for a child to agree that the taped statement is true than to repeat all of the details in court.

Accused Prohibited from Cross-Examination An accused acting as his own counsel is not usually permitted to cross-examine a witness under fourteen in cases involving sexual offences or violent crimes. The court will appoint a lawyer to undertake that task. There is some discretion granted to the judge, who can permit an unrepresented defendant to conduct the cross-examination if "the proper administration of justice" requires it.

CURRENT CHALLENGES FOR THE COURTS

Canada's criminal courts are encountering a number of challenges. Discussed next are some of the more significant issues relating to today's courts.

Delay and Backlog in the Criminal Court Process

Case delay is a problem in many jurisdictions and may compromise the administration of justice by:

- causing criminal charges to be stayed on constitutional grounds;
- jeopardizing the right of accused persons to a speedy trial;
- creating a stressful situation for crime victims;
- disrupting the lives of jurors and witnesses; *and/or*
- increasing the costs of criminal justice (Attorney General of Ontario, 1999: 2).

In a landmark case, *R. v. Askov* ([1990] 2 S.C.R. 1199), the court ruled that a twenty-three-month pretrial delay violated the defendant's right to trial within a reasonable time. This decision resulted in tens of thousands of pending cases being dropped, many of which involved serious offences; it also prompted efforts to speed the resolution of cases in the court system [see also *R. v. MacDougall* ([1998] 3 S.C.R. 45) and *Blencoe v. British Columbia (Human Rights Commission)* ([2000] 2 S.C.R. 307) for a discussion of the meaning of reasonable delay].

Recent court statistics indicate the following:

- The average elapsed time from first to last appearance for cases is just over seven months.
- Sixteen percent of cases take more than one year to resolve.
- The average number of court appearances per case is 5.9—a significant increase from the average of 4.1 a decade ago (Thomas, 2004).

Statistics on case delay indicate that some offences take longer to prosecute than others. The median elapsed time from first to last court appearance in cases of prostitution is 350 days—the longest average time, longer than the average time for cases involving sexual assault (331 days), fraud (315 days), and being unlawfully at large (106 days, the shortest average elapsed time) (Thomas, 2004: 4).

Especially in urban areas, where there are often serious backlogs of cases, the operations of the criminal courts easily bring to mind an overworked assembly line. Box 5.7 offers as accurate description of an Ontario courtroom.

It is important to note that the courts are not solely responsible for case delay. A number of other factors may slow the progress of a given case, including the complexity of the case, the type of offence being prosecuted, lawyers' decisions, and (sometimes) the failure of the accused person to appear in court. Defence lawyers may attempt to lengthen the pretrial process as long as possible, hoping that witnesses' memories will fade (and perhaps the notoriety of the crime) and that evidence will be misplaced. It is interesting that accused persons who represent themselves have not been found to slow the pace and efficiency of criminal courts (Currie, 2004).

There is also the practice of judge shopping. The defence counsel and Crown may seek out a judge who will be amenable to accepting a plea bargain

focus

BOX 5.7

A Day at the Justice Factory

A monotonous buzz fills the air of Courtroom 111, a crowded chamber where the misadventures of daily life are aired and swiftly judged. Suddenly the murmuring stops. Heads snap upright. Judge Tony Charlton has raised his voice.

"Look at me!" he barks at a 28-year-old construction worker, whom he has just sentenced for assaulting his wife.

"You can't hit her! Drunk or not, the police will put you in one of those stinking jails again.

"Good luck to you."

Conditional discharge, six months probation. Next!

As the man exits, trailed by his wife, Courtroom 111 returns to its former pace and purpose—the steady dispensation of trial dates, and the sentencing of those who have pleaded guilty to [summary conviction] offences. Here at Toronto's Old City Hall, one of the busiest justice mills in North America, the pace is numbing and the subject matter is wearying.

By the end of the afternoon, in Courtrooms 111, 115, and 101, hundreds of defendants will have appeared before a beleaguered corps of judges and justices of the peace. In other chambers, dozens more will have preliminary hearings or trials.

To the uninitiated, it's a bewildering scene. To the veterans, it's a daily grind.

The tension is present 24 hours a day, but for Ted Kelly it begins at 7:30 a.m., when he arrives at the cells in the building's basement to begin interviewing those arrested the night before. Mr. Kelly is a duty counsel in the bail courts, paid for by Ontario's legal-aid system to act on behalf of those who have no lawyer, or whose lawyer is not able to appear.

"We interview them through plexiglass and have to shout at them," he says. "Sometimes there are five or six lawyers in the interview area in these little stalls, all shouting at [their clients]. It's just not an effective environment to interview people."

He sees them again before their bail hearings; generally, he gets about a minute with each one. Amid the swirl of Courtroom 115, the drug court, Mr. Kelly is almost constantly in motion— slim, blond, almost deferential in manner.

He is the only really sympathetic face an accused person will encounter. "The hardest part of the job is not having enough time for people."

In the southeast corner of the building, outside Courtrooms 115, 116, and 117, the corridors form a bazaar of justice with police, prosecutors, and defence lawyers bargaining pleas and sentences. In theory, everyone charged by the police with an offence has the right to plead guilty or not guilty. In practice, if there is good evidence against a defendant, the bazaar takes over.

The Crown wants a conviction, without the expense to taxpayers of a preliminary hearing and trial. The defence wants the charge reduced, and the lightest possible sentence.

The most common bargain, in drug cases, is the reduction of a trafficking charge to the lesser one of possession. A trafficker can be sentenced to life; a typical sentence for cocaine possession is six months.

In Courtroom 115, a steady stream of slouching defendants emerges from the custody room at the rear. The public benches teem with chatting family members, children, defendants on bail waiting for a trial date, the odd lawyer.

At 10 a.m., in Courtroom 111, Judge Charlton is hearing bail requests, trial dates and guilty pleas. He decides it's time to speed things up.

"Everybody listen. Keep still," he says, as the hubbub in the spectators' gallery fades. "The name of the game is to get out of here.

"Go to that side, form a line, come up and tell me your name. The only people I want to see are the people who have been here since 9."

A line of about fifteen people forms at one side of the room. Within half an hour, Judge Charlton has dealt with each one, but more have arrived. An hour later, there is still a line-up.

Regularly, tempers boil over. During a trial that shows signs of dragging on well into the afternoon, Judge John Kerr is letting off steam because the room next door is empty and his own work is far from done.

"There's a court that's down before 1 o'clock," he says. "And we're still dealing with this case, and we've got one to follow. Where's the equity in the system?"

By the noon break in Courtroom 115, the prosecutor has grown noticeably more impatient with each bail hearing, more brusque with each drug charge defendant. He's opposing bail for almost all of them.

It's not uncommon for prosecutors to see several drug defendants go free because police witnesses are double-booked and don't show up.

Prosecutors "do drug court one day a week," comments one Crown counsel. "We rotate through. You couldn't do it full time because it would drive you nuts."

Then he offers a blunt assessment that might apply to many—perhaps too many—who are involved in the administration of summary justice. "You get pretty callous doing the job. You just sort of do it."

Source: D. Shoalts, "A Day at the Justice Factory," *The Globe and Mail* (29 January, 1992), A1, A7. Reprinted with permission from *The Globe and Mail.*

that has been struck. In the words of an Ontario provincial court judge: "Lawyers judge-shop as part of their plea bargain. They are seeking to find a judge who will agree in advance to commit him or herself to a range of sentence" (in Makin, 2003: F6). A defence lawyer will often attempt to get a case heard by a judge who is viewed as more lenient toward the particular offence the accused is alleged to have committed. Another common strategy among defence lawyers (noted earlier) is to prolong the process as long as possible. One result is that cases involving serious offences tend to drag on for years. There is also the very serious problem of lawyers "double and triple booking" court appearance times as legal aid rates fail to keep pace with the costs of operating a legal practice.

To speed up the flow of cases through the system, case flow management techniques have been introduced in many jurisdictions. For example, British Columbia is developing "paperless" courtrooms. Computer technology is being used to schedule cases, track cases through the court process, e-file court documents, and facilitate communication among court personnel. Innovations being introduced include video conferencing to increase access to the courts for those in isolated regions and video links to detention centres for use in bail hearings and adjournments.

Some argue that one way to reduce the backlog of cases would be to eliminate preliminary hearings. Proponents of this reform point out that charge screening, mandatory Crown disclosure of evidence to the defence, and pretrial conferences serve much the same function as these hearings.

The Emergence of Megatrials

A key theme in this book is the increasing complexity of the cases confronting the criminal justice system. This has been due, in part, to the rise of global criminal networks and to the increasing sophistication of criminal activity. Stronger enforcement efforts against outlaw motorcycle gangs and criminal syndicates have resulted in criminal trials involving multiple defendants, lengthy witness lists, and thousands of pages (and in many instances thousands of pieces) of evidence. Also, these types of cases are expensive. It is estimated, for example, that the convictions of four associates of the Rock Machine motorcycle gang in Quebec in 2001 for drug trafficking under the anti-gang legislation cost taxpayers $5.5 million. And in 2003, four Hells Angels pleaded guilty to similar charges in Montreal after the province constructed a special, high-tech courtroom at the cost of $16.5 million.

The case of Air India Flight 182 (see Box 5.8) illustrates the demands that a megatrial can place on governments, the police, the courts, and taxpayers. This crime devastated hundreds of families in Canada and abroad.

As of 2005, the federal government was developing guidelines for the management of megatrials.

focus   BOX 5.8

Air India Flight 182: The Worst Terrorist Act in Canadian History

On June 23, 1985, Air India Flight 182 exploded and crashed into the Atlantic Ocean off the west coast of Ireland while on a flight from Montreal to London. All 329 passengers on board, most of whom were Canadian citizens, were killed. This coincided with a suitcase explosion in the baggage terminal at Narita Aiport in Tokyo, Japan, just after the case was unloaded from a Canadian Pacific airlines plane that had arrived from Vancouver. The investigation into the bombings centred on certain individuals in British Columbia's Sikh community, who were involved in the struggle for an independent Khalistan in India. An Air India Task Force, led by the RCMP and working alongside police agencies in Europe, India, the United States, and Asia, spent fifteen years investigating the case, at one point offering a $1 million reward for evidence that would help convict the perpetrators.

In 2000, two B.C. residents were charged with multiple offences under the Criminal Code relating to the deaths of the passengers and crew on Air India 182. The charges included first degree murder, conspiracy to commit murder, and attempted murder. In 2001, a third defendant was charged with the same offences; two years later, he pleaded guilty to manslaughter for his part in the bombing and received a sentence of five years. He had previously been convicted for his role in the Tokyo explosion, which killed a baggage handler and seriously injured several others.

The Crown proceeded by direct indictment against the remaining two defendants. The trial began in April 2003 and went on for 19 months and 232 court days until December 2004. It was held in the B.C. Supreme Court in Courtroom 20, which had been built specifically for the Air India trial and for future megatrials. This courtroom (estimated cost: $7.2 million) is loaded with state-of-the-art technology, including security screening devices, video monitors for presenting evidence, voice-activated video cameras, cameras for broadcasting proceedings directly to other locations in the courthouse, digital recording and storage facilities, and videoconferencing technology. (See photo on page 209.)

In March 2005, the presiding judge found the two defendants not guilty on all charges, stating: "Despite what appears to have been the best and most earnest efforts by the police and the Crown, the evidence has fallen markedly short." The judge also commented on what he found to be "unacceptable negligence" on the part of the Canadian Security Intelligence Service (CSIS) for destroying evidence, including tapes of wiretaps. After the verdict, the families of the victims and others called for a public inquiry into the Air India investigation. The RCMP is still investigating what remains the worst terrorist incident in Canadian history, which led to the longest and most complex criminal trial in the annals of Canadian criminal justice.

It is estimated that, as of 2005, the Air India case had cost the federal government and the Government of British Columbia a total of nearly $60 million. As well:

- More than 300 RCMP officers have worked on the case over the past twenty years.
- More than a thousand witnesses and experts have been interviewed, some of them multiple times.
- Prior to and during the trial, more than thirty lawyers worked on the case on the defence and prosecution sides.
- There are more than one million documents of evidence.

After the trial, former Ontario premier Bob Rae was appointed by the federal government as an independent consultant regarding the case. In late 2005, he issued a report calling for an official inquiry. The federal government accepted this recommendation, and Mr. Rae will lead a policy-oriented public inquiry. As of early 2006, the terms of reference for the inquiry had not been released; however, it will be limited in scope.

Sources: K. Bolan, (March 11, 2005), "Evidence Falls Short: Judge," *National Post,* A9; Government of British Columbia, Ministry of Attorney General, "Air India" (2005). Online at http://www.ag.gov.bc.ca/airindia/index.htm; "On Trial: Air India Trial." Retrieved from http://www.lawcourtsed.ca; http://www.airindiatrial.ca

Wrongful Convictions

The criminal justice system operates within a legal and procedural framework that is designed to ensure that the rights of those accused of criminal offences are protected and that their guilt must be proved "beyond a reasonable doubt." Despite this, there has emerged in recent years an increasing concern about wrongful convictions (also referred to as miscarriages of justice)—cases in which individuals were convicted who were later found to be innocent. In many instances, these people served time in prison for crimes they did not commit (see Campbell and Denov, 2004). Many of these cases have been taken up by the Association in Defence of the Wrongfully Convicted (http://www.aidwyc.org).

Sections 696.1 to 696.6 of the Criminal Code—"Applications for Ministerial Review—Miscarriages of Justice"—give the federal Minister of Justice the power to review criminal cases to determine whether there has been a

Redman/AP/CP

A door from Air India Flight 182 floats in the Atlantic off the coast of Ireland after a bomb exploded causing the plane to crash.

miscarriage of justice. These regulations set out the requirements for an application for a criminal conviction review. Completed applications are forwarded to the Criminal Conviction Review Group (CCRG); lawyers on that body review and investigate the applications and make recommendations to the minister (see http://canada.justice.gc.ca/en/ps/ccr).

Box 5.9 lists several of the more high-profile wrongful conviction cases.

Just since 2000, there have been at least six government inquiries into wrongful convictions. A recent report on wrongful convictions found that they rarely occur as the result of a single mistake or event; they are almost always a consequence of a series of events (Federal/Provincial/Territorial Heads of Prosecutions Committee Working Group, 2004). The same report identified a number of factors that often contribute to innocent people being found guilty of crimes they did not commit, including:

- "tunnel vision" on the part of police and the Crown (that is, the focus of the investigation was too narrow);
- mistaken eyewitness identification and testimony;
- false confessions;
- the testimony of in-custody informers; *and*
- defective, unreliable, and unsubstantiated expert testimony.

For example, even though Simon Marshall had been diagnosed with schizophrenia in 1992, a psychiatrist declared him fit to stand trial in 1997 after conducting a ten-minute assessment. His defence lawyer presented no expert testimony by psychiatrists to rebut the Crown's witness, nor did his lawyer raise any arguments about his mental disorder. A commission of inquiry into the wrongful conviction of Guy Paul Morin (Kaufman, 1998) found that this miscarriage of justice had been the result of a flawed justice system and various systemic errors. In particular, the police, prosecutors, and forensic scientists had been guilty of tunnel vision of "staggering" proportions.

Accountability and Review of Judicial Conduct

Of all of the major components of the criminal justice system, the courts and the judiciary are probably the least known by the public. This is due in part to the difficulties of securing access to the judiciary, the tendency of judges to

focus

BOX 5.9

Selected Cases of Wrongful Conviction

- James Driskell was convicted of first degree murder in 1991 on the basis of perjured testimony from two witnesses and poor forensic work. He spent twelve years in prison before being exonerated by DNA evidence. It was later revealed that Manitoba prosecutors concealed evidence for a decade which indicated that the two witnesses had been paid to testify and had also been granted immunity (Tyler, 2005).

- Simon Marshall, a mentally disordered man, spent five years in prison after pleading guilty in 1997 to crimes that he did not commit. The federal justice minister ordered a new trial; however, the Province of Manitoba (where he was convicted) decided that Marshall would not be retried and stayed the charges against him. As of 2005, he was under psychiatric care in hospital.

- In 2000, DNA evidence cleared Thomas Sophonow of the killing of a shop clerk in Winnipeg in 1981. Tried three times before being convicted, he spent nearly four years behind bars before the Manitoba Court of Appeal acquitted him in 1985. As compensation for his wrongful conviction, the City of Winnipeg, the government of Manitoba, and the federal government contributed to a $2.6 million settlement.

- David Milgaard was convicted and given a life sentence in 1970 for the murder of a

A photograph of David Milgaard in a Manitoba prison in 1991

Gerard Kwiatkawski/CP Photo Archive

Saskatoon nursing aide. He spent twenty-three years in prison before the Supreme Court of Canada set aside his conviction in 1992. Five years later, he was exonerated by DNA evidence. In 1999, Larry Fisher was found guilty of the murder. Milgaard received a $10 million settlement for his wrongful imprisonment.

• Guy Paul Morin was charged in 1985 for the murder of a nine-year-old girl. He was acquitted by a jury at trial, but the Ontario Court of Appeal ordered a new trial and he was found guilty of the crime in 1992. Morin served fourteen months in custody before being released; he was subsequently declared innocent based on DNA evidence.

• Donald Marshall, a Mi'kmaq, was sentenced to life imprisonment in 1971 and spent eleven years in prison before being acquitted by the Nova Scotia Court of Appeal in 1983. A Royal Commission of Inquiry (Hickman, 1989) concluded that incompetence on the part of the police and the judiciary contributed

to his wrongful conviction; so did the fact that he was an Aboriginal person (Sack, 2005).

• Steven Truscott was convicted and sentenced to death in 1959 for the murder of a twelve-year-old girl near Clinton, Ontario. In 1960, the Ontario Court of Appeal dismissed his conviction appeal but commuted his sentence to life imprisonment. Truscott served ten years in prison (1959 to 1969) before being released on parole, which he successfully completed in 1974. In 2001, with the media again taking interest in the case, an application for a retrial was filed by the Association in Defence of the Wrongfully Convicted, based on new documents that were not presented at trial. In 2004, the federal Minster of Justice declined to order a new trial and referred the case to the Ontario Court of Appeal. There, as of early 2006, a decision was pending (Makarenko, 2005). See also Julian Schur's book, *'Until You Are Dead': Steven Truscott's Long Ride into History* (2002).

avoid publicity, and the reluctance of judges to express their opinions lest they be accused of harbouring biases. Impartiality and independence from government interference are hallmarks of the Canadian judiciary, and the Supreme Court of Canada has reaffirmed on several occasions the need for judges to be free from bias (see, for example, *R. v. Catcheway* [2000] 1 S.C.R. 838).

The judiciary's self-imposed silence has not spared it from a variety of criticisms, many of which concern its ethnic makeup, the sentences it metes out, and the extent to which judges are held accountable for their actions.

Where Do Judges Come From?

Judges at the provincial court level are appointed by provincial governments, while judges of the superior courts are appointed by the federal government. Appointments are for life so that once on the bench, judges need not consider the career implications when making controversial decisions. Each province/territory has in place a Judicial Advisory Committee composed of laypeople and lawyers. These screening committees forward nominations to the justice minister, who makes the final appointments. Most provinces have a parallel process for provincially appointed judges, and these referrals are made to the attorney general.

Elderly, white, Anglo males are overrepresented in the judiciary. Increasingly, efforts are being made to address the judiciary's failure to reflect Canada's cultural diversity. A recent study of federal, non-Supreme Court appointments found that more than half of appointees were supporters of the federal Liberal Party (Forcese and Freeman, 2005). The authors cross-referenced the records of donations to political parties between 1997 and 2003 and found that 53 percent of the appointees were probable or possible supporters of the Liberal Party. Only 4 percent were probable supporters of an opposition party. These findings raise concerns about the part played by political affiliation when federal judges are selected and the possible influence of patronage in the process for selecting judges.

Structures of Judicial Accountability

Criminal justice personnel are finding themselves increasingly accountable. The primary structure of accountability for federally appointed judges is the Canadian Judicial Council (http://www.cjc-ccm.gc.ca), which is chaired by the Chief Justice of Canada and is composed of judges. In keeping with its mandate— which is set out in the Judges Act—this council provides continuing education for judges, addresses issues concerning the administration of justice, and makes recommendations on judicial salaries and benefits. To exercise judicial accountability, the council investigates complaints made about federally appointed judges, be they from litigants, lawyers, complainants, accused persons, interest groups, or even other judges. Most complaints are made by the general public.

Complaints about judges arise from intemperate remarks and/or inappropriate conduct either on or off the bench. Displays of gender bias, racial bias, religious bias, conflict of interest, and cultural insensitivity are grounds for complaint, as is undue delay in rendering a decision (which should usually take no more than six months). In 2002, for example, the Supreme Court of Canada upheld the decision of the New Brunswick Judicial Council to remove a provincial court judge from the bench (*Moreau-Berube v. New Brunswick (Judicial Council)*, [2002] 1. S.C.R. 249). After hearing a number of break-and-enter cases in court, the judge had stated: "If a survey were taken in the Acadian Peninsula, the honest people against the dishonest people, I have the impression the dishonest people would win ... The honest people in the Peninsula, they are very few and far between" (Chwialkowska, 2002b). This was the first time a Canadian judge had ever been removed for comments made from the bench, although there have been investigations resulting in several judges "being allowed" to resign. An alleged inability to execute the functions of a judge because of mental infirmity is also grounds for complaint.

A review of the record indicates that few complaints ultimately result in the removal of a judge from the bench. As well, since the disciplinary procedure was established in 1971, there have been (as of 2004) only twelve inquiries by the council into the behaviour of a federal judge. Most complaints (there were a total of 194 files in 2004–05) are handled by the chairperson of the council. Sanctions range from removal from the bench (an extremely rare occurrence) to a leave of absence with pay or a letter of reprimand. Alternatives to these include counselling, educational workshops, or the requirement that the judge apologize to the complainant. In more serious cases, judges often choose to resign before the council can complete its inquiry. As of 2005, fewer than five judges had been removed from office on the vote of the council.

The conduct of provincial court judges is held to the same standards, but is monitored by local bodies rather than by the federal council. In Ontario and Quebec respectively, the Ontario Judicial Council and the Conseil de la magistrature du Québec review complaints made against provincially appointed judges. In jurisdictions that do not have stand-alone judicial councils, procedures have been established for receiving complaints. Some observers have questioned the adequacy of the structures for judicial accountability, especially in view of the fact that judges are generally appointed for life. Of particular concern is that in contrast to the accountability mechanisms for the police (see Chapter 4), those for judges are composed of judges and other members of the legal profession.

It is likely that some instances of judicial misconduct never come to the attention of the federal and provincial councils, owing to potential complainants feeling intimidated by the judge in question, the justice system, and

focus

BOX 5.10

The Case of Judge David Ramsay

On June 1, 2004, former B.C. provincial court judge David Ramsay was sentenced to seven years in prison for sexually assaulting several teenage Aboriginal girls in Prince George. He pleaded guilty to one count of sexual assault causing bodily harm, three counts of buying sex from minors, and one count of breach of trust. The sentence was two years longer than Crown counsel had asked for. During the sentencing hearing Ramsay apologized to four of his victims who were in court.

Evidence presented to the court indicated that Ramsay had sexually abused the young women, who were involved in the sex trade, over a ten-year period, intimidating them into remaining silent about his violent attacks on them. The girls, some as young as twelve, had appeared in court before Judge Ramsay, who was aware of their life circumstances and their vulnerabilities.

The RCMP authorities in Prince George were criticized for their slow response to the allegations against Ramsay; reports of his abuses had been circulating in the city for several years. After Ramsay was sentenced, two RCMP officers who had at one time been stationed in Prince George were themselves investigated for misconduct amidst allegations that they had covered up Ramsay's exploits. It was alleged that one of these officers had had sex with underage prostitutes in Prince George as well; he was suspended with pay while his case was being investigated by the RCMP Major Crime section. As of early 2006, no additional information on this investigation was available.

The Assembly of First Nations and the Native Women's Association of Canada called for an inquiry into the administration of justice in cases involving sexual assault against Aboriginal women and young women. The government did not act on this suggestion.

Source: "Judge David Ramsay." Retrieved from http://www.fathers.ca/judge_david_ramsay.htm

the complaint process. This may be particularly the case for people in vulnerable groups. And there may be times when judicial misconduct is shielded by the justice system itself.

It is possible that this occurred in the case of Judge David Ramsay, a provincial court judge in British Columbia (see Box 5.10). This case highlights the power that judges wield in the community, as well as the vulnerability of women, especially young women—in this case, Aboriginal young women.

Summary

This chapter has provided an overview of the criminal court system and the flow of cases through the court process. Most cases are heard and disposed of at the lowest level of the court hierarchy, the provincial/territorial courts. The Supreme Court of Canada—the federal "court of last resort"—hears cases that reflect ongoing efforts to balance the rights of individuals with the protection of society.

As a result of diversion, plea bargaining, and pre-charge screening by Crown counsel, not all incidents lead to trials. In those cases which do go to trial, the accused may offer a defence ranging from claims of innocence to various justifications and excuses for the behaviour. Once a case has been decided, the defendant can appeal the verdict or the sentence, or both. Among the challenges facing the criminal courts are case delay/backlog, the emergence of megatrials, the wrongfully convicted, and judicial accountability.

Key Points Review

1. The courts that deal with criminal cases in Canada are the provincial/territorial courts, courts of appeal, the provincial superior courts, and the Supreme Court of Canada.
2. The decisions of the Supreme Court of Canada are final and cannot be appealed.
3. Several important principles govern the processing of criminal cases, including presumption of innocence, burden of proof on the Crown, *mens rea*, and *actus reus*.
4. Criminal offences are classified as summary, indictable, or hybrid.
5. Crown counsel are responsible for prosecuting cases and are involved in a wide range of activities, including screening cases, negotiating pleas, and recommending sentences to the court.
6. Diversion programs are designed to keep offenders from being processed further into the criminal justice system.
7. Many factors influence the decision of the police and/or the Crown as to whether to lay a charge.
8. There are a variety of ways to ensure that an accused person will appear in court, including issuing an appearance notice, issuing a summons, and remanding the person into custody.
9. Security certificates are processes whereby non-residents of Canada who are deemed a threat to the country can be held, without charge, for an indefinite period of time with the objective of deportation.

10. All adult accused persons have the right to retain legal counsel but are required to pay for their own lawyer unless they qualify for legal aid.
11. There are various models by which legal aid is provided to those who qualify.
12. Although plea bargaining is not formally acknowledged in law, it plays a critical role in case processing.
13. Defence counsel typically use the "you've got the wrong person" defence, excuse-based defences, or procedural defences.
14. The courts are using a variety of measures to address the needs and rights of crime victims in the court process.
15. Case delays and backlogs are serious issues in the criminal courts and are caused by a variety of factors.
16. Megatrials are the consequence of the increasing complexity of cases entering the criminal justice system.
17. Increasing attention is being given to the wrongfully convicted and to the activities and decisions of the police, prosecutors, and judges that contribute to miscarriages of justice.
18. Structures of judicial accountability exist but have rarely been used to discipline judges.

Key Term Questions

1. What role does the principle of *stare decisis* play in the decision making of criminal courts?
2. What is a **preliminary hearing** and why is it an important part of the criminal court process?
3. What is a **hybrid** (or **elective**) **offence**?
4. Discuss the objectives of **diversion.**
5. What role does **judicial interim release** (or **bail**) play in the court process and in what circumstances might it be denied?
6. What are **security certificates** and why are they controversial?
7. Define **plea bargaining** and discuss the issues surrounding this practice.
8. What is meant by a **stay of proceedings**?

References

Attorney General of Ontario. (1999). *Report of the Criminal Justice Review Committee: Executive Summary.* Retrieved from http://www.attorneygeneral.jus.gov.on.ca/english/about/pubs/crimjr/

Bonta, J. (1998). "Adult Offender Diversion Programs." *Research Summary* 3(1): *Corrections Research and Development.* Ottawa: Solicitor General Canada.

Campbell, K.M., and M.S. Denov. (2004). "Wrongful Conviction: Perspectives, Experiences and Implications for Justice." *Canadian Journal of Criminology and Criminal Justice,* 46(2):101–208. Special Issue.

Canadian Bar Association. (2005). "Options for the Delivery of Legal Aid Services." Retrieved from http://www.cba.org/CBA/Advocacy/legalAid/options.aspx

Chiodo, A.L. (2001). "Sentencing Drug-Addicted Offender and the Toronto Drug Court." *Criminal Law Quarterly,* 45(1/2):53–100.

Choudhry, S., and C.E. Hunter. (2003). "Measuring Judicial Activism on the Supreme Court of Canada: A Comment on *Newfoundland (Treasury Board) v. NAPE*." *McGill Law Journal*, 48(3):525–62.

Chwialkowska, L. (2002a, February 2). "Supreme Court Split on Police Powers." *National Post*. Retrieved from http://www.nationalpost.com
———. (2002b, February 8). "Firing of Judge Who Insulted Acadians." *National Post*, A4.

Currie, A. (2004). "A Burden on the Court? Self-Representing Accused in Canadian Criminal Courts." *JustResearch*, 11. Ottawa: Department of Justice Canada.

Department of Justice Canada. (2005). *Canada's Court System*. Retrieved from http://canada.justice.gc.ca/en/dept/pub/trib

Ehman, A.J. (2005). "A People's Justice." Retrieved from http://www.cba.org/CBA/National/Cover2002/justice.asp

Federal/Provincial/Territorial Heads of Prosecutions Committee Working Group. (2004). *Report on the Prevention of Miscarriages of Justice*. Ottawa: Department of Justice Canada.

Forcese, C., and A. Freeman. (2005). *The Laws of Government: The Legal Foundations of Canadian Democracy*. Toronto: Irwin Law.

Gomme, I., and M.P. Hall. (1995). "Prosecutors at Work: Role Overload and Strain." *Journal of Criminal Justice*, 23(2):191–200.

Hickman, The Honourable T.A. (1986). *Royal Commission on the Donald Marshall, Jr. Inquiry*. Halifax: Government of Nova Scotia.

Horsey, J. (2005). "Security Certificate Inmates the Victims of Jurisdictional Limbo: Critics." Retrieved from http://cnews.canoe.ca/CNEWS/Canada/2005/09/27/1238081-cp.html

Johnson, S. (2004). "Adult Correctional Services in Canada, 2002/03." Catalogue no. 85-002-XPE. *Juristat*, 24(10). Ottawa: Canadian Centre for Justice Statistics, Statistics Canada.

Kaufman, The Honourable F. (1998). *Report of the Kaufman Commission on Proceedings Involving Guy Paul Morin*. Toronto: Ministry of the Attorney General. Retrieved from http://www.attorneygeneral.jus.gov.on.ca/english/about/pubs/morin

Layton, D. (2002). "The Prosecutorial Charging Decision." *Criminal Law Quarterly*, 46(1/2):447–82.

Legal Aid Ontario. (2004). *Legal Aid Ontario Business Plan, 2004–2006*. Toronto.

Makarenko. J. (2005). "The Steven Truscott Case." Retrieved from http://www.mapleleafweb.com/education/spotlight/issue_59/index.html

Makin, K. (2003, April 26). "In the Back Halls of Justice." *The Globe and Mail*, F6, F7.

Monahan, P.J. (2004, April 4). "Top Court Now Favours Charter Challengers." *Ottawa Citizen*, A5.

Palmer, T. (1979). "Juvenile Diversion: When and for Whom?" *California Youth Authority Journal*, 32:14–20.

Pron, N. (1995). *Lethal Marriage: The Unspeakable Crimes of Paul Bernardo and Karla Homolka*. Toronto: Seal.

Ratushny, The Honourable L. (1997). *Self-Defence Review: Women in Custody—Final Report*. Ottawa: Department of Justice Canada.

Sack, M. (2005). "The Donald Marshall, Jr. Story." *Issues in the Mi'kmaq Community*. Retrieved from http://home.rushcomm.ca/~hsack/ issues01.html

School Net. (2000). "The Jury in Criminal Cases. Basic Information." Retrieved from http://www.acjnet.org/youthfaq/basic.html

Schur, J. (2002). *"Until You Are Dead": Steven Truscott's Long Ride into History*. Toronto: Vintage Canada.

Tang, K.-L. (2003). "Battered Women Syndrome Testimony in Canada: Its Development and Lingering Issues." *International Journal of Offender Therapy and Comparative Criminology*, 4(6):618–29.

Thomas, M. (2004). "Adult Criminal Court Statistics, 2003/04." Catalogue no. 85-002-XPE. *Juristat*, 24(12). Ottawa: Canadian Centre for Justice Statistics, Statistics Canada.

Tyler, R. (2005, March 4). "Another 'Miscarriage of Justice.'" *Toronto Star*. Retrieved from http://www.prisontalk.com/forums/showthread. php?t=110120

University of Alberta Law Students. (2000). *A Guide to the Law Regarding How to Run Your Own Criminal Trial in Alberta*. Edmonton: Student Legal Services of Edmonton. Retrieved from http://www.acjnet.org

Verdun-Jones, S.N., and A.A. Tijerino. (2005). "Victim Participation in the Plea Negotiation Process: An Idea Whose Time Has Come?" *Criminal Law Quarterly* 50(1/2):190–212.

Webster, C.M., and A.N. Doob. (2003). "The Superior/Provincial Court Distinction: Historical Anachronism or Empirical Reality?" *Criminal Law Quarterly* 48(1):77–109.

Williams, S. (2003). *Karla: A Pact with the Devil*. Toronto: Cantos International.

York University. (2000). "Supreme Court Now Major Focus for Interest Group Activity, New Study Shows." Media Release. Retrieved from http://www. yorku.ca/mediar/releases_1996_2000/archive/040400.htm

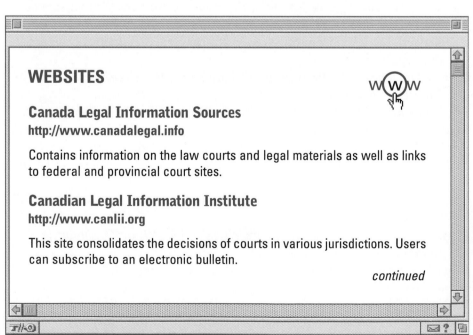

WEBSITES

Canada Legal Information Sources
http://www.canadalegal.info

Contains information on the law courts and legal materials as well as links to federal and provincial court sites.

Canadian Legal Information Institute
http://www.canlii.org

This site consolidates the decisions of courts in various jurisdictions. Users can subscribe to an electronic bulletin.

continued

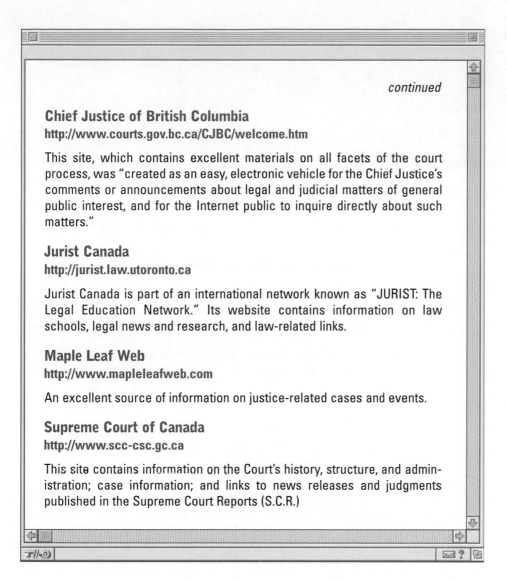

continued

Chief Justice of British Columbia
http://www.courts.gov.bc.ca/CJBC/welcome.htm

This site, which contains excellent materials on all facets of the court process, was "created as an easy, electronic vehicle for the Chief Justice's comments or announcements about legal and judicial matters of general public interest, and for the Internet public to inquire directly about such matters."

Jurist Canada
http://jurist.law.utoronto.ca

Jurist Canada is part of an international network known as "JURIST: The Legal Education Network." Its website contains information on law schools, legal news and research, and law-related links.

Maple Leaf Web
http://www.mapleleafweb.com

An excellent source of information on justice-related cases and events.

Supreme Court of Canada
http://www.scc-csc.gc.ca

This site contains information on the Court's history, structure, and administration; case information; and links to news releases and judgments published in the Supreme Court Reports (S.C.R.)

Sentencing is among the most difficult tasks that judges have to perform, and probably the most controversial. This is because Canada is a diverse and open society that encompasses a broad range of religious, social, cultural, and moral values and views; thus Canadians have widely disparate opinions on what constitutes a fit penalty for a particular offence (http://www.courts.gov.bc.ca).

For those convicted of a criminal offence, the next phase of the justice process is sentencing. A sentence represents far more than punishment for an offence. Offenders may well feel they are being punished; but sentencing has more important goals than this, including helping the convicted not to offend again in the future. Note that there are punitive aspects to a conviction in addition to the sentence: public humiliation or embarrassment; loss of employment; family breakup; possible exclusion from certain professions such as the law and policing; and, for those without Canadian citizenship, deportation from the country.

Recall from Chapter 2 that there are high levels of public dissatisfaction with the criminal courts, centring primarily on what Canadians perceive as overly lenient sentences imposed on the convicted. These negative perceptions probably owe much to media sensationalism and to a lack of understanding of the sentencing process. However, as the following discussion suggests, some aspects of the sentencing process do in fact warrant closer examination.

THE PURPOSE AND PRINCIPLES OF SENTENCING

Section 718 of the Criminal Code sets out the purpose and principles of sentencing:

> The fundamental purpose of sentencing is to contribute, along with crime prevention activities, a respect for the law and the maintenance of a just, peaceful and safe society by imposing just sanctions that have one or more of the following objectives:
>
> (a) to denounce the unlawful conduct;
> (b) to deter the offender and other persons from committing offences;
> (c) to separate offenders from society, where necessary;
> (d) to assist in rehabilitating offenders;
> (e) to provide reparations for harm done to victims or to the community; *and*
> (f) to promote a sense of responsibility in offenders, and acknowledgement of the harm done to victims and to the community.

Section 718.1 states that a sentence must be proportionate to the gravity of the offence and to the degree of responsibility of the offender. Section 718.2 states that a court that imposes a sentence shall also take into consideration the following principles:

(a) a sentence should be increased or reduced to account for any relevant aggravating or mitigating circumstances relating to the offence or the offender, and, without limiting the generality of the foregoing,

(i)   evidence that the offence was motivated by bias, prejudice or hate based on race, national or ethnic origin, language, colour, religion, sex, age, mental or physical disability, sexual orientation, or any other factor, *or*

(ii) evidence that the offender, in committing the offence, abused the offender's spouse or child, *or*

(iii) evidence that the offender, in committing the offence, abused a position of trust or authority in relation to a victim shall be deemed to be aggravating circumstances;

(iv) evidence that the offence was committed for the benefit of, at the direction of or in association with a criminal organization.

(b) a sentence should be similar to sentences imposed on similar offenders for similar offences committed in similar circumstances;

(c) where consecutive sentences are imposed, the combined sentence should not be unduly long or harsh;

(d) an offender should not be deprived of liberty, if less restrictive sanctions may be appropriate in the circumstances; *and*

(e) all available sanctions other than imprisonment that are reasonable in the circumstances should be considered for all offenders, with particular attention to the circumstances of Aboriginal offenders.

The Goals of Sentencing: The Cases of Mr. Smith and Mr. Jones

There are three main groups of sentencing goals in the criminal courts: utilitarian, retributive, and restorative. The real-life cases of "Mr. Smith" and "Mr. Jones" (not their real names) will be used to illustrate how these sentencing goals are applied. Mr. Smith was a Quebec-based police chief and swimming coach who was convicted of four counts of sexual assault for fondling two girls aged twelve and thirteen. Mr. Jones, a computer engineer based in British Columbia, was convicted of sexual assault for fondling his young stepdaughter over a two-year period. The cases of Mr. Smith and Mr. Jones—neither of whom had a prior criminal record—were heavily publicized in their respective communities, and both men eventually lost their jobs.

Utilitarian Goals

Utilitarian sentencing goals focus on the *future* conduct of Mr. Smith, Mr. Jones, and others who might commit similar offences. The sentence is designed to protect the public from future crimes in the following ways:

- By discouraging potential Mr. Smiths and Mr. Joneses from crime. This is *general* deterrence.

- By discouraging Mr. Smith and Mr. Jones from doing it again. This is *specific* deterrence.

- By curing Mr. Smith and Mr. Jones of what made them do it. This is *rehabilitation.*

- By keeping Mr. Smith and Mr. Jones in jail to protect society. This is *incapacitation.*

Retributive Goals

The past, rather than the future, is the focus of retributive sentencing goals, which include the following:

- *Denunciation*—that is, expressing society's disapproval of the behaviour of Mr. Smith and Mr. Jones.
- *Retribution*—that is, making the two men "pay" for their offences, based on the philosophy "an eye for an eye."

A key concept in retributive sentencing is *proportionality*–that is, the sentence handed down should be proportionate to the gravity of the offence and to the convicted person's degree of responsibility.

Restorative Goals

The most widely used restorative approaches are victim–offender reconciliation, circle sentencing, and family group conferencing. As noted in Chapter 1, restorative justice is based on the principle that criminal behaviour injures not only the victim, but communities and offenders as well. It follows that efforts to resolve the problems that the criminal behaviour has created should involve all three parties.

Regarding Mr. Smith and Mr. Jones, their victims were children, who because of their age would be excluded from any restorative justice forum. However, the victims' families would have the opportunity to discuss the impact of the crimes, and Mr. Smith and Mr. Jones would be held accountable for their criminal behaviour.

What Sentences Did Mr. Smith and Mr. Jones Receive?

The offence of sexual assault carries a maximum penalty of ten years' imprisonment. Neither Mr. Smith nor Mr. Jones had a prior criminal record, and both had a good job history; on the other hand, their offences were serious and had a significant impact on the victims. One of Mr. Smith's victims suffered long-term emotional and academic problems; Mr. Jones's former spouse and children experienced considerable emotional difficulties. In both cases, the children had been young and vulnerable. Mr. Smith had been an authority figure in the community, and parents had trusted him to supervise their children—a trust he violated. Similarly, Mr. Jones had violated the trust of his stepdaughter and most likely would have continued sexually abusing her had she not told her mother about his improper behaviour.

Mr. Smith was sentenced to three years' probation (the maximum) and 180 hours of community service. The Crown appealed the sentence on the grounds that it was too lenient. However, the Quebec Court of Appeal upheld the sentence, in part because Mr. Smith had been fired from his job as police chief and so had already experienced a severe sanction. The appeal court acknowledged that child abuse typically demands a denunciatory sentence for the protection of society, but noted that each case must be judged on its merits.

Mr. Jones was not so fortunate. He was sentenced to eighteen months' confinement in a provincial correctional facility and three years' probation (the maximum). In explaining the sentence, the presiding judge cited the objectives of denunciation and general and specific deterrence.

SENTENCING OPTIONS

The large majority of people convicted of criminal offences are not sent to prison but rather are placed under some form of supervision in the community. A number of traditional and contemporary dispositions are available to the courts; these range from probation—which was developed more than a century ago—to high-tech strategies such as electronic monitoring (EM). Falling somewhere between the sentencing extremes of absolute discharge and imprisonment are **intermediate sanctions.** These often involve the use of community programs and resources, and include EM, fines, community service, intensive supervision probation (ISP), and conditional sentencing.

> **intermediate sanctions**
> dispositions designed as alternatives to incarceration that provide control and supervision over offenders

The various alternatives to incarceration can reduce the numbers of adult and young offenders sent to correctional institutions, thereby reducing the costs of managing offenders and allowing offenders to avoid the negative consequences of incarceration (see Chapter 7). These programs also allow offenders to remain in the community, with their families and at their jobs, while receiving supervision and access to treatment programs and resources.

A variety of sentencing options are available to the courts. Some of these can be "mixed and matched." For example, a judge may impose a fine in conjunction with a sentence of incarceration, or a period of probationary supervision in conjunction with a sentence of two years less a day.

The case outcomes for 2003–04 are presented in Figure 6.1. Note that a conviction was recorded in 58 percent of the cases, while the charges were stayed or withdrawn in 36 percent of the cases. Also in 2003–04, approximately one-fifth (22 percent) of those found guilty received an absolute discharge, a conditional discharge, or a suspended sentence (Thomas, 2004).

Absolute Discharge

> **absolute discharge**
> a sentence wherein the accused is found guilty but does not gain a criminal record and is given no sentence

Adult and young offenders given an **absolute discharge** are free to leave the courtroom with no penalty. In fact, they are not technically convicted and can rightly claim to have no criminal record (although police information systems will continue to include discharges until the record is automatically removed after six months). In choosing this option, the court must be satisfied that it is "in the best interests of the accused and not contrary to the public interest."

Conditional Discharge

A conditional discharge is similar to an absolute discharge except that the offender is placed on probation, with various conditions, one of which is "to keep the peace and be of good behaviour." If the offender satisfies all the conditions within the specified period, he or she is discharged and deemed never to have been convicted. Those who fail to abide by the conditions, or who

FIGURE 6.1   **Adult Court Processing of Federal Statute Cases in Provincial and Selected Superior Courts, Provinces and Territories, 2003–04**

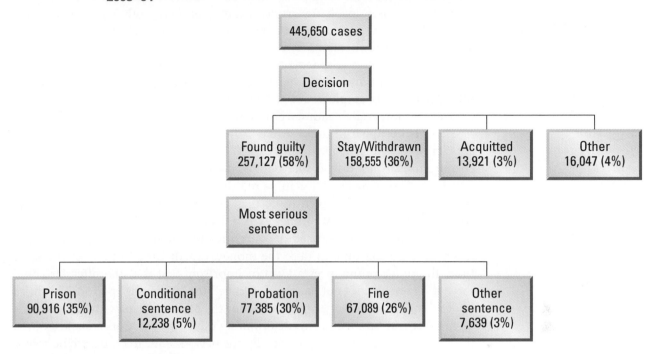

Source: Adapted from the Statistics Canada publication, *Juristat*, Adult Criminal Court Statistics, 2003/04, vol. 24, no.12, Catalogue 85-002, December 10, 2004, p. 5.

Notes: 1. "Found Guilty" decisions include absolute and conditional discharges. 2. No data on conditional sentences were gathered from Quebec during 2003–04, so there was an undercount for this disposition. 3. Differences in how cases are classified among the various jurisdictions may slightly affect the overall totals. 4. Where there are convictions for multiple offences, a disposition is recorded for only the most serious offence. This results in undercounting.

commit a new offence, can be brought back to court. One of the options available is to revoke the discharge, convict the offender of the original offence, and impose a new sentence. This disposition is also available to youth court judges.

The Fine

The fine is one of the most frequently used sentences imposed by criminal courts in Canada, second only to probation. It is a straightforward penalty that involves the payment of a specific amount of money within a specified period. The amount is determined by the judge although there are statutory maximums set out in the Criminal Code. The amount of the fine that is levied generally reflects the severity of the crime, tempered with consideration for the offender's ability to pay. Indeed, a judge cannot order a fine unless the offender is able to pay it or work it off in a **fine option program.** Note that a fine is not compensation to the victim. This is covered by a different sentencing option called restitution, discussed below.

fine option program
a program that provides offenders who cannot pay a fine with the opportunity to discharge, through community service work, all or part of the fine

In 2003–04, fines were most frequently imposed for traffic offences (in nearly 80 percent of Criminal Code traffic offences and in 86 percent of impaired driving offences) and for drug possession (52 percent). Fines were less frequently imposed in cases involving crimes against property (18.5 percent) and—predictably—crimes against the person (10.5 percent) (Thomas, 2004). Fines are a popular sentencing option for the following reasons: they generate revenue instead of costing money; they save offenders the harmful consequences of imprisonment, including the stigma of being incarcerated; they allow the offender to retain employment and maintain family contact; and they may be the only option available for corporate offenders.

There is no upper limit for a fine if the offence is indictable. For summary conviction offences the maximum fine is $2,000 for individuals and $25,000 for corporations. A few offences have mandatory minimum fines—for example, a fine for a first conviction for impaired driving must be at least $300. Young offenders can never be fined more than $1,000. In provincial and territorial courts, just over 5 percent of fines exceed $500.

When a fine is ordered in adult court, the judge may define a specific period to give the person time to raise the money, called "time to pay." At the same time, the judge pronounces a term of imprisonment that the offender could serve if he or she does not pay the fine, called the "in lieu of" period. This has a twofold intent: to provide an incentive for the offender to pay, and to make the fine enforceable. A formula for determining these two amounts is spelled out in the Criminal Code (s. 734.5). Generally, the period of incarceration to be served if the fine is not paid is never longer than the prison sentence that would have been given in the first place.

Enforcement of Fines and the Problem of Default

The vast majority of offenders pay their fines. Those who do not pay by the deadline are deemed to be in default. Until recently, fine defaulters were routinely sent to jail for the "in lieu of" period. As well, some offenders preferred to serve the time rather than pay the fine. Both these situations resulted in large numbers of fine defaulters serving terms of custody in provincial/territorial institutions for relatively minor offences that did not warrant prison sentences in the first place.

Fines have also been criticized for leading to incarceration of the poor; the affluent, meanwhile, can avoid prison by making the payment. In some jurisdictions, Aboriginal people—women in particular—are adversely affected by their inability to pay fines and avoid incarceration. In *R. v. Wu* ([2003] 3 S.C.R. 530), the Supreme Court of Canada reaffirmed the principle that if an offender does not have the means to pay a fine immediately upon conviction, he or she should be given a reasonable time to pay, and that every attempt should be made to avoid sending the offender to jail.

To address the problem of default, several provinces/territories have developed fine option programs that allow defaulters to work off fines in the community. Each hour the defaulter works is good for credit toward the payment of the fine. The work can be extended over as long as two years.

Once in prison, the defaulter can pay part or all of the fine in order to hasten release. A partial payment will reduce the period of imprisonment on a pro rata basis, as long as the amount of money is large enough to "buy" at least one day. However, in addition to the original fine, the incarcerated defaulter must also pay the costs and charges incurred in being admitted to confinement. Fine defaulters who do not pay but serve the "in lieu of" period are considered to have discharged the fine.

Suspended Sentence

Upon a finding of guilt, the judge can suspend the passing of a sentence altogether. The offender is convicted of the offence and has a criminal record, but is given no sentence. This is not technically true, since a **suspended sentence** must be accompanied by a probation order. If the offender successfully completes the period of probation, a sentence is not imposed. However, offenders who reoffend during the probation period can be brought back to court, and the judge can revoke the original probation order and "impose any sentence that could have been imposed if the passing of the sentence had not been suspended." Alternatively, the court can extend the period of probation for up to one year. This sentencing option is not available in the youth courts.

> **suspended sentence**
> a sentencing option whereby the judge convicts the accused but technically gives no sentence and instead places the offender on probation, which, if successfully completed, results in no sentence being given

Probation

Section 731 of the Criminal Code provides that in those cases in which there is no minimum penalty prescribed, the sentencing judge may place the offender on probation. Once ordered, probation falls under the authority of the provincial/territorial correctional systems. It can be ordered alone or in conjunction with a period of provincial/territorial imprisonment. The only exceptions to the general rule that probation is not available for federal offenders involve an offender who is sentenced to *exactly* two years of confinement. In such cases the judge may attach a probation order of up to three years, to be completed following the custodial sentence in a federal correctional facility.

About 100,000 adult offenders are under probation supervision across the country. This makes probation the most widely used sanction for controlling and supervising offenders. In 2003–04, probation was imposed in 46 percent of all cases with a conviction (as opposed to prison in 35 percent of cases, and fines in 32 percent of all cases). Probation was imposed, often in conjunction with a term of imprisonment, for 28 percent of offenders convicted of a crime against the person (Thomas, 2004). Note that probation is mandatory where the accused is given a conditional discharge or a suspended sentence.

The popularity of probation is due in large measure to its versatility. The length and conditions of a probation order can be tailored to the needs and circumstances of the individual offender, to a maximum of three years for adults and two years for young offenders. The average length of adult probation orders is one year.

Adult offenders can be on probation in one of six ways:

- as part of a conditional discharge;
- as a condition of a suspended sentence;
- as part of an intermittent sentence;
- as a sentence on its own (the most common);

© James Shaffer/PhotoEdit Inc.

A probation officer meets with an offender on probation.

- following a prison term of less than two years (provincial/territorial sentence); *or*
- following a prison term of exactly two years (federal sentence). (Note: This is the only instance in which an offender who has served federal time will be on provincially supervised probation.)

When probation follows a term of confinement, probation supervision begins either on release from custody or on the expiration of parole, if the offender was released on parole.

Probation is also the most common disposition in youth courts. It can be a stand-alone disposition, or it can follow a period of confinement in a youth detention facility.

Conditions of Probation

There are three *compulsory* conditions set out in Section 732.1 of the Criminal Code, with which every adult probationer must comply:

(a) keep the peace and be of good behaviour;
(b) appear before the court when required to do so by the court; *and*
(c) notify the court or the probation officer in advance of any change of name or address, and promptly notify the court or the probation officer of any change of employment or occupation.

The judge can also impose *optional* probation conditions. These are tailored to meet the offender's specific needs and circumstances and include the following:

- Report as required to a probation officer.
- Abstain from drugs and/or alcohol.

- Abstain from owning, possessing, or carrying a weapon.
- Provide for the support and care of dependants.
- Perform up to 240 hours of community service work over a period not to exceed eighteen months.
- Participate in a treatment program (if agreed to by the offender and if the program director accepts the offender).
- Comply with any other conditions set out by the court.

This last condition allows the sentencing judge to specifically tailor the order to the individual offender; this provides flexibility in sentencing.

Other optional conditions include the following: restitution to the victim; non-association with co-accused (if any); non-association with the victim; participation in a victim–offender reconciliation program; area and travel restrictions; and curfews. Where available, a probation order may contain the provision that the offender be placed under home confinement and electronically monitored for compliance. Common optional conditions for young offenders include attendance at school, job seeking, maintenance of employment, and living with parents or another adult the court deems appropriate.

Probation conditions must be stated clearly; they must also be reasonable so that the probationer knows what is expected and is not saddled with conditions that are impossible to meet. Ideally, the conditions should relate to the offence and be designed to prevent the offender from committing further crimes. During the period of supervision, the probation officer may ask the sentencing judge to increase or decrease the number of optional conditions, eliminate all optional conditions, or reduce (but not lengthen) the total period of the probation order.

An adult probationer who, "without reasonable excuse, fails or refuses to comply" with a condition or who commits a new offence may be charged with what is commonly referred to as *breach of probation*. A breach of probation is a hybrid offence and can carry a maximum penalty of two years' incarceration (if pursued as an indictable offence) or eighteen months' imprisonment and a fine not to exceed $2,000 (if prosecuted as a summary offence). Young offenders on probation can also be charged with failing to adhere to the conditions of their probation order.

Dual Role of Probation Officers

Probation officers play two roles: they provide assistance and support for offenders, and they enforce the general and specific conditions of the probation order. In the support role, the probation officer helps the offender address the issues that likely contributed to the offence and identifies for the offender resources in the community (such as alcohol and drug treatment programs, education-upgrading courses, and mental health services). According to Johnson (2004), the length of probation terms has been increasing, and more and more people sentenced to probation (nearly 50 percent in 2002–03) were convicted of violent offences (Johnson, 2004). This further challenges probation officers to ensure that sufficient supervision is provided to those offenders who are "high need, high risk."

The dual role of probation officers can be a barrier to effective case management. For example, a probationer with a history of drug addiction who has started "using" again may want to ask his or her probation officer for help in finding a treatment program. Disclosing the illegal drug use could trigger a charge of breach of probation; yet failing to disclose the relapse could result in the probationer committing additional criminal acts to support the addiction. A probation officer who becomes aware that probation conditions have not been met may enjoy considerable discretion in deciding whether or not to revoke probation. The decision often turns on the severity of the alleged breach. Similar challenges of balancing the helping and enforcement roles are encountered by parole officers; this is discussed in Chapter 8.

Electronic Monitoring

electronic monitoring (EM)

a sentencing option involving the use of high technology to ensure that offenders remain in their residences except while working or for other authorized activities

Electronic monitoring (EM) is the first widespread application of high technology in non-institutional corrections. In the coming decades, the majority of offenders may be supervised in this way. Under electronically monitored home confinement, offenders who would otherwise have been sentenced to prison must remain in their residences except while working or for other authorized activities. As a sentencing condition, EM is available to provincial/territorial offenders in some jurisdictions but not to federal offenders. The National Parole Board does not use EM as a condition of release from correctional institutions.

There are several ways that offenders under home incarceration can be monitored electronically. The most widely used system involves the offender wearing a tamper-proof transmitting device (usually an anklet) that sends radio signals to a receiver attached to the offender's telephone, which relays those signals to a central computer at the probation or parole office. The preprogrammed computer calls the offender's home at random times; the offender then has a limited amount of time to place the anklet or bracelet in the verifier attached to the telephone. Second-generation EM systems use cellular telephone systems to constantly monitor the offender's location.

Participation in EM programs is generally restricted to offenders who have been convicted of less serious, non-violent offences and who have a stable residence and a telephone. However, EM can also be attached as a condition of parole to increase the capacity of parole officers to monitor the activities of offenders who have committed serious crimes. The program may require the offender to pay a fee to assist in offsetting the costs of equipment and monitoring. EM programs were originally designed as a "true" alternative to confinement—that is, in the absence of the EM program, the offender would have served a period of incarceration in a correctional facility. However, their application has since expanded to include provincial/territorial offenders who are released on parole.

Each province has developed its own unique way of applying EM technology. EM can be used for accused persons who are out on bail awaiting trial; as a condition of probation; for offenders who are on intermittent sentences during the periods when they are not in custody; and as a condition of day parole or full parole.

Saskatchewan, for example, uses EM as a "front end" sentencing option. The decision to place an inmate on EM is made by the judge at the time of sentencing. Newfoundland uses "back end" EM, whereby offenders who are placed on EM must comply with an ambitious schedule of community programming, which can include sessions on anger management and treatment for alcohol abuse. These programs are operated by the John Howard Society. Home visits are also made, and sobriety may be verified with the use of an "alco-meter." Participation in the program is restricted to low-risk/low-need offenders; offenders convicted of sex crimes and men with records of family violence may not be eligible. Offenders who are eligible may be required to sign a consent form to allow victims to be notified of their early return to the community.

Another example of "back end" EM is found in British Columbia, where EM is used only as a condition of day parole or full parole. The decision to place an inmate on EM is made by the B.C. Board of Parole at the parole hearing.

Supporters of EM argue that it is less costly than incarceration, provides a humane alternative to confinement, and allows offenders to remain in the community, at work, and with their families. There are, however, legal and ethical concerns surrounding the use of EM, and these will surely increase as the capacity for monitoring offenders through the use of high technology increases. In the not-too-distant future, monitoring systems may be developed that are able to determine not only the location of an offender, but also what the offender is doing and (possibly) thinking. As the technology becomes more sophisticated and the costs of incarcerating offenders rise, corrections systems will likely be tempted to place even more offenders under some type of electronic surveillance. This will pose challenges for the criminal justice system and for Canadian society as a whole.

Restitution

Several types of sentences involve compensation to victims, commonly known as restitution. This type of sentencing is often used in cases of mischief, fraud, and theft. Although there are minor differences between adult and youth court, these sentences typically involve cash payments in compensation for stolen or damaged property or for property seized from someone who innocently purchased it without knowing it had been stolen. The Youth Criminal Justice Act also permits "compensation in kind," which means that the young offender can do work or perform a service for the victim that is equivalent to the value of the stolen or damaged property.

Restitution is an important concept in the restorative justice paradigm because it helps restore victims to their pre-offence financial condition and because it holds the offender accountable to the parties he or she has wronged. This measure is encouraged and supported by victims' groups. A judge can make the order unilaterally or at the request of the Crown.

The main problem with restitution orders is that they are difficult to enforce. There is very little that can be done if the offender does not pay. Embedding restitution orders in probation orders is one option, such that non-payment becomes a breach of probation. Beneficiaries of the restitution order can use the civil courts to enforce the order. Restitution orders can also be part of a conditional sentence order—a novel sentencing option that will be discussed later. This route may well increase the rate of compliance, in that imprisonment may result if payment is not forthcoming.

A controversial provision in the Criminal Code extends the concept of restitution to include expenses associated with bodily harm offences. For example, a victim of assault might lose income while recuperating, require costly dental work, need to replace broken eyeglasses, or require physiotherapy. All specific and ascertainable costs such as these can be included in the restitution order. Many of these expenses are the same as those covered by criminal injury compensation programs (see Chapter 2).

The Criminal Code also provides that in cases of spousal or child abuse where victims are members of the same household, the offender may be ordered to pay expenses incurred by the victim as a result of moving out of the offender's household (for example, for temporary housing, food, childcare, and transportation).

Community Service Order

Community service work can be a condition of an adult probation order or an order by a youth court judge. This work may include sorting material at a recycling depot, picking up litter in parks, or scrubbing graffiti from highway structures. The judge sets the number of hours of work and defines a specific period during which the hours should be completed.

Imprisonment

The rarest—and most onerous—of all sentences, imprisonment is generally reserved for serious crimes or for offenders who have long criminal records. Statistics on adult imprisonment for 2003–04 indicate the following:

- Imprisonment was imposed in cases involving conviction for crimes against property (40.6 percent), crimes against the person (35 percent), drug trafficking (43.1 percent), and impaired driving (12.4 percent). Note that when minor assaults are removed from the crimes-against-person category, the percentage of cases resulting in incarceration in this category rises to 44 percent.

- Incarceration rates and lengths of incarceration vary according to the type of offence.

- Incarceration rates vary greatly across jurisdictions, from Prince Edward Island (58 percent), to Ontario (41 percent) and British Columbia (40 percent), to New Brunswick (25 percent) and Saskatchewan (24 percent). This is for a number of reasons, including the offence mix in each jurisdiction and the sentencing styles of judges.

- Most prison terms are short: 57 percent of sentences are one month or less; an additional 31 percent are one to six months (Thomas, 2004) (see Figure 6.2).

FIGURE 6.2 Length of Provincial/Territorial and Federal Sentences, 2003–04

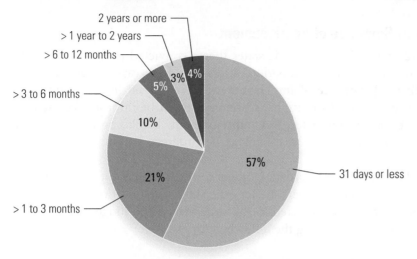

Federal Sentence Length (I)	percent
2 years < 3 years	53
3 years < 4 years	20
4 years < 5 years	8
5 years < 10 years	12
10 years or more (but not life)	3
Life	4

Sources: Adapted from the Statistics Canada publication, *Juristat*, Adult Criminal Court Statistics, 2003/04, vol. 24, no. 12, Catalogue 85-002, December 10, 2004, p. 8, and from *Juristat*, Adult Correctional Services in Canada, 2002/03, vol. 24, no.10, Catalogue 85-002, October 27, 2004, p. 12.

A sentence of two years or more means the offender will most likely serve the term in a federal institution operated by the Correctional Service of Canada. Provincial sentences are those which total two years less a day. Whether or not offenders serve their time in a correctional institution depends on the type of sentence they receive. Among the sentences provincial offenders can serve are intermittent sentences and conditional sentences.

Intermittent Sentence

If the prison sentence is ninety days or less, the judge has the option of allowing the offender to serve it on an intermittent basis. The most common arrangement is that the offender lives at home and works or attends school during the week but spends the weekends (specifically, the period from Friday evening to Monday morning) in jail. During the periods outside the institution, the offender is technically on probation and must abide by the conditions of the order. Fine defaulters who are sent to confinement may also serve their time intermittently.

Conditional Sentence of Imprisonment

Section 742 of the Criminal Code states that a convicted person who would otherwise be incarcerated for a period of less than two years can be sentenced to a **conditional sentence of imprisonment** to be served in the community rather than in custody. Conditional sentences cannot be imposed for offences punishable by a minimum term of imprisonment. Unlike probation orders, conditional sentences allow treatment to be ordered without the offender's consent.

All conditional sentence orders contain compulsory conditions similar to those found in probation orders. Besides the standard conditions (such as the requirement to report as directed to the supervisor), the judge can prescribe optional conditions, involving the following, for example:

- abstaining from alcohol or other intoxicants;
- abstaining from owning, possessing, or carrying a weapon;
- providing support or care for dependants;
- performing up to 240 hours of community service work; *and/or*
- attending a treatment program.

An offender who does not comply with the details of a conditional sentence order can be incarcerated. Once an allegation is made that a condition has been breached, the offender may have to appear in court to prove that the allegation is not true. In this reverse onus situation, it is up to the offender to prove that the breach did not occur. A judge who is satisfied, on a balance of probabilities, that a condition has been breached has four options:

- take no action.
- add or eliminate optional conditions.
- suspend the conditional sentence order, commit the offender to custody to serve a portion of the unexpired sentence, and resume the conditional sentence when the offender is released from confinement.
- terminate the conditional sentence order and commit the offender to custody to serve until the release on parole or the *discharge possible date* (that is, the date at which an offender has served two-thirds of a provincial prison sentence).

Offenders serving conditional sentences are supervised in the community by probation officers; however, they are not technically on probation. In *R. v. Proulx* ([2000] 1 S.C.R. 61), the Supreme Court of Canada expressly defined the differences between conditional sentences and probation. The main difference is that probation focuses on rehabilitation, whereas conditional sentences are intended to provide both denunciation and rehabilitation. In practical terms, this means that the conditions attached to a conditional sentence are generally

conditional sentence of imprisonment

a sentence for offenders who receive a sentence or sentences totalling less than two years whereby the offender serves his or her time in the community under the supervision of a probation officer

more onerous than those attached to a probation order. The Supreme Court also ruled that no offences—including serious and violent crimes—could be excluded from consideration for a conditional sentence.

Research findings and statistics on the use of conditional sentences indicate the following:

- There has been a steady increase in the use of conditional sentences, with 8 percent of the adult correctional population in 2002–03 on conditional sentences—up nearly 70 percent from 1998–99.
- Over the past five years, the length of conditional sentences has increased, as has the severity of the attached conditions.
- Conditional sentences are imposed by sentencing judges in more cases involving crimes against property (6.6 percent) than crimes against the person (6.3 percent). However, they are imposed in nearly 18 percent of cases involving sexual assault, 12 percent of cases involving robbery, and nearly 35 percent of cases involving drug trafficking.
- There is considerable variation in the use of conditional sentences across jurisdictions. Judges in Quebec, Ontario, and British Columbia use this sanction most frequently.
- Crime victims are not uniformly opposed to the use of conditional sentences; some feel that these sentences can be effective if properly structured and supervised.
- Reducing the use of imprisonment is the most frequently stated objective noted by judges who impose conditional sentences.
- Judges would impose conditional sentences more frequently if there were sufficient community resources.
- Judges believe that the general public's understanding of conditional sentences is limited but that people who do understand them support their use (Johnson, 2004; North, 2001; Roberts and LaPrairie, 2000; Roberts and Roach, 2004; Thomas, 2004).

The Continuing Controversy over Conditional Sentences

Conditional sentences are a popular option among judges; they are also controversial. Defence lawyers regularly ask the courts to impose conditional sentences on their clients instead of sending them to prison. Critics cite numerous cases in which an offender who should have been incarcerated was instead granted a conditional sentence. This debate is fuelled by newspaper stories about people who committed serious offences while serving a conditional sentence:

- "School Clerk Gets No Jail for Killing Principal" (*The Province* [Vancouver], August 30, 2005);
- "House Arrest for Killer: Victim's Family Appalled" (*Calgary Herald*, March 17, 2005);
- "House Arrest for Pedophile" (*Windsor Star*, March 10, 2005); *and*
- "Sex Offender Sent Home" (*Daily News* [Halifax], February 16, 2005).

A study of sentencing in family violence cases found that nearly 25 percent of spousal offenders convicted of sexual assault received a conditional sentence (Gannon and Brzozowski, 2004: 57). Observers contend that there was not sufficient consultation between the federal and provincial/territorial governments before conditional sentences were implemented and that there are often insufficient supervisory resources in the community for offenders (Mazey, 2002). Also, the original intent of conditional sentences was to reduce overcrowding in correctional institutions (and the associated expense) while at the same time helping offenders reintegrate into the community (see *R. v. Proulx* [2000] 1 S.C.R. 61); yet it is not certain that conditional sentences have achieved these objectives (see below).

The Supreme Court of Canada has strongly endorsed the use of conditional sentences and has rejected efforts by some provinces, including Ontario, to prohibit judges from using conditional sentences for violent offences. And appellate court judges strongly oppose proposals to amend the Criminal Code to restrict the ability of judges to impose conditional sentences in certain types of cases. In the view of these judges, such restrictions would constitute an encroachment of Parliament into the exercise of discretion by members of the judiciary (Roberts and Manson, 2004).

The general public seems to misunderstand conditional sentences. In a national survey, only 43 percent of the respondents were able to identify the correct definition of "conditional sentence"; 38 percent selected the definition of "parole." The survey also found that community residents would support the use of conditional sentences even in cases of assault so long as sufficient restrictive conditions were attached to the sentence. Respondents were nearly unanimous, however, in the belief that conditional sentences should not be used in cases of sexual violence (Sanders and Roberts, 2000).

Selected Sentencing Trends

The sentencing practices of judges in specific types of cases have come under increasing scrutiny.

Impaired Driving Causing Death or Bodily Harm

Recent years have seen a stronger focus on drinking and driving; this has included scrutiny of sentencing in cases of impaired driving involving death or bodily harm. Mothers Against Drunk Driving (MADD; see Chapter 1) has been a vocal critic of what it considers the lenient treatment of impaired drivers by the courts.

MADD has accused B.C.'s justice system of being "soft" on drunk drivers. To support this contention, it presented the following statistics:

- Criminal charges were laid by Crown counsel in roughly 7,000 of 11,000 impaired driving incidents in B.C., compared to 19,000 of 20,000 incidents in Ontario.
- Eight percent of offenders convicted of impaired driving in 2001–02 went to prison in B.C., compared to 91 percent in Prince Edward Island (Baron, 2004).

A number of provincial justice ministers have supported MADD in calling on the federal government to eliminate the use of conditional sentences for those convicted of impaired driving causing death or bodily harm. However, an analysis of fifty-five cases of impaired driving causing death found that only nine resulted in a conditional sentence. And in 339 convictions for impaired driving involving death or bodily harm, only 84 (25 percent) resulted in a conditional sentence (Paciocco and Roberts, 2005).

The study also found that those offenders whose blood alcohol levels were far beyond the .08 limit rarely received conditional sentences; the same for offenders who were found to have driven recklessly. Also, first-time offenders were more likely to receive conditional sentences than those with prior convictions. Nationally, in cases of impaired driving causing death or injury, conditional sentences are imposed far less often than media headlines might lead one to assume.

Marijuana Grow-Ops

In Chapter 5, it was noted that a significant percentage of marijuana "grow-op" seizures in British Columbia do not result in further investigation or charges. Statistics on sentencing indicate that 14 percent of those convicted of growing marijuana in the province receive a jail term, 27 percent a fine, and 59 percent neither. In Vancouver, just under 8 percent of convicted growers receive jail terms, nearly 10 percent a fine, and 83 percent neither. The "neither" category includes probation or conditional sentences (Skelton, 2005).

Additional research by economist Stephen Easton (2004) on those convicted of running marijuana grow-ops indicates that for those who do receive jail time, the sentences are short: most received a sentence of ninety days or less in a provincial correctional facility. As well—and significantly—70 percent of these people had previous convictions for marijuana cultivation; this indicates that the sentences the courts are imposing are not serving as a deterrent. Easton concludes from this that "whatever it is that these people are doing, they are continuing to do it! From the point of view of an ongoing business, court time, or a charge, are simply part of the costs of doing business" (ibid., 25).

These findings raise a number of issues regarding the effectiveness of sentencing as one component of the criminal justice response to marijuana cultivation.

Family Violence

Over the past decade a wide range of programs and initiatives have been implemented in order to prevent, reduce, and respond more effectively to family violence. For example, pro-prosecution policies have been developed that are designed to reduce the number of cases of family violence in which charges are withdrawn or stayed; and domestic violence courts have been established (Gannon and Brzozowski, 2004; and see Chapter 5).

A recent review of sentencing in cases of family violence reported the following:

- Probation is the most common sentence in cases involving spousal violence, and most probation orders are in the range of six to twelve months.

- Spouses who inflict major injury are more likely to be sent to prison, although over 50 percent of the sentences are for one month or less.
- Husbands convicted of spousal violence are three times more likely than wives to receive a prison term.
- Family members convicted of abuse against children are less likely to receive a prison term than non-family members. Probation is imposed in two-thirds of cases of physical violence, and in just over one-third of cases involving sexual violence against children and youth.
- Most prison sentences (67 percent) in cases involving family violence against children were three months or less; 8 percent of offenders received a sentence of more than two years.
- Family members (22 percent) are less likely than non-family members (34 percent) to receive a prison sentence for violence against seniors. Family members are much more likely to receive probation (Gannon and Brzozowski, 2004; see also Ursel, 2003).

Concurrent and Consecutive Sentences

Whenever an offender is convicted on more than one charge, it is important to note whether the sentences imposed are to be served concurrently or consecutively. **Concurrent sentences** are rolled into one time period and served simultaneously. For example, an offender who is sentenced to two concurrent terms of imprisonment of eight months each will serve eight months, not sixteen. In contrast, **consecutive sentences** run separately and one after the other. An offender who is sentenced to two prison terms of three years each will serve a six-year sentence. In Canada, most sentences meted out by judges are to be served concurrently.

Judicial Determination

Federal inmates typically can apply for release on full parole after serving one-third of their sentence. However, Section 743.6 of the Criminal Code gives sentencing judges the authority to order certain federal inmates to serve half of the sentence before becoming eligible for parole. This **judicial determination** may be imposed on offenders who have been convicted of one or more Schedule I offences (specific crimes against the person) and Schedule II offences (drug offences specified in the Corrections and Conditional Release Act). The primary objectives of judicial determination are denunciation of the criminal behaviour and general and specific deterrence.

Judicial determination is used in less than 5 percent of federal cases, and Aboriginal offenders are overrepresented among those offenders who receive it. Offenders who receive judicial determination are more likely than other offenders to serve their entire sentence in prison (Solicitor General Canada, 1998).

Judicial Restraint Order

Under Section 810 of the Criminal Code, you may lay an information before a justice of the peace if you have reasonable grounds to believe that another person will injure you, your spouse, your children, or your property. The person

What do you think?

What is your response to the sentencing patterns for cases involving family violence? What patterns emerge from these findings, and what do these patterns tell us about sentencing practices and family violence?

concurrent sentences

sentences that are amalgamated and served simultaneously

consecutive sentences

sentences that run separately and are completed one after the other

judicial determination

order that a federal inmate serve half of the sentence before becoming eligible for parole

Adults and children protest outside a British Columbia provincial court the day of the first appearance in court of an alleged child murderer.

need not have a criminal history at the time of the application. Other sections—810.01(1), fear of a criminal organization offence; 810.1(1), fear of a sexual offence; and 810.2, fear of serious personal injury—require an information to be laid before a provincial court judge. Section 810 has withstood Charter challenges.

If the JP or the judge is satisfied that there are reasonable grounds for the threat, the defendant is required to enter into a recognizance to keep the peace and be of good behaviour for a period not to exceed twelve months. The court may also impose conditions on the defendant—for example, to abstain from possessing a firearm, to avoid contact with persons under fourteen, or to stay away from places frequented by children (such as school or daycare grounds). Violation of the conditions of an 810 order is an offence and can result in imprisonment. A defendant can also be imprisoned for refusing to agree to an 810 order. Critics of Section 810 argue that the conditions are too broad in their application in that no crime need have been committed in order for them to be imposed.

Section 810 orders can also be imposed by judges when an offender is released from custody following the completion of his or her sentence (see Chapter 8).

HOW DO JUDGES DECIDE?

Sentencing is a very human process. Most attempts to describe the proper judicial approach to sentencing are as close to the actual process as a paint-by-numbers landscape is to the real thing.—Ontario Court of Appeal Judge David Doherty in *R. v. Hamilton and Mason* (2004: 17)

One of the main themes of this book is that the criminal justice process is a human enterprise involving people making decisions about people. Nowhere is this better illustrated than in the sentencing process in the criminal courts. Within the framework of the Criminal Code, Canadian judges have considerable discretion in selecting a sentence. Section 718.3(1) of that code states: "Where an enactment prescribes different degrees or kinds of punishment in respect of an offence, the punishment to be imposed is, subject to the limitations prescribed in the enactment, in the discretion of the court that convicts a person who commits the offence."

For most offences, the Criminal Code prescribes only the maximum sentence that may be imposed by the sentencing judge. Any number of sentencing options and sentencing goals can be applied to any specific case. This can lead to disparities in the specific penalties imposed, even for similar types of crimes. In any given case, five different judges could easily hand down five different sentences. Also, judges may have personal opinions and philosophies that guide their sentencing decisions.

There are some limits to judges' discretion. A key principle of sentencing—set out in Section 718.2(b) of the Criminal Code—is that two similar crimes committed by two similar offenders in similar circumstances should draw similar sentences. In their deliberations, judges consider the sentences handed down by other judges; and as just noted, the Criminal Code gives some guidance by setting maximum sentence limits. The appellate courts defer to the sentencing decisions of lower court judges and are reluctant to overturn these unless the sentence is found to be "demonstrably unfit [due to] an error of principle, failure to consider a relevant factor, or overemphasis of the appropriate factors" (*R. v. McDonnell* [1997] 1 S.C.R. 948).

In making a sentencing decision, a judge may sometimes seek to impose a sentence that not only fits the crime and reflects the "going rate" for similar offences but also takes into account the offender's particular circumstances. In other cases, the sentence may reflect only the severity of the crime, with no consideration to the situation of the offender. In still other cases, judges are confronted with difficult issues that generate considerable media attention and public and political debate (see the Robert Latimer case, Box 6.2).

You Be the Judge

To gain an appreciation of the challenges judges face in making sentencing decisions, review the summaries of actual cases presented in Box 6.1 and place yourself in the position of the sentencing judge. The purposes of sentencing and the various sentencing options available to judges were presented earlier in this chapter. Note that you can mix and match options—that is, you can sentence the offender to a period of custody in a provincial correctional facility and, as well, add on a period of probation of up to three years. Probation cannot be used in conjunction with a sentence of more than two years, which places the offender under the jurisdiction of federal corrections. As well, the various objectives of sentencing were discussed earlier.

Record your sentencing decisions and the reasons why you selected each particular sentence. Once you have completed all five cases, check at the end of

focus

BOX 6.1

You Be the Judge

Read each of the following case summaries. Then decide on a sentence and note the purpose of your sentence.

Case 1. On March 28, 2005, a twenty-three-year-old who was high on drugs and speeding in a stolen SUV ran a stop sign and collided with a car driven by a church pastor. The victim died on impact. Today the offender appears before you in court, pleads guilty to the offence of criminal negligence causing death, and expresses remorse for his actions. He has a lengthy criminal record dating back to 1996 and was serving a conditional sentence at the time he committed the offence. The maximum sentence for this offence, as set out in the Criminal Code, is life imprisonment. His defence lawyer asks for a sentence of three to four years in jail; the Crown counsel is seeking a sentence of five to seven years, with the recommendation that you, the judge, consider the higher end of that range. As the presiding judge, what do you decide?

Case 2. A police investigation reveals that two parents had subjected their two adopted sons to thirteen years of abuse. The abuse occurred between January 1988 and June 2001 and included locking the boys in cages fashioned from baby cribs wrapped in chicken wire. Sometimes they were tied to a bed in the basement; both were beaten with slippers and a shoe horn. The boys were also forced to wear diapers when not at school and were required to spread their buttocks to show their parents they had not defecated. The boys sometimes ate their own feces to avoid being punished for accidents.

The parents appear before you and plead guilty to two counts each of forcible confinement, assault with a weapon, and failure to provide the necessities of life. Each of these is a hybrid offence (see Chapter 5). As indictable offences, assault with a weapon and forcible confinement each carry a maximum sentence of ten years; on summary conviction, each carries a maximum of eighteen months. Crown counsel requests a sentence of four to eight years in federal prison; the defence lawyers have argued for a conditional sentence. Evidence is presented during the sentencing hearing that the boys may have suffered from attention deficit disorder and fetal alcohol syndrome. However, each was at the appropriate grade level and had no behaviour problems in school.

As the judge, what sentence would you impose on the adoptive parents?

Case 3. In January 2004, a lawyer in private practice (a former Crown counsel) enters a plea of guilty for assaulting his former girlfriend. A review of his record indicates that he has twice before been convicted of assaulting former girlfriends. On both previous occasions, he was granted a conditional discharge and did not incur a criminal record. The current assault involved a violent attack on his former girlfriend, which included kicking and punching, two weeks after she broke up with him. Evidence submitted to the court indicates that the accused's long-term battle with alcoholism had been aggravated by the breakup.

The defence lawyer points out that since being charged, his client has been attending Alcoholics Anonymous regularly and that this and other efforts at rehabilitation make him a candidate for another conditional discharge. Furthermore, this lawyer continues, there is nothing to be gained by imposing a more severe sentence. Crown counsel contends that since the accused has been convicted twice before of the same offence, a conditional discharge would not deter future assaults. The maximum punishment for assault as an indictable offence is five years in prison; as a summary conviction, the maximum is eighteen months. What do you decide?

Case 4. A Montreal advertising executive appears before you and pleads guilty to fifteen counts of fraud. The criminal offences involved defrauding taxpayers of $1.5 million as part of what has become known as the Liberal Party Sponsorship Scandal. The executive admits that he filed false invoices, inflated costs, billed for meetings he did not attend, and billed the government for work that was done by nonexistent employees. The executive before you is the first person charged in the $250 million scandal. Crown counsel is asking for a thirty-four-month term in federal prison based on the severity of the breach of trust. Defence counsel is arguing for a conditional sentence, to include a series of speeches on ethics at Canadian business schools. Since the amount involved is over $5,000, the maximum penalty that can be imposed under Section 280(1) of the Criminal Code is ten years' imprisonment. As the presiding judge, what sentence do you impose?

Case 5. A twenty-one-year-old appears before you having pleaded guilty to criminal negligence causing death and leaving the scene of an accident. Two years earlier, in the early morning hours, he had driven his Honda Civic through a red light and T-boned an RCMP police cruiser. The police investigation revealed that the Civic was travelling at more than 130 km/h at the time of the impact, and that it hit the cruiser with such force that the officer was thrown through the back window, killing him instantly. While the officer was lying on the roadway, the injured Civic driver was picked up by his friend in another vehicle and the two sped from the scene. Witnesses later stated that the two friends had been street racing prior to the accident.

The defence lawyer argues for a conditional sentence, noting that his client is remorseful, has become a productive member of society, and has already spent forty-five days in jail and almost two years under house arrest. Crown counsel argues that a three-year prison sentence would be appropriate, given the high-risk nature of the behaviour that led to the crash and the death of the RCMP constable. The mother and father of the deceased constable read victim impact statements to you. He was their only son.

The maximum penalty for criminal negligence causing death is life imprisonment. For leaving the scene of an accident, the maximum penalty is five years in prison, or punishable upon summary conviction with a term of custody generally not longer than six months. As the presiding judge, what sentence would you impose?

Case 6. Appearing before you for sentencing is a twenty-one-year-old University of Manitoba student, whom you have found guilty of dangerous driving causing death. Two years earlier, the young man ran a stop sign on the way to Bible camp and killed three people in the ensuing crash. Evidence presented at trial indicated that speed and alcohol were not factors in the crash and that the man had no prior driving infractions. The man had just come from a church service and had become lost on his way to the Bible camp prior to running the stop sign. He has expressed deep remorse for the accident and for the deaths of the people in the other vehicle.

At trial, the man pleaded not guilty to the criminal charge. His defence lawyer sought to have his client found guilty of careless driving under the provincial Highway Traffic Act, which carries a penalty of a fine. However, you have determined that the evidence supports a conviction for dangerous driving causing death. The maximum penalty for this offence is fourteen years in prison. The Crown is seeking an eighteen-month jail term. What is your decision?

this chapter and see the actual sentences imposed by the judges for these cases. Then ask yourself these questions:

- Did my sentence match the sentence of the judge?
- Was it more lenient or more harsh?
- Did the judge in the actual case make a good decision?

For each decision, be prepared to discuss why you selected that sentencing option, what purpose it was supposed to serve, and whether you agree or disagree with the actual sentence that was imposed by the presiding judge.

Statutory Guidance

Increasingly in recent years, judges looking for guidance in sentencing can find direction from Parliament in some statutes. However, Section 718 of the Criminal Code, reproduced earlier in the chapter, is merely a list of the sentencing rationales typically presented in textbooks such as this one. The fundamental principle of sentencing, as stated in Section 718.1 of the Criminal Code, is that of proportionality: a sentence must be proportionate to the gravity of the offence and to the degree of responsibility of the offender.

In what may well be the beginning of a trend, Parliament has specified factors that judges should consider when sentencing drug cases under the Controlled Drugs and Substances Act. According to Section 10(2) of that act, an offender may deserve a harsher sentence when he or she carried, used, or threatened to use a weapon; used or threatened to use violence; trafficked in one of the specified substances or possessed such a substance for the purposes of trafficking, in or near a school, on or near school grounds, or in or near any other public place usually frequented by people under eighteen; or trafficked one of the specified substances or possessed such a substance for the purpose of trafficking to a person under eighteen. These provisions join the principles and purposes of sentencing set out in Section 718 of the Criminal Code and indicate that Parliament is willing to give sentencing judges some guidance by designating certain types of crime as deserving of greater punishment.

Maximum Sentences

Every offence has a maximum sentence that a judge cannot exceed. However, these maximums are so high as to provide little practical guidance. For example, life imprisonment is the maximum sentence for manslaughter. Life imprisonment is also a possible (but not probable) sentence for offences such as piracy (s. 74), breaking and entering a dwelling house (s. 306), and stopping a mail truck to rob or search it (s. 345). If no maximum sentence is specified for an indictable offence, the maximum allowable is five years. A maximum sentence is rarely applied for an indictable offence.

For summary conviction offences, the maximum sentence is six months in prison and/or a $2,000 fine (except for sexual assault, where the maximum sentence is eighteen months). The same maximum sentence applies when the Crown prosecutor elects to proceed summarily on a hybrid offence. However, these limits do not apply when the defendant elects trial in a provincial court on an indictable offence (this is a common misunderstanding). Judges cannot exceed the statutory maximum sentence, even when they disagree with the decision to proceed summarily. Prosecutorial election for summary proceedings is one way that Crown counsel can limit the severity of the sentence. Election can, therefore, be used as a bargaining chip in sentencing negotiations (plea bargaining).

Mandatory Minimum Sentences (MMSs)

Several offences carry a mandatory minimum sentence—for example, murder (both first and second degree). Another is use of a firearm during the commission of an offence, which carries a minimum of one year in prison for the first conviction and three years for subsequent ones (both to be consecutive to any term of imprisonment imposed for the offence itself). Yet another is a second conviction for impaired driving.

Pressure is growing on the federal government to lengthen the list of offences subject to mandatory minimum sentences. Those in favour contend that MMSs serve as a general and specific deterrent; that they prevent crime by removing offenders from the community; that they serve as symbolic denunciations of certain behaviours; and that they reduce sentencing disparities. Those opposed counter that MMSs have little or no deterrent value; that they limit judicial discretion (with a resulting impact on individual cases); that they have significant cost implications; and that they may lead to unfair sentencing practices (for an example, read about Robert Latimer in Box 6.2).

There has been little research about MMSs in Canada. However, a review of mainly American and British studies found that MMSs:

What do you think?

1. Some observers (including groups representing children with disabilities) have condemned Latimer's actions (Box 6.2), while others feel a ten-year sentence was not warranted, notwithstanding the requirements of the Criminal Code. What is your own view?
2. Should Latimer have been given an exemption from the mandatory penalty? Explain.

- have only a modest impact on crime;

- have no effect on rates of impaired driving (mainly because repeat offenders with substance abuse problems contribute disproportionately to the offence rate and are not generally deterred by specific sanctions);

- have no effect on drug consumption or drug-related crime;

- do not reduce sentencing disparities, mainly because prosecutors apply more discretion in MMS situations; *and*

- result in more "not guilty" pleas and in more crowded prisons; both of these increase the costs to the justice system (Gabor and Crutcher, 2002).

Appellate Decisions and Legal Precedents

One limitation on the discretion of judges is that sentences can be changed by appellate courts if deemed too lenient or too severe. Appellate courts can also set guidelines to assist judges in the trial courts. The general range of acceptability can be discovered by reviewing judicial precedents (decisions of other trial court judges) and decisions of courts of appeal. Also important is the designation of factors which indicate that deviation from the norm is warranted.

focus

BOX 6.2

The Robert Latimer Case

In 1994, a Saskatchewan farmer named Robert Latimer was found guilty of second degree murder. Following is a chronology of the case, which was concluded in 2000.

October 1993: Latimer kills his intellectually and physically disabled twelve-year-old daughter, Tracy, by placing her in his pickup truck (from which she cannot escape due to her disability) and asphyxiating her with exhaust fumes. He initially tries to hide his actions, but later admits to poisoning his daughter and is charged with murder.

November 16, 1994: At trial, Latimer admits killing Tracy but suggests that his actions were justified and not criminal because she suffered chronic pain. A jury convicts him of second degree murder. The judge has no choice in sentencing because under the Criminal Code, murder carries a mandatory sentence of life in prison. The second degree murder conviction means that Latimer will be eligible for parole after serving ten years—the minimum sentence available to the judge. His lawyers appeal the conviction.

July 18, 1995: In a two-to-one decision, the Saskatchewan Court of Appeal upholds the conviction.

October 25, 1995: It is revealed that the Crown counsel interfered with the jury by questioning them about religion, abortion, and mercy killing.

November 27, 1996: The Supreme Court of Canada hears the case.

February 6, 1997: The Supreme Court orders a new trial due to jury interference, but upholds the admissibility of Latimer's confession.

October 27, 1997: Latimer's second trial begins.

November 5, 1997: The jury finds Latimer guilty of second degree murder and recommends that he be eligible for parole after one year, contrary to the mandatory minimum sentence of ten years before eligibility for parole.

December 1, 1997: The trial judge gives Latimer a "constitutional exemption" to the mandatory sentence of ten years and imposes a sentence of less than two years, with one year to be spent in the community.

November 23, 1998: The Saskatchewan Court of Appeal sets aside the constitutional exemption and upholds the mandatory minimum sentence of ten years of parole ineligibility.

February, 1999: Latimer appeals to the Supreme Court of Canada.

May 6, 1999: The Supreme Court announces it will hear the appeal.

June 14, 2000: The Supreme Court hears the appeal.

January 18, 2001: The Supreme Court upholds Latimer's life sentence, with no possibility of parole for ten years (*R. v. Latimer* [2001] 1 S.C.R. 3).

Aggravating Factors

Certain factors in a particular offence can result in a more severe sentence than would normally be imposed. These "aggravating factors" can include:

- the abuse of a spouse or child (s. 718.2);
- a previous criminal record;

What do you think?

1. What is your opinion of Section 718.2(e)?
2. Review the points/ counterpoints in Box 6.3. What are the strongest arguments made by each side in the debate?

- breach of trust or a position of authority (s. 718.2);
- premeditation;
- the use of force or a weapon;
- injury to the victim;
- the high value of stolen or damaged property; *and/or*
- a victim who is either a youth or a vulnerable adult (such as a senior citizen).

Membership in or association with a criminal gang is also an aggravating factor. In 2005, in a landmark case, an Ontario judge sentenced two members of the Woodbridge chapter of Hells Angels to additional time in prison for committing extortion against a Barrie businessman. One Hells Angel received an additional two years on his four-year sentence, the other an additional year on his two-year sentence. The additional sentences were to be served consecutively (Avery, 2005).

An offender with a lengthy criminal record can expect a more severe penalty than a first offender would receive for the same offence. If the offender's last conviction was quite recent—especially if the present offence was committed while the offender was out on bail or some form of conditional release—this can also add time to a sentence (this is sometimes referred to as the "gap principle"). Hindering a police investigation—for example, by evading lawful arrest or by giving a false identity—can also weigh against the offender.

Mitigating Factors

Factors that point toward a more lenient sentence than usual are called mitigating factors. Characteristics of the offence that can mitigate a sentence include

"As a mitigating circumstance, may I say that my client's getaway car was a hybrid."

intoxication, lack of premeditation, being provoked into acting in self-defence, and acting out of financial need rather than for greed or profit. Also, a number of characteristics of the offender may be considered in mitigation, including:

- psychological problems;
- the age of the offender;
- being Aboriginal (see Box 6.3);

focus   BOX 6.3

Point/Counterpoint: Section 718.2(e)—Valuable Sentencing Option or Misguided Reform?

Point

- Section 718.2(e) requires only that judges *consider* sanctions other than confinement when sentencing.

- Non-Aboriginal offenders are not precluded from being considered for non-incarcerative sentences.

- Section 718.2(e) represents enlightened sentencing policy and is only one component of a broader effort to address the overrepresentation of Aboriginal people in correctional institutions.

- Aboriginal communities become more involved in assisting offenders.

- The development and use of restorative justice and traditional healing programs and practices is encouraged.

- Aboriginal offenders have the opportunity to participate in culturally based programs, which may be more effective than institutional programs.

Counterpoint

- Special sentencing provisions for Aboriginal people discriminate against non-Aboriginal offenders and are based on faulty assumptions.

- Complex historical and contemporary factors, rather than sentencing practices, are the main reason for the high rates of Aboriginal incarceration.

- Research evidence shows that Aboriginal people generally receive shorter sentences than non-Aboriginal offenders who have been convicted for comparable offences.

- Section 718.2(e) does not protect Aboriginal victims of crime.

- There is no evidence that judges systematically discriminate against Aboriginal offenders during sentencing.

- Section 718.2(e) is more a product of political correctness than of research findings.

- Defence lawyers will try to use the provisions in Section 718(2)(e) to mitigate the severity of sentencing for non-Aboriginal offenders, including visible/cultural minority offenders.

Sources: S. Haslip, "Aboriginal Sentencing Reform in Canada: Prospects for Success—Standing Tall with Both Feet Planted Firmly in the Air," *Murdoch University Electronic Journal of Law,* 7(1) [2000]; P. Stenning and J.V. Roberts, "Empty Promises: Parliament, the Supreme Court and the Sentencing of Aboriginal Offenders," *Saskatchewan Law Review,* 64(1) [2001]; M.E. Turpel-Lafond, "Sentencing within a Restorative Justice Paradigm: Procedural Implications of *R. v. Gladue*," *Criminal Law Quarterly,* 43(1), 34–50 [2000].

- absence of a prior criminal record;
- addiction to alcohol or drugs; *and/or*
- having been abused by a spouse or family member (Gannon and Brzozowski, 2004, 65).

Sentencing judges can also consider the general good character of the offender and a stable employment record; the defendant's conduct at arrest and during the legal proceedings (cooperation with the police, for example, can be seen as mitigating); and post-offence efforts at rehabilitation, such as treatment for alcoholism. Time spent in custody on pretrial remand (called "dead time" because it cannot be credited against the sentence the offender receives on conviction) may also be considered. A common example of mitigation is the "guilty plea discount." Defendants who plead guilty may expect a lower sentence because a guilty plea is an acknowledgment of blame, indicates remorse, saves the taxpayers the cost of a trial, and saves the victim the inconvenience and/or trauma of testifying. The courts must maintain a delicate balance in such cases; that is, they must justify a guilty plea as a mitigating factor but at the same time affirm that a not guilty plea does not work to aggravate a sentence.

Sentencing Aboriginal Offenders

Section 718.2(e) was created in an attempt to reduce the overrepresentation of Aboriginal people in correctional institutions. Perhaps the most controversial sentencing section of the Criminal Code, it was reaffirmed by the Supreme Court of Canada in *R. v. Gladue* ([1999] 1 S.C.R. 688). In that landmark case, the Court held that where a term of incarceration would normally be imposed, judges must consider the unique circumstances of Aboriginal people. More specifically, judges must consider (1) the unique systemic or background factors that may have contributed to the criminal behaviour of the Aboriginal person before the court, and (2) specific sentencing procedures and sanctions (including restorative justice and traditional healing practices) that may be more appropriate for the individual Aboriginal offender.

However, in subsequent cases, such as *R. v. Wells* ([2000] 1 S.C.R. 207), the Supreme Court of Canada has held that Section 718.2(e) does not alter the fundamental duty of sentencing judges to impose a sentence that is appropriate for the offence and the offender. Furthermore, there is no requirement that the sentencing judge, in every circumstance, give the greatest weight to restorative justice and less weight to other possible goals of sentencing (see Carter, 2001).

The extent to which Section 718.2(e) may apply to other offenders remains to be seen. A recent case involving the sentencing of two Black Canadian women convicted of drug trafficking raises the issue of the extent to which Section 718.2(e) may apply to other offenders (see Box 6.4).

Pre-Sentence Reports

Judges may consider information and recommendations contained in reports that are prepared for the sentencing hearing. A **pre-sentence report** (PSR) may be prepared by a probation officer at the request of the judge. This document

pre-sentence report
a document, prepared by a probation officer for the sentencing judge, that contains sociobiographical and offence-related information about the convicted offender and may include a recommendation for a specific sentence

focus

BOX 6.4

Extending *Gladue*? Race and Sentencing

In November 2000, Marsha Hamilton, a Canadian citizen and a resident of Toronto, was caught returning from a trip to Jamaica with ninety-three cocaine pellets in her stomach. The cocaine was nearly 80 percent pure and had a street value of $70,000. In May 2001, Donna Mason, also a Toronto resident, returned from holidays in Jamaica and was arrested following an investigation that revealed she had eighty-three pellets of cocaine of 90 percent purity in her stomach. Both women were acting as drug couriers and had no financial interest in the cocaine. Neither had a criminal record. The two women also had in common that they were Black single mothers living in poverty. Ms. Hamilton stated that she had committed the crime due to financial hardship. Ms. Mason offered no explanation.

At trial, Crown counsel requested a sentence of two to five years for each woman; the defence lawyers sought conditional sentences for both. The judge decided to impose conditional sentences on both. Among the reasons he gave was that systemic racism and gender bias had played a role in the commission of the offences and thus should mitigate the penalties imposed. During the court process, he introduced 700 pages of materials relating to racism and gender bias in the criminal justice system.

The federal justice department appealed his decisions to the Ontario Court of Appeal. This appeal argued (in part) that the judge had compromised his impartiality by producing evidence on issues related to racism and gender bias and that "the trial judge effectively took on the combined role of advocate, witness, and judge, thereby losing the appearance of a neutral arbiter" (p. 7). The Ontario Court of Appeal concurred with the Crown and held that the trial judge had exceeded the proper role of a judge in sentencing by considering broader social issues:

> Sentencing is not based on group characteristics, but on the facts relating to the *specific* offence and *specific* offender as revealed by the evidence adduced in the proceedings. A sentencing proceeding is also not the forum in which to right perceived societal wrongs, allocate responsibility for criminal conduct as between the offender and society, or "make up" for perceived social injustices by the imposition of sentences that do not reflect the seriousness of the crime. (*R. v. Hamilton and Mason*, 2004: 2).

In the decision, Justice David Doherty reaffirmed that Section 718.2(e) applied to all offenders, but that the restraint against the use of imprisonment is to be used particularly in cases of Aboriginal offenders owing to the potential ineffectiveness of imprisonment. He noted that no evidence was presented at trial to indicate that poor Black women share a similar situation with Aboriginals nor similar notions of punishment. Aboriginal Legal Services of Toronto had intervenor status in the case and argued that Section 718(2)(e) should be applied in determining the sentences for the women.

The three appellate court judges concurred that more appropriate sentences would have been twenty months in prison for Ms. Hamilton and two years less a day for Ms. Mason. The court decided not to send the women to jail, however, since they had satisfactorily completed most of their conditional sentences and felt it would impose a hardship on them.

It is likely that the issue of race and sentencing will continue to resound in the criminal justice system (see Ives, 2004).

What do you think?

What role should race play in sentencing? Should it be a mitigating factor? Do you agree with the decision of the trial judge or with the decision of the Ontario Court of Appeal in Box 6.4?

(known as a pre-disposition report in youth court) contains information on the offender's life history as well as details of the current offence and prior offences. It generally concludes with an assessment of the offender's suitability for various sentencing options and may include a recommendation for a specific sentence.

A recent study found that most judges (87 percent) were satisfied with the pre-sentence reports submitted to them. However, far fewer probation officers (40 percent) were satisfied with the PSRs they had prepared, citing inadequate training, the requirement to include information in the PSR they considered of little value, and a lack of resources to prepare adequate PSRs. The same study found that PSRs had an impact on sentencing; cases in which a PSR was presented were more likely to result in a community sentence rather than a custodial sentence (Bonta et al., 2005).

Plea Bargains

Plea bargaining also affects sentencing (see Chapter 5). Despite the name of this practice, the focus from the defendant's point of view is clearly on the sentence rather than the plea. As noted earlier, there appears to be a "guilty plea discount" whereby those accused persons who plead guilty can expect a more lenient sentence than would have been the case had they pleaded not guilty and then been convicted at trial. A guilty plea can also be offered to the Crown in exchange for various other benefits that can affect the sentence. These benefits include:

- the dropping of charges from the indictment;
- a reduction of the charge to a lesser but included offence (for example, from assault causing bodily harm to common assault);
- summary proceedings rather than an indictment (for hybrid offences); *and/or*
- the promise of a joint submission about a sentence from the Crown and defence counsel.

Unlike their counterparts in many American jurisdictions, Canadian crime victims have no veto power over the "deals" resulting from plea negotiation between the prosecution and the defence. This has led many observers to argue that plea bargains "trivialize" the role of crime victims in that these people are excluded from the negotiations between Crown and defence (Henneberry, 2005).

SENTENCING IN YOUTH COURT

The Youth Criminal Justice Act contains a number of provisions relating to the sentencing of young offenders. As in adult criminal courts, the primary principle of sentencing is proportionality: the sentence a youth receives should be in proportion to the seriousness of the offence. The purpose of sentencing in youth courts is to hold young offenders accountable for their behaviour. This is reflected in a number of provisions, including the following:

- The use of intensive custody and supervision sentences for high-risk youth, including treatment plans to deal with psychological, emotional, and mental difficulties.
- The possibility of an adult sentence for any youth fourteen or older who is convicted of an offence that is punishable by more than two years in jail, upon successful application by the Crown to the court.
- An expansion of the offences for which a youth can be subjected to an adult sentence to include a pattern of convictions for serious violent offences.
- The publication of the name of any youth convicted of a crime who receives an adult sentence.
- Scope for victim impact statements to be presented in youth court.

For less serious offences and for offenders who are not high-risk, the Youth Criminal Justice Act encourages youth courts to use community-based sentences that involve probation, community service, and the offender paying restitution or compensation to the crime victims.

ADDITIONAL SENTENCING OPTIONS

When handing down a sentence, the judge can attach to it one or more other dispositions contained in the Criminal Code, including specified prohibitions and forfeitures.

Prohibitions

The Criminal Code outlines prohibitions that can be ordered at the time of sentencing. These include the following:

- *Prohibition from driving.* Those convicted of impaired driving may be prohibited from operating a motor vehicle, vessel, aircraft, or railway equipment for a set period of at least three months and possibly as long as three years.
- *Prohibition from attending places frequented by children.* Those convicted of a sexual offence involving a complainant under fourteen may be prohibited (for a period that can be as long as life) from attending specified places such as school grounds where those under fourteen might be present.
- *Prohibition from working with children.* Those convicted of sexual offences with children under fourteen may also be prohibited from seeking, obtaining, or continuing employment or volunteer work that involves being in a position of trust or authority toward those under fourteen.
- *Prohibition from possessing firearms, etc.* Those convicted of an indictable offence involving violence (including attempts or threats), or who used a firearm during an offence, may be prohibited from possessing firearms, ammunition, or explosives for a set period—at least ten years and possibly life.

- *Prohibition from owning or caring for animals or birds.* Those convicted of causing the unnecessary suffering of animals or birds can be subject to this prohibition for up to two years.

These prohibitions are usually imposed in addition to other penalties. An offender who violates any of the prohibitions may be charged with breach of probation. Charter challenges to these prohibitions have not found them to be cruel and unusual punishment.

Forfeitures

Convicted offenders may be required to forfeit goods to the Crown. For example, those found in the possession of counterfeit money, narcotics, illegal pornography, hate propaganda, or some types of explosives and weapons may be required to hand the goods over to the government. Forfeited items are either destroyed, or sold with the proceeds going to the government.

Proceeds of Crime

A special kind of forfeiture is described at length in Part XII.2 of the Criminal Code. Section 462.3 defines the "proceeds of crime" as any property, benefit or advantage, within or outside Canada, obtained or derived directly or indirectly as a result of:

(a) the commission in Canada of a designated offence, *or*
(b) an act or omission anywhere that, if it had occurred in Canada, would have constituted a designated offence.

If an offender accumulated money, property, or goods as a result of their crimes, those items may be seized by the government and sold or otherwise liquidated.

EXTRAORDINARY MEASURES

There are two dispositions that are quite different from the sentences discussed so far in that they are not time limited and are used only in the most serious and unusual cases. These dispositions involve declaring offenders either dangerous offenders or long-term offenders.

Dangerous Offender (DO) Designation

Section 752 of the Criminal Code contains procedures and criteria for declaring someone a "dangerous offender." That section defines a dangerous offender (DO) as a person who is given an indeterminate sentence upon conviction for a particularly violent crime and/or who has demonstrated a pattern of committing serious violent offences. In the judgment of the court, the offender's behaviour is unlikely to be controlled or prevented by normal approaches to behavioural restraint. The purpose of the section is to identify those persons with unacceptable propensities for violence and to incapacitate them in order to protect the public interest.

A person can be declared a DO by a sentencing judge only if the Crown makes a formal application after conviction but before sentencing. The provincial attorney general must approve such an application beforehand.

If the Crown proves the case, the judge *may* order detention for an indeterminate period. If this happens, the offender is detained in a federal prison, but there is no set length on the sentence. The offender can be released by the National Parole Board the following year, the following decade, or never (see Chapter 8). These applications are rare, and there is a high burden of proof on the Crown. Two elements are considered in making this determination: *past* offence history, and the likelihood of serious offences in the *future*.

Past Offences

The first threshold is that the current offences of conviction must involve at least one "serious personal injury offence"—that is, an indictable offence for which the possible sentence is at least ten years and which involved:

- the use or attempted use of violence against another person; *or*
- conduct endangering or likely to endanger the life or safety of another person, *or* conduct inflicting or likely to inflict severe psychological damage on another person.

Four offences are specifically excluded: treason, high treason, and first and second degree murder. All four carry a mandatory life sentence, so the indeterminate sentence would be redundant.

The second threshold involving past behaviour is that the offender must have a history characterized by at least one of these phrases (found in Section 753):

- a pattern of repetitive behaviour showing a failure to restrain his behaviour;
- a pattern of persistent aggressive behaviour showing a substantial degree of indifference on the part of the offender respecting the reasonably foreseeable consequences to other persons of that behaviour;
- any behaviour of a brutal nature; *or*
- a failure to control his sexual impulses.

The current offence(s) must form a part of this pattern. Past sexual offences that did not go to court can be considered.

Predictions about Future Offences

The indeterminate sentencing option is unique in that judges are explicitly called upon to predict, based on patterns of past behaviour, the likelihood of serious offences in the future. Specifically, the Crown must prove (beyond a reasonable doubt) that the offender "constitutes a threat to the life, safety or physical or mental well-being of other persons." The Crown must show that the historical patterns are proof that:

- there is a likelihood of death or injury to other people, or severe psychological damage to other people, through the convicted person's future failure to restrain his behaviour;

- the convicted person's future behaviour is unlikely to be inhibited by normal standards of behaviour or restraint; *or*
- there is a likelihood of injury, pain, or other evil to other people through the convicted person's future failure to control his sexual impulses.

Expert witnesses are often called to help the court make these determinations. At least two psychiatrists—one nominated by the defence, the other by the prosecution—must testify. Other experts may be called, and the offender can call witnesses to testify to his character and reputation.

The number of offenders given the DO designation has increased sharply in recent years, largely because of an increase in proactive efforts of Crown counsel and public concerns about violent offenders. As of May 2005, 336 offenders had been designated as DOs (none of whom were female). On average, 24 DOs are admitted to federal correctional facilities each year. The majority of DO designations have been made in Ontario and British Columbia. Aboriginal offenders account for nearly 20 percent of DOs. Seventeen of the active DOs have been granted some type of release from incarceration: fifteen supervised, one temporarily detained, and one deported. The remaining 319 are in custody. The DOs in the community will be under parole supervision for the remainder of their lives (Public Safety and Emergency Preparedness Canada, 2005).

Long-Term Offender (LTO) Designation

Also contained in the Criminal Code (Section 753) are provisions for declaring someone a "long-term offender" (LTO). Crown counsel may use this option when the case falls short of the stringent criteria for filing a DO application. As with dangerous offenders, evidence must be presented to indicate that there is substantial risk that the offender will commit a serious personal offence after release from prison.

The LTO designation is available only for offenders who have received a sentence of more than two years and who are, therefore, under the jurisdiction of the federal correctional system. At sentencing, the judge sets the length of the Long-Term Supervision Order. After imprisonment and any post-release supervision, the Long-Term Supervision Order comes into effect. The offender is supervised by a parole officer for the duration of the order, which can be for a period of up to ten years. As of May 2005, there were 309 active LTO offenders across the country, 188 of whom were in prison; the other 121 were under community supervision. The majority of LTO designations are for sexual offences (Public Safety and Emergency Preparedness Canada, 2005). Female LTOs are rare. Most offenders with Long-Term Supervision Orders have a ten-year supervision period (Solicitor General Canada, 2001).

Long-Term Offenders and Long-Term Supervision Orders: What's the Difference?

The LTO designation is imposed by the sentencing judge and is the actual sentence of the court under Section 753.1 of the Criminal Code. A Long-Term Supervision Order refers to the administration of the sentence and is the responsibility of the National Parole Board under the Corrections and Conditional Release Act.

SENTENCING AND CRIME VICTIMS

It was noted in Chapter 2 that there have been increased efforts to involve victims in the criminal justice process and to ensure that the interests of crime victims are addressed. These efforts have included allowing victim impact statements, collecting "fine surcharges" from offenders for use in supporting victim services programs, and establishing mediation programs as forums for victim–offender reconciliation.

Victim Impact Statements

> There are not enough words to express our sadness, loneliness, loss, and anger . . . This incident has changed my life forever. He was my only son. Who's going to look after us when we're old? Who will bury us if we both die at the same time? Nothing is important anymore. I wonder why I have to wake up every day and face the situation. — Portion of a victim impact statement read to a provincial court judge by the mother of a police constable killed by a speeding car that ran a red light (in Beutel, 2004). [See Case 5 in Box 6.1.]

Section 722.1 of the Criminal Code provides that at the sentencing stage, a crime victim can submit to the court a **victim impact statement (VIS)** explaining:

- his or her personal/emotional reaction to being victimized;
- any physical injuries caused by the victimization; *and*
- the financial impact of the victimization.

victim impact statement (VIS)
submission to a sentencing court explaining the emotional, physical, and financial impact of the crime

FIGURE 6.3 Sample Victim Impact Statement Form

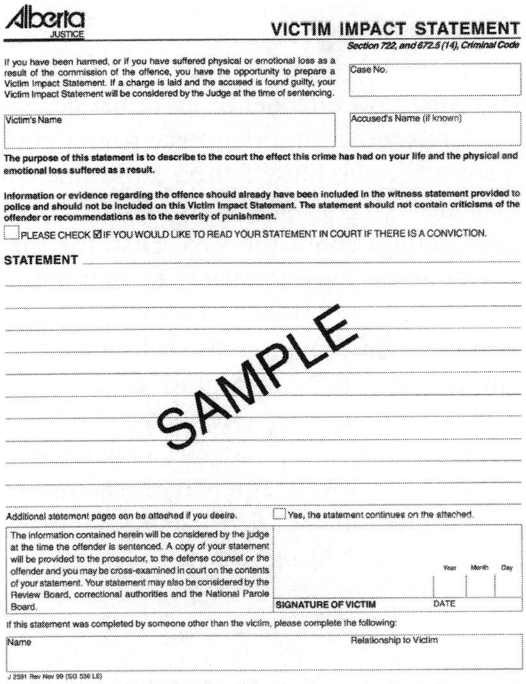
Alberta JUSTICE

VICTIM IMPACT STATEMENT

Section 722, and 672.5 (14), Criminal Code

If you have been harmed, or if you have suffered physical or emotional loss as a result of the commission of the offence, you have the opportunity to prepare a Victim Impact Statement. If a charge is laid and the accused is found guilty, your Victim Impact Statement will be considered by the Judge at the time of sentencing.

Case No.

Victim's Name

Accused's Name (if known)

The purpose of this statement is to describe to the court the effect this crime has had on your life and the physical and emotional loss suffered as a result.

Information or evidence regarding the offence should already have been included in the witness statement provided to police and should not be included on this Victim Impact Statement. The statement should not contain criticisms of the offender or recommendations as to the severity of punishment.

☐ PLEASE CHECK ☑ IF YOU WOULD LIKE TO READ YOUR STATEMENT IN COURT IF THERE IS A CONVICTION.

STATEMENT

SAMPLE

Additional statement pages can be attached if you desire. ☐ Yes, the statement continues on the attached.

The information contained herein will be considered by the judge at the time the offender is sentenced. A copy of your statement will be provided to the prosecutor, to the defense counsel or the offender and you may be cross-examined in court on the contents of your statement. Your statement may also be considered by the Review Board, correctional authorities and the National Parole Board.

Year Month Day

SIGNATURE OF VICTIM DATE

If this statement was completed by someone other than the victim, please complete the following:

Name Relationship to Victim

J 2591 Rev Nov 99 (SG 536 LE)

Source: Retrieved through the Alberta Justice/Solicitor General website (http://www.gov.ab.ca/just). Reprinted by permission of the Alberta Solicitor General.

There are no limitations on the kinds of offences for which a VIS can be submitted. However, it is most commonly used for crimes against the person.

A VIS can take the form of a letter to the judge. Many provinces distribute standard forms, which typically ask the victim to itemize physical injuries and any permanent disability, as well as the dollar value of financial losses, such as property loss or damage, lost wages, or medical expenses not covered by insurance. There is also space to express personal reactions to the crime, including any need for counselling. At the discretion of the judge, victims may read their VIS aloud in court or testify about the impact of the crime; they are not allowed to request specific penalties or directly address the issue of sentencing.

Opponents of VISs argue that they are emotionally charged and thus undermine the objectivity of the justice process. (In fact, research studies indicate that VISs have little influence on the sentence a convicted offender receives.) Advocates contend that VISs ensure that victims are involved in the justice process, make the justice system more accountable, help the victim recover from the victimization, and educate both offenders and judges about the real-life consequences of crime.

The Canadian criminologist Julian Roberts (2003) points out that although VISs are enshrined in the Criminal Code, no guidance is provided as to how the courts should utilize the information they contain. Rather, how the information is used is left to the discretion of individual judges. He also notes that even though crime victims who submit a VIS expect it to affect the sentence, there is no evidence that it does. This may result in disillusionment among crime victims due to unmet expectations: "It is not clear to date that victim impact statements have been implemented in ways that optimize the benefits for the victim and criminal justice professionals" (ibid., 395).

Victim Fine Surcharge

Unless it would constitute undue hardship to the offender or his or her dependants, a sentencing judge in adult court must order the offender to pay a victim fine surcharge (VFS) equal to 15 percent of any fine. If there is no fine, an amount of up to $10,000 is set by the judge.

There are two common misunderstandings about the VFS. First, the surcharge is *not* a sentence in its own right and is always ordered in addition to another disposition. Second, the money is *not* paid to the victim. It goes into a provincial fund to pay for victim services. Some provinces also collect the VFS for *provincial* offences. On Prince Edward Island, the surcharge for each provincial offence is $10. In Saskatchewan, provincial offenders must pay a surcharge of about 10 percent of any fine, or a flat fee of $20. The Saskatchewan government raises about $2 million every year through these surcharges. The rate of non-payment of VFSs is unknown, although it can be anticipated that for many offenders, even a small amount may be beyond their means.

Victim–Offender Mediation

victim–offender mediation

a restorative justice approach in which the victim and the offender, with the assistance of a mediator, work to resolve the conflict and consequences of the offence

Victim–offender mediation (VOM) programs (also known as victim–offender reconciliation programs, or VORPs) are designed to address the needs of crime victims while ensuring that offenders are held accountable for their actions.

focus

BOX 6.5

Benefits of Victim–Offender Mediation

Listed below are some of the potential benefits of victim–offender mediation.

Benefits for Victims

- The victim is given the opportunity to express his or her views directly to the offender.
- The victim has the opportunity to obtain realistic compensation for losses incurred as a result of the incident.
- The victim receives answers to questions about the offence that only the offender can provide.
- The victim has the opportunity to be involved in the sentence of the offender.
- Victims are more likely to receive restitution through VORPs than through the courts. The collection rates for court-ordered restitution are low—around 58 percent. For VORPs, rates are generally over 80 percent.

Benefits for Offenders

- The offender is able to take direct and personal accountability for his or her actions.

- Offenders have the opportunity to learn about the consequences of their actions, apologize, express regrets, and make amends directly to the victim.
- The offender is given the opportunity to participate in a process through which the stigma of a criminal record can be avoided.

Benefits for the Community

- VORPs contribute to the peace of the community by helping people reach resolutions that address the causes of the conflict.
- VORPs save society money: it costs several hundred dollars for an offender to be placed in a VORP; it costs thousands for an offender to be on probation, in custody, or supervised on parole.

Source: John Howard Society of Alberta (1998), "Victim Offender Reconciliation Programs." Retrieved from http://www.johnhoward.ab.ca. Reprinted by permission of the John Howard Society of Alberta.

The first VOM program was established by Community Justice Initiatives in Kitchener, Ontario, in 1974 and has served as a model for many subsequent programs developed across Canada and internationally. VOM programs were among the earliest restorative justice initiatives. The benefits of these programs for victims, offenders, and the community are outlined in Box 6.5.

VOM programs are operated by provincial agencies and also by not-for-profit organizations. They are generally restricted to cases involving less serious offences and are used at the pre-charge, post-charge, and pretrial stages. Referrals come from the police, the courts, and probation offices. Three basic conditions must exist before a VOM program is used:

1. The offender must accept responsibility for the crime.
2. Both the victim and the offender must be willing to participate.
3. Both the victim and the offender must consider it safe to be involved in the process.

focus

BOX 6.6

Restorative Resolutions

Based in Winnipeg, Restorative Resolutions is an intensive supervision program that uses victim–offender mediation to achieve restorative justice. The program was designed as an alternative to incarceration for offenders willing to take responsibility for their behaviour and to provide compensation to their victims. Operated by the John Howard Society of Manitoba, and staffed by workers trained in probation practices and restorative justice, the program provides counselling, anger management programs, and a wide range of other services.

A Case Study

A thirty-two-year-old man with a lengthy criminal record was charged with four new counts of theft and break and enter. The Crown counsel wanted the offender to serve a prison term. Restorative Resolutions staff prepared an alternative plan in which they recommended a suspended sentence, with supervision to be carried out by Restorative Resolutions.

Under the plan, the offender would be required to complete an interpersonal communication skills course; complete an Addictions Foundation of Manitoba assessment and attend Alcoholics Anonymous (AA) meetings regularly; adhere to the conditions outlined in the mediation agreement; and receive literacy training. The plan was accepted by the judge.

Source: *Satisfying Justice: Safe Community Options That Attempt to Repair Harm from Crime and Reduce the Use or Length of Imprisonment*, p. 5. Correctional Service of Canada, 1996. Reproduced with the permission of the Minister of Public Works and Government Services Canada, 2006.

The mediator assists the victim and the offender in arriving at a settlement that addresses the needs of both parties and that provides a resolution to the conflict. A conciliation agreement mediated between the offender and the victim before the sentence is passed is included in the probation order. A VOM program called Restorative Resolutions is profiled in Box 6.6.

SENTENCING AND RESTORATIVE JUSTICE

The emergence of restorative justice was identified in Chapter 1 as one of the key trends in Canadian criminal justice. As Table 6.1 illustrates, the underlying principles of this alternative approach are in sharp contrast to those on which the adversarial system of criminal justice is based. Other key differences between restorative justice and retributive justice are outlined in Figure 6.4.

Circle Sentencing

Circle sentencing is an example of the application of restorative justice, a key principle of which is that the sentence is less important than the process used

circle sentencing
a restorative justice strategy that involves collaboration and consensual decision making by community residents, the victim, the offender, and justice system personnel to resolve conflicts and sanction offenders

TABLE 6.1 Comparison of Retributive and Restorative Justice Principles	
Retributive Justice	**Restorative Justice**
Crime violates the state and its laws.	Crime violates people and relationships.
Justice focuses on establishing guilt so that doses of pain can be meted out.	Justice aims to identify needs/obligations so that things can be made right.
Justice is sought through conflict between adversaries in which the offender is pitted against the state.	Justice encourages dialogue and mutual agreement and gives victims and offenders central roles.
Rules and intentions outweigh outcomes, one side wins while the other loses.	The outcome is judged by the extent to which responsibilities are assumed, needs are met, and healing (of individuals and relationships) is encouraged.

Source: H. Zehr, *Changing Lenses: A New Focus for Crime and Justice* (Scottsdale, PA: Herald Press, 1990). Reprinted by permission of Herald Press.

FIGURE 6.4   Comparison of Criminal Justice and Restorative Justice Participants and Processes

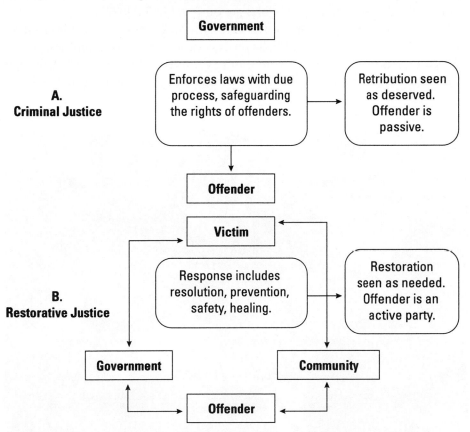

Source: C.G. Nicholl, *Community Policing, Community Justice, and Restorative Justice: Exploring the Links for the Delivery of a Balanced Approach to Public Safety* (Washington, DC: Office of Community-Oriented Policing Programs, U.S. Department of Justice, 1999), 113.

to select it. In this approach, traditional Aboriginal healing practices and processes of reconciliation, restitution, and reparation are used to address the needs of victims and offenders, their families, and the broader community. Circle sentencing originated in several Yukon communities as a collaborative initiative involving community residents and territorial justice personnel (mainly RCMP officers and judges from the Territorial Court of Yukon), but its principles and procedures can be adapted to other communities. The city of Minneapolis, Minnesota, for example, has a successful circle sentencing program, and the Island Community Justice Society in Nova Scotia operates sentencing circles for youth (see Island Community Justice Society at http://www.islandcommunityjustice.com).

In circle sentencing, all participants—the judge, the defence lawyer and prosecutor, the police officer, and community residents, as well as the victim and offender and their families—sit facing one another in a circle. Discussion is aimed at reaching a consensus about the best way to dispose of the case, taking into account both the need to protect the community and the rehabilitation and punishment of the offender. There are a number of important stages in the circle sentencing process, each of which is critical to its overall success. Table 6.2 compares the attributes of formal criminal courts with those of restorative justice as represented by circle sentencing. Further differences between criminal court and circle sentencing principles are outlined in Table 6.3.

Circle sentencing is generally available only to offenders who plead guilty (see s. 717 of the Criminal Code). Its process varies across communities; generally, though, it relies heavily on community volunteers (Stuart, 1996). Although offenders who have their cases heard through circle sentencing may still be

At Aboriginal Ganootamaage Justice Services of Winnipeg, Manitoba, participants take part in a healing circle for a 20-year-old shoplifter (not shown in photo).

TABLE 6.2 Comparison of Formal Criminal Courts and Restorative Justice

Attribute	Court	Restorative Justice
Participants	experts	local people non-residents
Process	adversarial state vs. offender	consensus community vs. problem
Central issue	laws broken	relationship broken
Focus	guilt	identification of victim/offender/community needs as part of solution to problem
Tools	punishment control	healing/support
Procedure	fixed rules	flexible guidelines

TABLE 6.3 Differences between Criminal Court and Circle Sentencing Principles

Criminal Court	Circle Sentencing
The conflict is the crime.	Crime is a small part of a larger conflict.
The sentence resolves the conflict.	The sentence is a small part of the solution.
The focus is on past conduct.	The focus is on present and future conduct.
Takes a narrow view of behaviour.	Takes a larger, holistic view of behaviour.
Not concerned with social conflict.	Focuses on social conflict.
The sentence is the most important part of the process.	The sentence is the least important part of the process; most important is the process itself, which shapes the relationship among all parties.

Source: Reprinted by permission of Judge Barry D. Stuart.

required to serve a period of incarceration, a wide range of other sanctions are available, including banishment (generally to a wilderness location), house arrest, and community service. Although circle sentencing has thus far mainly involved adult offenders, the approach is being applied more and more often to young offenders. Circle sentencing has spawned a number of variations, including community sentence advisory committees, sentencing panels, and community mediation panels.

In contrast with the adversarial approach to justice, circle sentencing has the potential to:

- reacquaint individuals, families, and communities with problem-solving skills;
- rebuild relationships within communities;
- promote awareness and respect for values and the lives of others;
- address the needs and interests of all parties, including the victim;
- focus on causes, not just symptoms, of problems;
- recognize existing healing resources and create new ones;
- co-ordinate the use of local and government resources; *and*
- generate preventive measures.

It should be pointed out that circle sentencing is not appropriate for all offenders or for all crimes. Moreover, the success of a given circle will depend on the extent to which all of its participants are committed to the principles of restorative justice (see Dickson-Gilmore and LaPrairie, 2005). Concerns have been raised, for example, as to whether crime victims—especially Aboriginal women who have been the victims of sexual assault and domestic abuse—may be pressured into participating in circle sentencing.

Collaborative Justice

A high-profile program that is premised on restorative justice principles is the Collaborative Justice Project, which operates within the criminal justice system and is designed to provide an alternative to the traditional criminal justice process. The program is a joint initiative of the Church Council on Justice and Corrections and the Ottawa Crown attorney's office and is supported by a number of federal departments. It is unique in that it takes cases involving serious criminal offences and operates at the pre-sentence stage. The objective is to facilitate a dialogue between the victim and the offender that can be presented to the court at sentencing.

Box 6.7 presents a case from the Collaborative Justice Project.

An evaluation of the Collaborative Justice Project examined the levels of satisfaction among victims and offenders with the collaborative justice process, whether the needs of clients were met, and whether clients were satisfied with a restorative approach as opposed to the traditional criminal justice process (Rugge, Bonta, and Wallace-Capretta, 2005). This study involved comparing victims and offenders who participated in the Collaborative Justice Project with a control group. The results of the evaluation revealed high levels of satisfaction with the process among victims and offenders; furthermore, there was a shared sense of empowerment that was not found in criminal justice case processing. This led the authors to conclude that a restorative justice approach can succeed at the pre-sentence stage in cases involving serious offences.

Restorative justice has its critics. It has been pointed out that very few evaluations of restorative justice have been conducted and that its principles are difficult to transform into practice. An extended critique of restorative justice is *Compulsory Compassion* (2004) by Annalise Acorn, a law professor at the University of Alberta. Her book raises a number of issues surrounding the challenges of reconciling the core principles of restorative justice with notions of justice and accountability.

focus

BOX 6.7

Impaired Driving Causing Death: A Collaborative Justice Approach

An impaired driver ended up driving down the wrong side of a multilane highway. He collided with a van, killing the driver and trau- matizing the driver's wife. The accused was also seriously injured. The son of the deceased wished to meet the offender to talk about the impact on his family and to determine whether the offender was genuinely remorseful. They met on three occasions. The offender got to know something about the deceased and what he meant to his family. The son of the deceased saw something of the character of the accused and his sincere remorse. They agreed to go together to speak to high school classes about the dangers of drinking and driving. They also participated in a video produced by the Law Commission on restorative community initia- tives. The offender received a sentence of two years less a day, three years' probation, and a seven-year driving prohibition. Later the son of the victim wrote:

I believe that the importance of victims being given the opportunity to explain their feelings and posi- tion, directly or indirectly, to the person responsible for the harm done to them is inadequately under- stood. In my experience, victims have an over- whelming need to know that the person responsible for their suffering understands the nature and extent of what has been inflicted on them. They need to see some comprehension by the perpetrator of the degree and scope of the harm done. For many, recognition and an indication of remorse is far more important than punishment. Being able to communicate in such a fashion undoubtedly can speed and facilitate the healing process and, thereby, reduce the overall harm suffered. From my perspec- tive, the Project made it possible to create positive outcomes from my father's tragic death.

Source: J. Scott, "Restorative Justice and the Criminal Justice System." Retrieved from http://www.ccjc.ca/ collaborativejustice.htm. Reprinted by permission of the Church Council on Justice and Corrections.

MEASURING THE EFFECTIVENESS OF SENTENCING OPTIONS

Measuring the effectiveness of the various sentencing options discussed in this chapter is no easy task. There are several reasons why:

- Although judges may include recommendations for treatment in their sentencing orders, once an offender is convicted, he or she becomes the responsibility of the federal or provincial/territorial correctional system. There are few (if any) formal channels of communication between the judiciary and corrections once a case has been decided in the courts. A number of factors beyond the control of the court—including limited opportunities for treat- ment, overcrowding in correctional facilities, and the heavy caseloads of pro- bation and parole officers—may undermine the objectives of the sentence.

- Matching specific sentencing options with the needs of offenders (and the risks they pose) is an inexact science at best. No body of scientific knowledge is available to help judges select sanctions that would have the maximum positive impact on the offender.

• Studies about the impact of criminal court sanctions are inconclusive; whatever the specific objectives of a sentence are, there is no way to know whether the sanction will have the intended effect.

These concerns have prompted many jurisdictions to explore alternative approaches to sentencing, including restorative justice.

The Effectiveness of Alternatives to Imprisonment

Research on the effectiveness of alternatives to imprisonment presents a mixed picture, as the following discussion suggests.

Diversion Programs

Few formal evaluations of diversion programs have been conducted. Because they tend to target low-risk, first-time offenders convicted of minor offences, diversion programs may "widen the net," resulting in offenders being placed under supervision who otherwise would not have been subjected to any sanction. There is no evidence that diversion has any impact on recidivism or on correctional populations. Diversion may increase the justice system's workload and costs.

Fine Option Programs

Fine option programs have not been especially effective, as evidenced by the high numbers of fine defaulters admitted to custody in a number of provinces.

Probation

Probation is the most widely used offender supervision program, yet few evaluations have been carried out on its effectiveness. Studies conducted in the United States have found that probation has failed to protect the community and to assist offenders (see Reinventing Probation Council, 2000).

Home Confinement/Electronic Monitoring Programs

There is no evidence that EM programs reduce prison admissions, lower costs for correctional systems, or reduce the likelihood of future criminal behaviour (Bonta, Wallace-Capretta, and Rooney, 1999).

Conditional Sentences

Conditional sentences may not reduce admissions to custody, which supposedly is their purpose. Studies indicate that the number of violent and sexual offenders being given conditional sentences is increasing (Johnson, 2004). One study found that 40 percent of offenders in three locations across Canada had breached the conditions of their conditional sentences (North, 2001). Many jurisdictions lack adequate resources to supervise offenders on conditional sentences (Mazey, 2002).

Restorative Community Justice

A review of research findings on restorative justice published over the past twenty-five years found lower rates of reoffending among participants in restorative justice programs relative to those who participated in traditional correctional approaches such as probation (Latimer, Dowden, and Muise, 2001). As well, victim–offender mediation programs receive high marks from victims and offenders in terms of the fairness of the process and the outcomes of mediation sessions (Umbreit, 1995).

Perhaps the longest running and most successful community-based restorative justice program is the Hollow Water First Nation Community Holistic Circle Healing (CHCH) process in Manitoba. Evaluations of this program for sexual abuse victims and offenders indicate that it has:

- significantly reduced the prevalence of alcohol abuse in the community;
- improved educational standards;
- increased services for infants, children, and youth;
- increased community awareness of sexual abuse and family violence and the rates of disclosure by offenders; *and*
- been more cost effective than the traditional criminal justice process (Couture et al., 2001; Lajeunesse and Associates, 1996; Rugge, Bonta, and Wallace-Capretta, 2005).

The Effectiveness of Incarceration

Supporters of incarceration argue that it is an effective general and specific deterrent to crime. However, there is some evidence that imprisonment does not have a significant impact on crime rates. For example, Nova Scotia and Saskatchewan are similar in terms of population size and per capita income, yet Saskatchewan's incarceration rate (161 per 100,000) is four times higher than Nova Scotia's (40 per 100,000). And Saskatchewan continues to have higher levels of violent and property crime (Sauve, 2005). Box 6.8 provides further information on the effectiveness of incarceration.

focus   BOX 6.8

How Effective Is Incarceration?

Research studies on the use of incarceration reveal the following:

- Except when used for violent, high-risk offenders, incarceration is not an effective individual deterrent. There are nearly identical rates of reoffending for those sentenced to probation supervision and those sent to prison. All else equal, offenders sent to prison are more likely to recidivate.

- It is unlikely that incarceration is effective as a general deterrent. Rather, the evidence suggests that the likelihood of being caught and punished is the most effective general deterrent.

- For some offenders, incarceration may not be effective as a specific deterrent. Offenders who have spent much of their lives in institutions, and who as a result have few if any family or other ties to the community, may be more comfortable in prison than in the community.

- To be effective, incarceration must be used selectively. However, it is difficult to determine which offenders should be incarcerated and for how long in order for imprisonment to be effective.

- Longer terms of incarceration are ineffective. There is no evidence that increasing the length of a prison term reduces either the overall crime rate or the likelihood that an offender will recidivate once released from an institution. In fact, low-risk offenders who receive longer sentences in custody have higher rates of reoffending because they have been exposed to more serious offenders.

- There is no relationship between recidivism rates and the type or severity of sanctions. There is, however, conclusive evidence that treatment approaches are more effective than punishment in reducing reoffending. As well, community-based programs have higher overall rates of success than institutional programs.

- Incarceration is not necessarily the most severe punishment that can be inflicted on offenders. Research on the perceptions of sanctions held by offenders in several American prisons found that many inmates prefer incarceration to probation. Two-thirds of the inmates indicated that they would rather spend one year in prison than ten years on probation. Nearly half the sample expressed a preference for prison when the probationary period was reduced to five years, while one-third would have preferred confinement to three years on probation. For some of the offender population—especially offenders in provincial correctional institutions—incarceration may be part of a life pattern and of no significant consequence. For drug-addicted offenders, a period of incarceration may offer the opportunity to get over being "drug sick" and regain their health.

Sources: D. Andrews, "The Psychology of Criminal Conduct and Effective Treatment," in J. McGuire (ed.), *What Works: Reducing Reoffending—Guidelines from Research and Practice* (Chichester, UK: John Wiley and Sons, 1995), 35–62; B.M. Crouch, "Is Incarceration Really Worse? An Analysis of Offenders' Preferences for Prison over Probation," *Justice Quarterly,* 10(1) [1993], 67–68; National Crime Prevention Council, *Incarceration in Canada* (Ottawa, 1997); P. Smith, C. Goggin, and P. Gendreau, *The Effects of Prison Sentences and Intermediate Sanctions on Recidivism: General Effects and Individual Differences* (Ottawa: Solicitor General of Canada; Public Safety and Emergency Preparedness Canada, *Fast Facts 4.* Retrieved from http://www.psepc-sppcc.gc.ca/prg/cor/acc/_fl/ff4-en.pdf

Summary

This chapter focused on sentencing in the criminal justice process. The purpose, principles and goals of sentencing were identified and discussed, as were the various sentencing options available to Canadian judges. These options include conditional sentences, which have been the focus of controversy. The factors that influence the sentencing decisions of judges were then discussed. After this, an exercise was provided wherein you, the student-reader, were given an opportunity to make decisions based on actual case scenarios and then compare your sentencing decisions with those of the presiding judges. The involvement of victims in the sentencing process was then considered, as well as the influence (or lack thereof) of victim impact statements on sentences imposed by judges. The chapter concluded with a discussion of the use of various restorative justice approaches and programs in addressing the needs of victims and offenders, and of the contribution of these approaches to sentencing.

Key Points Review

1. Among the statutory objectives of sentencing are denunciation, deterrence, the separation of offenders from society, rehabilitation, and reparation for harm done.
2. Section 718.2(e) of the Criminal Code, which states that sentencing judges must consider all available sanctions other than imprisonment—for Aboriginal offenders in particular—has generated considerable controversy.
3. The sentencing goals in the criminal courts fall into three main groups: utilitarian, retributive, and restorative.
4. Judges in the criminal courts can select from a range of sentencing options, which include various alternatives to confinement and varying terms of imprisonment in correctional institutions.
5. Falling between traditional probation and imprisonment are intermediate sanctions, which include electronic monitoring (EM), fines, community service, intensive supervision probation (ISP), and conditional sentences.
6. Probation is the most widely used sanction for controlling and supervising criminal offenders.
7. The discretion exercised by judges in choosing sentencing options is limited by the key sentencing principle that similar crimes committed by similar offenders should draw similar sentences.
8. Victim impact statements provide an opportunity for crime victims to have input at the sentencing stage, although the influence of these statements on judicial decision making is uncertain.
9. Circle sentencing is one example of how restorative justice principles can be used to develop an alternative forum for sentencing offenders.
10. Research evidence is for the most part inconclusive as to the effectiveness of the various sentencing options, although it appears that incarceration is not an effective general or specific deterrent.

Key Term Questions

1. What are **intermediate sanctions,** and what are some examples of this sentencing option?
2. Compare and contrast the sentencing options of **absolute discharge** and **suspended sentence.**
3. How do **fine option programs** work, and why are they a popular disposition in the criminal courts?
4. What is **electronic monitoring,** and what are some ethical concerns surrounding its use as an intermediate sanction?
5. What are **conditional sentences of imprisonment**, and why are they a popular but controversial sentencing option?
6. Compare and contrast **concurrent** and **consecutive** sentences.

7. What is **judicial determination** and what role does it play in sentencing?
8. What are **pre-sentence reports,** and what role do they play in sentencing?
9. What is a **victim impact statement (VIS),** and what role does it play in the criminal justice process?
10. How do **victim–offender mediation (VOM)** programs work?
11. In what ways does **circle sentencing** demonstrate how restorative justice principles are applied at the sentencing stage of the criminal justice process?

References

Acorn, A. (2004). *Compulsory Compassion: A Critique of Restorative Justice.* Vancouver: UBC Press.

Agrell, S. (2004, July 6). "Judge Says Abusers Meant Well." *National Post,* A1, A6.

Andrews, D. (1995). "The Psychology of Criminal Conduct and Effective Treatment." In J. McGuire (ed.), *What Works: Reducing Reoffending—Guidelines from Research and Practice.* Chichester, UK: John Wiley and Sons.

Avery, R. (2005, September 15). "Hells Angels Get Extra Jail Time for Membership in a Bike Gang." *The Barrie Advance.* Retrieved from http://www.simcoe.com/sc/barrie/law/story/3033344p-3517244c.html

Baron, E. (2004, December 19). "B.C. Soft on Drunk Drivers: MADD." *The Province,* B8.

Beutel, T. (2004, June 17). "Guan Gets Off with Just Probation." *Richmond (BC) News Online Edition.* Retrieved from http://www.richmond-news.com/issues04/063204/news/063204nn2.html

Bonta, J., G. Bourgon, R. Jesseman, and A.K. Yessine. (2005). *Presentence Reports in Canada.* Ottawa: Public Safety and Emergency Preparedness Canada.

Bonta, J., S. Wallace-Capretta, and J. Rooney. (1998). *Restorative Justice: An Evaluation of the Restorative Resolutions Project.* Ottawa: Solicitor General of Canada.

———— (1999). *Electronic Monitoring in Canada.* Ottawa: Solicitor General of Canada.

Carter, M. (2001). "Addressing Discrimination in the Sentencing Process: "C.C. S. 718.2(a)(i) in Historical and Theoretical Context." *Criminal Law Quarterly,* 44(3):399–436.

Couture, J., T. Parker, R. Couture, and P. Laboucane. (2001). *A Cost-Benefit Analysis of Hollow Water's Community Holistic Healing Circle Process.* Ottawa: Solicitor General of Canada and the Aboriginal Healing Foundation.

Crouch, B.M. (1993). "Is Incarceration Really Worse? Analysis of Offenders' Preferences for Prison over Probation." *Justice Quarterly,* 10(1):67–68.

Dickson-Gilmore, J., and C. LaPrairie. (2005). *Will the Circle Be Unbroken? Aboriginal Communities, Restorative Justice, and the Challenges of Conflict and Change.* Toronto: University of Toronto Press.

Easton, S.T. (2004). "Marijuana Growth in British Columbia." *Public Policy Sources*, 74. Vancouver: The Fraser Institute.

Gabor, T., and N. Crutcher. (2002). *Mandatory Minimum Penalties: Their Effects on Crime, Sentencing Disparities, and Justice System Expenditures.* Ottawa: Research and Statistics Division, Department of Justice Canada.

Gannon, M., and J.-A. Brzozowski. (2004). "Sentencing in Cases of Family Violence." In Kathy AuCoin (ed.), *Family Violence in Canada: A Statistical Profile of Cases.* Ottawa: Minister of Industry.

Globe and Mail (2005, April 9). "Driver in Deadly Crash Gets Community Work," A9.

Hanes, A. (2005, September 20). "$1.5M in Ad Fraud, No Jail." *National Post*, A1, A6.

Hansen, D. (2004, October 12). "Ng's Killer Gets Two Years." *Richmond News Online Edition.* Retrieved from http://www.richmond-news.com/issues04/102104/news/102104nn1.html

Haslip, S. (2000). "Aboriginal Sentencing Reform in Canada: Prospects for Success—Standing Tall with Both Feet Planted Firmly in the Air." *Murdoch University Electronic Journal of Law*, 7(1). Retrieved from www.murdoch.edu.au/elaw/issues/v7n1/haslip71.html

Henneberry, J. (2005). "Retired Judge Says Plea Bargains are a Bad Deal for Victims." *National Justice Network Update* 10(1):2.

Ives, D.E. (2004). "Inequality, Crime, and Sentencing: Borde, Hamilson, and the Relevance of Social Disadvantage in Canadian Sentencing Law." *Queen's Law Journal*, 30(1):114–55.

Johnson, S. (2004). "Adult Correctional Services in Canada, 2002/03." Catalogue no. 85-002-XPE. *Juristat*, 24(10). Ottawa: Canadian Centre for Justice Statistics, Statistics Canada.

Lajeunesse, T., and Associates. (1996). *Evaluation of Community Holistic Circle Healing: Hollow Water First Nation. Vol. 1. Final Report.* Winnipeg: Manitoba Department of Justice.

Latimer, J., C. Dowden, and D. Muise. (2001). *The Effectiveness of Restorative Justice Practices: A Meta-Analysis.* Ottawa: Research and Statistics Division, Department of Justice Canada.

Law Society of British Columbia. (2005). "In the matter of the Legal Profession Act, SBC 1998, c. 9 and a hearing concerning Stephen Neville Suntok, Respondent. Decision of the Hearing Panel." Retrieved from http://www.lawsociety.bc.ca/professional_regulation/hearings/disc-report_suntok.html

Mazey, E. (2002). "Conditional Sentences under House Arrest." *Criminal Law Quarterly*, 46(2):246–64.

National Crime Prevention Council. (1997). *Incarceration in Canada.* Ottawa.

North, D. (2001). "The 'Catch-22' of Conditional Sentencing." *Criminal Law Quarterly*, 44(3):342–74.

Paciocco, D.M., and J. Roberts. (2005). *Sentencing in Cases of Impaired Driving Causing Bodily Harm or Impaired Driving Causing Death, With a Particular Emphasis on Conditional Sentencing.* Ottawa: Canada Safety Council.

Public Safety and Emergency Preparedness Canada. (2005). "Frequently Asked Questions: Release of Offenders." Ottawa.

R. v. Hamilton and Mason. (2004). O.J. NR. v.o. 3252 (C.A.). Court of Appeal of Ontario. Retrieved from http://www.ontariocourts.on.ca/decisions/2004/august/C39716.htm

Reinventing Probation Council. (2000). *Transforming Probation through Leadership: The "Broken Windows" Model.* New York: Center for Civic Innovation at the Manhattan Institute.

Roberts, J.V. (2003). "Victim Impact Statements and the Sentencing Process: Recent Developments and Research Findings." *Criminal Law Quarterly,* 47(3):365–96.

Roberts, J.V., and P. Healy. (2001). "The Future of Conditional Sentencing." *Criminal Law Quarterly,* 44(3):309–41.

Roberts, J.V., and C. LaPrairie. (2000). *Conditional Sentencing in Canada: An Overview of Research Findings.* Ottawa: Research and Statistics Division, Department of Justice.

Roberts, J.V., and A. Manson. (2004). *The Future of Conditional Sentencing: Perspectives of Appellate Court Judges.* Ottawa: Department of Justice Canada.

Roberts, J.V., and K. Roach. (2004). *Community-Based Sentencing: The Perspectives of Crime Victims.* Ottawa: Department of Justice Canada.

Rugge, T., J. Bonta, and S. Wallace-Capretta. (2005). *Evaluation of the Collaborative Justice Project: A Restorative Justice Program for Serious Crime.* Ottawa: Public Safety and Emergency Preparedness Canada.

Sanders, T., and J.V. Roberts. (2000). "Public Attitudes toward Conditional Sentencing: Results of a National Survey." *Canadian Journal of Behavioural Science,* 32(4):199–207.

Sauve, J. (2005). "Crime Statistics in Canada, 2004." Catalogue no. 85-002-XPE. *Juristat,* 25(5). Ottawa: Canadian Centre for Justice Statistics, Statistics Canada.

Sin, L. (2005, May 27). "Addict Gets 7 Years for Crash that Killed Pastor." *The Province,* A3.

Skelton, C. (2005, January 18). "B.C. Jails One in 7 Pot Growers; Vancouver Jails Only One in 13." *Vancouver Sun,* A1.

Sokora, G. (2005a, April 11). "Driver Who Killed RCMP Constable Ng Denied Parole." *Richmond News Online Edition.* Retrieved from http://www.richmond-news.com/issues05/042105/news.html

——— (2005b, June 5). "Mountie's Killer Granted Full Parole." *Richmond News Online Edition.* Retrieved from http://www.richmond-news.com/issues05/062105/news.html

Solicitor General Canada. (1998). *CCRA 5 Year Review: Judicial Determination.* Ottawa.

——— (2001). *High Risk Offenders: A Handbook for Criminal Justice Professionals.* Ottawa.

Stenning, P., and J.V. Roberts. (2001). "Empty Promises: Parliament, the Supreme Court, and the Sentencing of Aboriginal Offenders." *Saskatchewan Law Review,* 64(1):137–68.

Stuart, B. (1996). "Circle Sentencing in Yukon Territory, Canada: A Partnership of the Community and the Criminal Justice System." *International Journal of Comparative and Applied Criminal Justice,* 20(2):291–309.

Thomas, M. (2004). "Adult Criminal Court Statistics, 2003/04." Catalogue no. 85-002-XPE. *Juristat,* 24(12). Ottawa: Canadian Centre for Justice Statistics, Statistics Canada.

Turpel-Lafond, M.E. (1999). "Sentencing within a Restorative Justice Paradigm: Procedural Implications of *R. v. Gladue.*" *Criminal Law Quarterly,* 43(1):34–50.

Umbreit, M.S. (1995). *Mediation of Criminal Conflict: An Assessment of Programs in Four Canadian Provinces.* Minneapolis: Center for Restorative Justice and Mediation, University of Minnesota.

Ursel, J. (2003). "Using the Justice System in Winnipeg." In H. Johnson and K. AuCoin (eds.), *Family Violence in Canada: A Statistical Profile.* Ottawa: Minister of Industry.

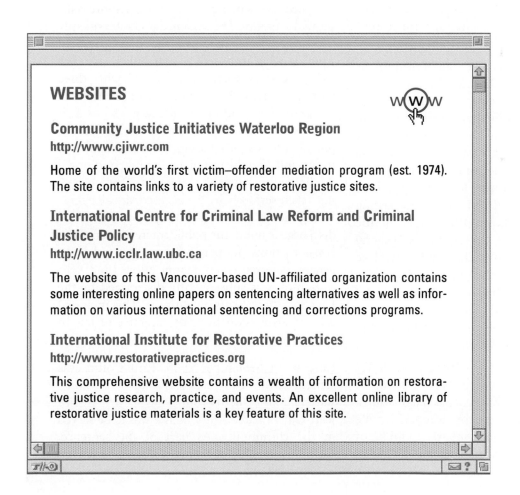

WEBSITES

Community Justice Initiatives Waterloo Region
http://www.cjiwr.com

Home of the world's first victim–offender mediation program (est. 1974). The site contains links to a variety of restorative justice sites.

International Centre for Criminal Law Reform and Criminal Justice Policy
http://www.icclr.law.ubc.ca

The website of this Vancouver-based UN-affiliated organization contains some interesting online papers on sentencing alternatives as well as information on various international sentencing and corrections programs.

International Institute for Restorative Practices
http://www.restorativepractices.org

This comprehensive website contains a wealth of information on restorative justice research, practice, and events. An excellent online library of restorative justice materials is a key feature of this site.

focus

Actual Sentencing Decisions

Case 1. The twenty-three-year-old addict who committed the offence was Benjamin Bleinis; the victim was church pastor Joseph Chan. On May 26, 2005, B.C. Provincial Court Judge R.D. Franklin sentenced Bleinis to seven years in prison. As well, Bleinis was required to provide a DNA sample for the National DNA Data Bank and given a lifetime driving prohibition. Also, he will be required to complete the 203 days remaining on his conditional sentence in jail following the seven-year sentence. In passing sentence, the judge stated: "If I can send a message to car thieves and those in possession of stolen property, here's a message for you: I impose seven years in prison" (Sin, 2005: A3).

Case 2. Justice Donald Halikowski of the provincial court in Oshawa, Ontario, imposed a sentence of nine months in a provincial correctional facility, to be followed by three years' probation. The latter was to include counselling, reporting to the police on a monthly basis, and not having any contact with anyone under the age of sixteen, or with the boys who had been placed in care by the child protection society. In passing sentence, the judge stated the actions of the adoptive parents were the result of "general good intentions," noting that the discipline was designed to "train" the children. The judge further referred to the abuse as "intrusive but not physically harming" and stated that there were "no long-term psychological effects for the children" (Agrell, 2004: A1, A6).

Case 3. In February 2004, the Honourable Judge W.J. William Diebolt of the B.C. Provincial Court imposed a conditional discharge with a three-year probation order on Stephen Neville Suntok. In addition to the standard conditions of probation, the order contained the following

requirements: Suntok was to take and complete counselling as directed by his probation officer; to have no direct or indirect contact with the victim; to complete 200 hours of community work service; and to not consume alcohol (cited in Law Society of British Columbia, 2005).

In passing sentence, the judge indicated that the public interest would be best served if Suntok were rehabilitated, noting that he had been publicly humiliated and embarrassed. The sentence was criticized by Vancouver Rape Relief and Women's Shelter and by the victim. The Crown appealed the sentence, requesting a more substantial sentence. The sentence was subsequently overturned by a judge of the B.C. Court of Appeal, who imposed a suspended sentence (which carries a criminal record) with three years probation. This judge stated: "In the circumstances a conditional discharge is not a fit sentence. I think that the nature of the assault, one akin to a spousal assault, the seriousness of the assault, and the fact of a previous conditional discharge for an assault, the terms of which were not taken seriously by the accused, are all aggravating factors . . . The granting of a conditional discharge is not in the public interest given those considerations, in particular as they relate to general deterrence and denunciation" (ibid.).

Suntok was subsequently disciplined by the Law Society of British Columbia. This included a brief suspension from the practice of law in the province.

Case 4. In September 2005, Paul Coffin was sentenced to a two-years-less-a-day conditional sentence by Quebec Superior Court Justice Jean-Guy Boilard. This sentence was to be served in the community. As part of the sentence, Coffin must abide by 9 p.m. to 7 a.m. curfew on weekdays and must deliver a series of lectures on business ethics to university students. Among the topics proposed by Coffin: "Never compromise your integrity, no matter what the perceived benefit" and "The only person who can rob you of your reputation, credibility, and good name is yourself" (Hanes, 2005: A1). The presiding judge indicated that the guilty plea entered by Coffin, his repayment of nearly $1 million to the government, and his remorse were key factors in the decision to impose a conditional sentence rather than jail time. As well, the judge stated: "Denunciation and deterrence were the prime objectives of the sentence that will be meted out" (Hanes, 2005: A6). In October 2005, the Quebec Court of Appeal gave the Crown permission to appeal the sentence imposed on Coffin. In April, 2006, the Quebec Court of Appeal overturned the conditional sentence and ordered Coffin to serve 18 months in jail for his role in the sponsorship scandal.

Case 5. In October 2004, Stuart Chan was sentenced in B.C. Supreme Court to two years less a day for criminal negligence causing death and to an additional six months in custody for leaving the scene of an accident involving bodily harm. Noting that Chan had already served twenty-one days in jail and, considering the time spent on house arrest, B.C. Supreme Court Justice Harvey Groberman reduced the total sentence to two years less a day in custody. In addition, the judge imposed a two-year period of probation and a three-year driving prohibition. Justice Groberman had previously discounted a conditional sentence as inappropriate.

In handing down the sentence, the judge indicated that a term of incarceration was necessary to deter others, but that the sanction should not be so severe as to negatively affect Chan for the rest of his life: "The accused is now a hardworking, productive member of the society and he is highly remorseful" (Hansen, 2004).

After serving one-sixth of his sentence, Chan applied for day parole, but the application was denied by the B.C. Board of Parole. In its decision, the Board indicated that Chan had demonstrated an inability to "control impulses" such as those which contributed to his poor decision making on the night of the crash (Sokora, 2005a). Chan subsequently applied and was granted full parole after serving one-third of his sentence. In its formal decision, the parole board noted that Chan had set out an acceptable parole plan and had performed well inside the institution: "He now takes responsibility for his actions and thinks about the consequences of his behaviour" (Sokora, 2005b). During the hearing, Chan apologized to Cst. Ng's parents, who were in attendance.

Case 6. On April 8, 2005, Madam Justice Colleen Suche of the Court of Queen's Bench imposed a two-year suspended sentence on Charles Manty that required him to perform 100 hours of community service work by speaking to young drivers about the importance of paying attention while driving (*The Globe and Mail*, 2005: A9).

chapter

7

Correctional Institutions

learning objectives

After reading this chapter, you should be able to

- Discuss the origins and development of correctional institutions.
- Describe the federal and provincial/territorial systems of corrections.
- Identify and discuss the types of challenges that confront prison managers.
- Describe the recruitment and training of correctional officers, their role and activities in the prison, and sources of occupational stress.
- Describe the dynamics of life inside prisons, including the inmate code, violence and victimization, and the experience of prison inmates.
- Discuss the classification of offenders, the role of risk/needs profiles, and case management.
- Discuss institutional treatment programs and their effectiveness.
- Discuss the unique treatment challenges to corrections personnel posed by sex offenders, female offenders, and Aboriginal offenders.
- Describe the various ways in which the success of treatment programs is measured.
- Describe the role of provincial and federal ombudsmen.
- Discuss current trends in institutional corrections.

key terms

Correctional institutions (or prisons, as they are more commonly called) have always been a source of curiosity and controversy. Perhaps it is because the prison is, for most Canadians, an unknown, mysterious world that few will enter. Perhaps it is because of how life inside prisons has been portrayed in movies and on television over the decades. Or perhaps it is due to the fact that the most notorious and dangerous criminals go to prison.

Canada's incarceration rate is often depicted as high relative to the rates in other Western countries (see Figure 7.1). However, extreme caution must be exercised when comparing the number of inmates per 100,000 population across international jurisdictions. Changes in legislation and policy can have a powerful effect on reported rates. For example, an offender in Canada who is on a conditional sentence is counted as incarcerated even though that person is serving the sentence under house arrest in the community. As noted in Chapter 6, more and more offenders are being given conditional sentences, and this skews the incarceration rate presented in Figure 7.1.

Correctional institutions account for nearly 75 percent of the approximately $2.7 billion spent annually by federal and provincial/territorial corrections systems even though only about 5 percent of convicted offenders are incarcerated. In contrast, the 95 percent of offenders who serve their sentence under

FIGURE 7.1   Comparative Incarceration Rates per 100,000 Population, 2003

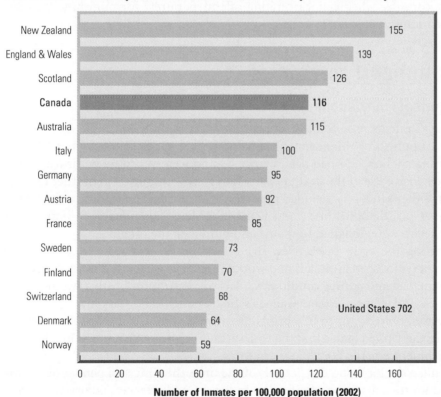

Number of Inmates per 100,000 population (2002)

Source: Public Safety and Emergency Preparedness Portfolio Corrections Statistics Committee, *Corrections and Conditional Release Statistical Overview* (Ottawa: Public Safety and Emergency Preparedness Canada, 2004), 5. Reproduced with the permission of the Minister of Public Works and Government Services Canada, 2006.

supervision in the community require only 13 percent of the overall corrections budget. The remaining monies are costs for headquarters and central services.

Although their numbers are small, incarcerated offenders are often those who have inflicted the greatest harm on the community. Many Canadian politicians and citizens' groups are calling for more punitive responses to criminal offenders; if those calls are heeded, more offenders will be sent to prison. You may feel that the best way to respond to crime in society is for judges to hand down more harsh sentences and for convicted offenders to be given longer periods of time in prison. Or you may feel that longer sentences in prison may not be the most effective way to protect society, to deter others from committing crime, or to reduce the likelihood that an offender will commit further crimes when released.

Whatever your personal feelings are on this issue, keep in mind one fact: 100 percent of provincial inmates and 95 percent of federal inmates, even those serving twenty-five-year minimum sentences, will one day be released back into society. Very few offenders will never be released from confinement. Offenders will return from prison to live in our neighbourhoods, attend school with us, and work alongside us, and they will be around our children in public settings. For this reason alone, we need to consider the role of correctional institutions in the criminal justice system, the dynamics of life inside prisons, and the impact that incarceration has on the offenders who experience it. Above all, we must consider whether incarcerating offenders purely for punishment makes our communities safer.

INSTITUTIONAL CORRECTIONS

Historical Foundations

In pre-Confederation Canada, imprisonment was not used for the specific purpose of punishing offenders. Local jails and lockups held offenders who were awaiting trial or who had been convicted and were yet to be punished. There was extensive use of the death penalty for crimes ranging from murder to petty theft. Other sanctions included branding, transportation (sending offenders to another jurisdiction), fines, and whipping. Many sanctions were imposed in public and were designed to shame and humiliate the offender.

During the early 1800s, New Brunswick, Nova Scotia, Lower and Upper Canada established local jails and workhouses. These facilities were plagued by poor management, foul conditions, a lack of separation of men, women, and youths, and general chaos. Canada's first penitentiary was built in 1835 in Kingston, Ontario, from blueprints provided by officials in New York state. It was in the United States that the widespread use of imprisonment as a form of punishment emerged in its modern form.

Unlike the local and district jails of the time, the Kingston penitentiary provided for the separation of offenders by gender and offence, and prisoners were given bedding, clothing, and adequate food. The cornerstones of the Kingston penitentiary were hard labour, silence, and strict adherence to prison regulations. Prisoners were required to walk in lock step and were forbidden to speak or even gesture to one another. All activities were conducted "by the bell"

(Simonds, 1996). This was the Auburn model, a "separate and silent" system first applied in a penitentiary in Auburn, New York. This system allowed prisoners to work and eat together during the day and provided housing in individual cells at night. A strict regime of silence, which forbade prisoners from communicating or even gesturing to one another, was enforced at all times. The Auburn system was the model on which most prisons in the United States and Canada were patterned. The other model for penitentiaries was the Pennsylvania system, in which inmates worked, ate, and slept in their cells. Prisoners were completely isolated from one another and even kept out of eyesight of one another. This became the model for prisons in Europe, South America, and Asia.

By the early 1840s concerns were being voiced about the excessive use of corporal punishment in the Kingston prison and the effectiveness of the prison regimen in reforming offenders. A commission of inquiry documented widespread abuses by the prison administration and raised questions as to whether the prison could be used to reform offenders; but the federal government ignored its findings and constructed a series of penitentiaries across the country, most of which still operate today.

In the 1900s, governments began passing legislation to improve conditions in correctional institutions. This effort continued throughout the twentieth century. In the early twenty-first century, however, the effectiveness of incarceration is still very much in doubt and many of the less positive features of institutional life remain.

Structure of Institutional Corrections

Corrections services and facilities are generally divided into "custodial" and "non-custodial." Non-custodial services include bail supervision and probation (discussed in Chapter 6), as well as temporary absences from institutions, provincial and federal parole, and statutory release (discussed in Chapter 8). Figure 7.2 shows the composition of the adult correctional population in Canada. Note that only about one-fifth of those who are the responsibility of correctional agencies are confined in a correctional institution.

In Canada, the responsibility for corrections is shared between the federal, provincial, and municipal levels of government. Actually, the federal and provincial levels are the most important (see s. 91.28 and s. 92.6 of the Constitution Act, 1867). The only facilities at the municipal level are local police lockups, which are used for short-term detention. The basis for the split in correctional jurisdiction is the **two-year rule:** an offender who receives a sentence (or consecutive sentences) totalling two or more years falls under the jurisdiction of federal corrections, whereas an offender whose prison sentence totals less than two years is the responsibility of provincial corrections services.

> **two-year rule**
> the basis for the split in correctional jurisdiction

The two-year rule was established at the time of Confederation and is a unique feature of Canadian corrections. The historical record indicates no clear rationale for it, but a number of explanations have been offered, including these:

- The federal government was interested in strengthening its powers.
- Only the federal government had the resources to establish and maintain long-term institutions.

FIGURE 7.2   Composition of the Adult Correctional Population, 2003–04

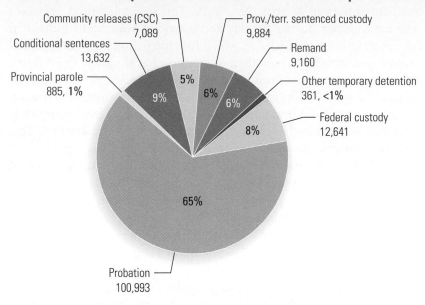

Community releases (CSC)
7,089

Conditional sentences
13,632

Provincial parole
885, 1%

Prov./terr. sentenced custody
9,884

Remand
9,160

Other temporary detention
361, <1%

Federal custody
12,641

5%   6%   6%   8%

9%

65%

Probation
100,993

Source: Adapted from the Statistics Canada publication, *Juristat,* Adult Correctional Services in Canada, 2002/03, vol. 24, no. 10, Catalogue 85-002, October 27, 2004, p. 4.

- Offenders receiving short sentences were seen as in need of guidance, whereas those receiving longer sentences were seen as more serious criminals who had to be separated from society for longer periods (Ouimet, 1969).

A General Profile of Custodial Populations

The total number of offenders admitted to federal and provincial/territorial custody has been declining for the past decade. Incarcerated offenders tend to have the following characteristics:

- *They are men.* Women constitute just under 10 percent of provincial/territorial inmate populations and about 5 percent of admissions to federal facilities.

- *They are in their thirties.* However, the average age of offenders in the federal system has been rising in recent years.

- *They have been convicted of a property offence.* Note, though, that offenders convicted of violent crimes are a higher percentage of the population in federal institutions.

- *They are single.* That is, they are neither married nor in a stable common-law relationship.

- *They are parents.* About 60 percent of male offenders and nearly 70 percent of female offenders have children or stepchildren.

- *They are marginally skilled.* Offenders typically have low levels of education and are usually unemployed at the time of conviction.

- *They are disproportionately Aboriginal.* Aboriginal offenders are significantly overrepresented in institutional populations (see Figure 7.3). This suggests that

FIGURE 7.3 **Aboriginal Adult Admissions to Custody, Provincial/Territorial and Federal Institutions, 1998–99**

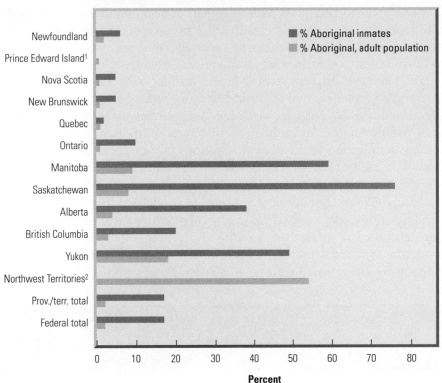

1. Number too small to be expressed: percentage, Aboriginal inmates.
2. Figure not available: percentage, Aboriginal inmates.

Source: Adapted from the Statistics Canada publication, "Canadian Centre for Justice Statistics profile series," Aboriginal People in Canada, 1999, Catalogue 85F0033MIE2001001, June 14, 2001, p. 10.

little progress has been made in reducing the numbers of Aboriginal people involved in the justice system and sentenced to custody.

- *They are disproportionately Black.* Although they account for only 0.02 percent of Canada's population and 0.03 of Ontario's, Black men make up 5 percent of the federal institution population and 11 percent of Ontario's institution population. Black women account for 7 percent and 9 percent of these respective populations (Canadian Centre on Substance Abuse, 2004; Johnson, 2004).

- *They are substance addicted.* High numbers of offenders are addicted to alcohol and/or drugs. Moreover, for the majority of offenders, drugs or alcohol were directly or indirectly involved in their offences.

- *They suffer mental impairment.* Many offenders have mental disorders or deficiencies in mental abilities, including damage caused by fetal alcohol spectrum disorder (FASD), the result of the birth mother drinking while pregnant.

Recall from Chapter 1 that special populations present challenges to the criminal justice system. This is evident in corrections. Special populations in correctional institutions include elderly inmates, violent offenders, sex offenders, and offenders serving lengthy sentences.

Provincial/Territorial Corrections

Provincial/territorial corrections agencies are responsible for offenders who are serving sentences totalling less than two years. The provinces and territories operate approximately 160 facilities with different levels of security. There are no uniform designations; these facilities are variously called jails, detention centres, correctional centres, reformatories, établissements, or correctional institutions.

Ontario's jails and detention centres house offenders who are awaiting trial or other court proceedings as well as offenders who are serving short sentences. Most jails are smaller, older facilities; detention centres tend to be larger, modern, regional facilities built to meet the needs of several municipalities. Correctional centres house offenders typically serving periods of incarceration of up to two years less a day. A profile of provincial/territorial inmates is presented in Box 7.1.

Provincial corrections systems operate institutions at all security levels (maximum, medium, and minimum) as well as specialized treatment facilities. Jails and detention centres are the point of entry into institutional corrections and are classified as maximum security. They hold offenders who are on **remand,** those who are sentenced to short terms of confinement, and sentenced offenders who are awaiting transfer to a federal penitentiary. There are also

remand
the status of accused persons in custody awaiting trial or sentencing

focus
BOX 7.1

Who Goes to Provincial/Territorial Institutions?

There has been a general decline in the number of offenders being sent to provincial/territorial institutions; even so, in many jurisdictions facilities are overcrowded. Following are some of the characteristics of offenders confined in provincial/territorial institutions:

- *They are on remand.* Nearly half of admissions to custody are on remand.
- *They have a prior criminal record.* The majority of inmates have previous convictions and have already served time in custody.
- *They are property offenders.* Most inmates have been convicted of property crimes or other non-violent offences, rather than crimes against the person. This has led many observers to argue that incarceration is being overused and that alternatives should be considered.
- *They are Aboriginal.* Aboriginals comprise 21 percent of admissions to provincial/

territorial custody. However, the levels are much higher in the Prairie provinces. In Saskatchewan, 78 percent of the custody population is Aboriginal; in Manitoba, the figure is 68 percent.

- *They are Black.* Black offenders, especially Black women, are overrepresented in Ontario's provincial institutions.
- *They are addicted to drugs and/or alcohol.* One study found that 75 percent of the women in the facility at Prince Albert, Saskatchewan, had been or still were injection drug users.
- *They are serving short sentences.* Just over 50 percent of sentences are for thirty-one days or less.
- *They have defaulted on a fine.* About 20 percent of inmates have been incarcerated for fine default—that is, for failing to pay a fine levied by a judge.

treatment centres that provide specialized and intensive treatment for special categories of offenders such as sex offenders. There is considerable variation across the provincial/territorial corrections systems in terms of the specific programs that are offered inside facilities.

Federal Corrections

The Correctional Service of Canada (CSC) has three levels: national, regional, and institutional (or district). The CSC is directed by a Commissioner of Corrections, who is appointed by the governor in council under provisions in the Corrections and Conditional Release Act. The CSC has a mission statement that reads, in part, as follows: "The Correctional Service of Canada, as part of the criminal justice system and respecting the rule of law, contributes to the protection of society by actively encouraging and assisting offenders to become law-abiding citizens while exercising reasonable, safe, secure and humane control" (Correctional Service of Canada, 1997: 4).

The CSC's national headquarters are in Ottawa. There are also five regional headquarters:

- Moncton, New Brunswick (Atlantic region, which includes Newfoundland and Labrador, Prince Edward Island, Nova Scotia, and New Brunswick)
- Laval, Quebec (Quebec region)
- Kingston, Ontario (Ontario region)
- Saskatoon, Saskatchewan (Prairie region, which includes Manitoba, northwestern Ontario, Saskatchewan, Alberta, and the Northwest Territories)
- Abbotsford, British Columbia (Pacific region, which includes B.C. and Yukon)

These regional headquarters administer the system's maximum, medium, and minimum security institutions, of which there are nearly fifty. They also run community corrections centres and forest work camps. The features of each of these levels of security are as follows:

- *Minimum security.* These environments allow unrestricted inmate movement, except during the night, and generally have no perimeter fencing.
- *Medium security.* In these environments, inmates have more freedom of movement than in maximum security but are surrounded by high-security perimeter fencing. Most federally incarcerated offenders are housed in medium security institutions.
- *Maximum security.* In these highly controlled environments, surrounded by high-security perimeter fencing, inmates' movements are strictly controlled and constantly monitored by video surveillance cameras.
- *Special Handling Units (SHU)*: These are high-security institutions for inmates who present such a high level of risk to staff and inmates that they cannot be housed in maximum security facilities. There is only one special handling unit, in Ste-Anne-des-Plaines, Quebec.
- The CSC also operates a number of regional health centres, which house violent offenders and offer treatment programs in violence and anger management.

Courtesy of the Correctional Service of Canada

Inside a maximum security correctional institution

- *Multilevel units.* These institutions contain one or more of the above security levels in the same facility or on the same grounds. In some cases there are distinct inmate populations within the same institution that, for security and safety reasons, are not allowed to co-mingle. For example, in Kent Institution, a federal maximum security institution east of Vancouver, there are three separate populations: closed custody, general population, and protective custody.

Federal female offenders are housed in a number of small, regional facilities. These were developed to replace the Kingston Prison for Women, which was closed in 2000. These facilities include the Nova Institution for Women (Truro, NS), Joliette Institution (Joliette, QC), Grand Valley (Kitchener, ON), Edmonton Institution for Women (Edmonton, AB), and Fraser Valley Institution for Women (Abbotsford, BC). In addition, there are separate maximum security units for women within the prisons for men at Springhill Institution (Springhill, NS) and the Saskatchewan Penitentiary (Prince Albert, SK). As well, the Regional Psychiatric Centre in Saskatoon contains a unit for women.

Facilities designed specifically for Aboriginal offenders include the following:

- *Pe Sakastew* (a Cree word pronounced "Bay Sah-ga-stay-o," meaning "new beginning"). Located near Hobbema, Alberta, this minimum-security facility is a joint initiative of the federal government and the Samson Cree Nation.
- *Aboriginal Men's Healing Lodge.* Operated by the Prince Albert (Saskatchewan) Grand Council under agreement with the federal government, the lodge houses federal and provincial offenders.
- *The Okimaw Ohci Healing Lodge.* Located on the Nekaneet Indian Reserve in Maple Creek, Saskatchewan, this facility for federally sentenced Aboriginal women offers a full range of correctional treatment programs. It operates under the direction of a governing council composed of elders and other representatives of the Nekaneet First Nation.

As of 2003, just under 13,000 offenders were confined in federal correctional facilities. This population is profiled in Box 7.2.

The correctional process through which federal offenders pass is set out in Figure 7.4. Refer to this chart when reading both this chapter and Chapter 8 on the release and re-entry of offenders.

---

focus   BOX 7.2

Who Goes to Federal Institutions?

Among those inmates confined in federal correctional institutions, 70 percent have been convicted of crimes against the person, and 17 percent have been convicted of first or second degree murder. Specific categories of federal inmates are profiled below.

Aboriginal Offenders

- 18 percent of all federal inmates are Aboriginal. This overrepresentation is most pronounced in the Prairie region, where Aboriginal males constitute 37 percent and Aboriginal females 51 percent of federal institutional populations.

  The number of Aboriginal women in federal custody increased nearly 75 percent over the seven years between 1996 and 2003.

  A higher percentage of Aboriginal offenders (82 percent) than non-Aboriginal offenders (67 percent) are serving sentences for a violent offence.

  80 percent of Aboriginal women are serving a sentence for a violent offence; for non-Aboriginal women, the figure is 50 percent.

Female Offenders

- About 70 percent are incarcerated for violence or murder.
- Nearly 50 percent are serving sentences of four years or less.
- 24 percent are serving sentences of more than ten years (including life).
- 25 percent are Aboriginal.

Sex Offenders

- This group accounts for 20 percent of the custodial population.

Elderly Offenders

- The federal inmate population is aging. Offenders over fifty now account for 16 percent of the federal institutional population, and this figure is expected to increase dramatically over the next twenty years.

Mentally Disordered

- 14 percent of admissions to federal prisons have had recent psychiatric or psychological treatment and a recent report of the federal Office of the Correctional Investigator (2005) found that the number of prison inmates with a diagnosed mental disorder on admission rose from 6.8 percent in 1997 to 11.1 percent in 2004—a 61 percent increase.

Source: B. Moloughney, "A Health Care Needs Assessment of Federal Inmates in Canada," *Canadian Journal of Public Health,* 95 (2004, Supp. 1), S1–S62; Public Safety and Emergency Preparedness Portfolio Corrections Statistics Committee, *Corrections and Conditional Release Statistical Overview* (Ottawa: Public Safety and Emergency Preparedness Canada, 2004; Office of the Correctional Investigator, 2005.) Retrieved online at http://www.oci-bec.gc/ca/release-20051104-e.asp

FIGURE 7.4 The Correctional Process for Federal Offenders

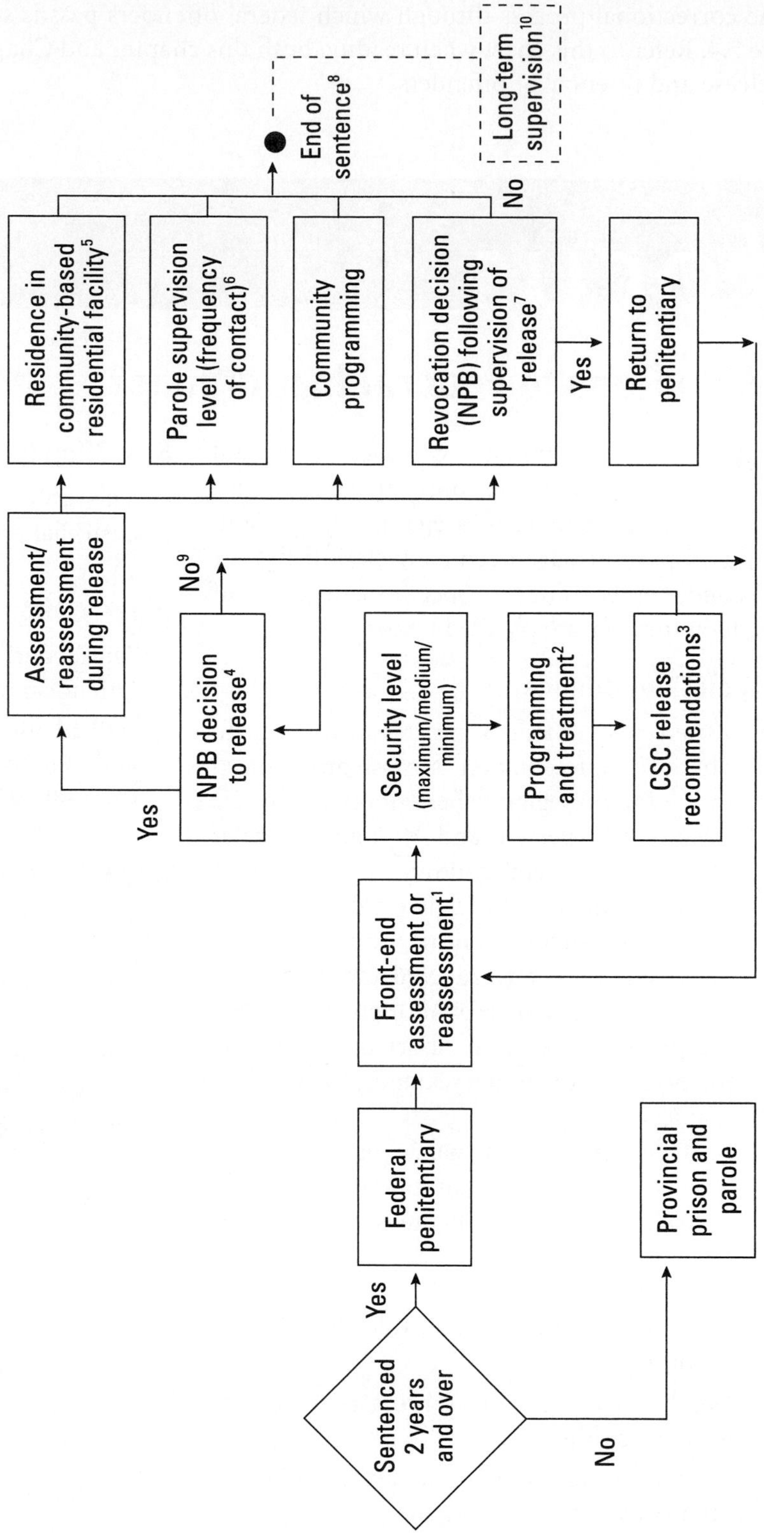

1. Front-end assessment is a comprehensive and integrated evaluation of the offender at the time of admission. It involves collecting and analyzing information about the offender's education, social situation, and criminal and mental health history, and about other factors relevant to determining risks and needs. This provides a basis for deciding on the offender's institutional placement and establishing his or her correctional plan.

2. During the incarceration period, inmates may periodically be outside an institution, either on a work release program or on escorted or unescorted temporary absences.

3. The CSC prepares the cases of inmates eligible for day parole and full parole for review and decision by the National Parole Board (NPB). The CSC's recommendation may be positive or negative.

4. The NPB may impose conditions on the release to control the risk of reoffending.

5. The residential facility may be operated by the CSC or by a private agency under contract.

6. The purpose of supervision is to monitor the offender's behaviour and adjustment (compliance with conditions) so as to minimize the risk of reoffending. Minimum frequency of contact ranges from four times per month to once per month according to assessed risk and need.

7. A designated NPB or CSC officer may suspend the release for a breach of conditions, to prevent a breach of conditions, or to protect society.

8. For offenders who have received a life sentence, the sentence never ends, although they can serve part of their sentence in the community. Also, offenders declared by the courts to be dangerous offenders (DOs; see Chapter 6) serve an indeterminate sentence, subject to NPB review three years after the declaration and every two years thereafter.

9. Subject to the detention provisions in the CCRA, an offender who is not conditionally released by the NPB is entitled to statutory release after serving two-thirds of the term of imprisonment.

10. A period of supervision of up to ten years following the warrant expiry date (end of sentence) may be imposed by the sentencing judge on offenders designated as "long-term offenders" (LTOs; see Chapter 6) due to the risk they present to the community.

Source: *Basic Facts About Corrections in Canada, 1994 Edition*, pp. 24–25, Correctional Service of Canada, 1995. Reproduced with the permission of the Minister of Public Works and Government Services Canada, 2006.

Profile of Federally Sentenced Women

The number of men admitted to federal custody has declined in recent years; the number of women admitted has risen. Women, however, still account for only 5 percent of the federal prison population. A woman serving a federal sentence is typically from a marginalized social group. Her past and current situation is likely to include poverty, a history of abuse, long-term drug and alcohol dependency, responsibility for the primary care of children, limited educational attainment, and few opportunities to obtain adequately paid work. For example, 70 percent of women in maximum security prisons are identified at intake as having alcohol abuse issues, and nearly 80 percent as having drug abuse problems. Also, the percentage of women who have had a prior mental health intervention or hospitalization is much higher than for male inmates (Moloughney, 2004).

Surveys of women serving federal sentences reveal that two-thirds have children and are single parents and that four-fifths have been abused sexually and/or physically. Most have limited education and skill levels. In recent years the female offender population has become increasingly diverse; in percentage terms, there are fewer Caucasian women than there used to be and more Aboriginal and Black women (Canadian Human Rights Commission, 2003).

Figure 7.5 gives the levels of involvement in federal custodial and community corrections for Asians, Blacks, Aboriginals, and Caucasians.

FIGURE 7.5   The Involvement of Visible Minorities in Federal Corrections

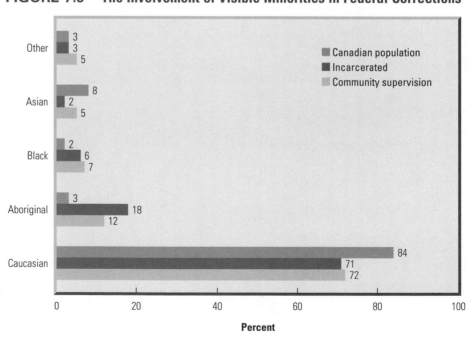

Source: Adapted from the Statistics Canada publication, *Juristat*, Adult Correctional Services in Canada, 2002/03, vol. 24, no. 10, Catalogue 85-002, October 27, 2004, p. 4.

LIFE INSIDE PRISON

A defining characteristic of prisons is that they are **total institutions.** This term was coined by sociologist Erving Goffman (1961) to describe life inside hospitals, military training camps, mental hospitals, and prisons. Total institutions hold specific groups of individuals who are cut off from the broader society and whose activities are controlled twenty-four hours a day. Although conditions inside Canadian correctional facilities are decidedly more humane than they used to be, there are still considerable controls on inmates' movements and activities. And much like their counterparts in earlier decades, they are still denied their freedom and are under constant control and surveillance.

All correctional institutions are total institutions; they differ, however, in how strictly they control inmates. Life for inmates in a minimum security work camp is considerably different from life in a maximum security penitentiary. Inmates in maximum security institutions require closer supervision than those in less secure facilities; they may also be more likely to engage in violence against staff and other inmates. Although offenders in provincial/territorial institutions are not confined for long periods of time, there may still be problems, including physical altercations, intimidation, and illicit drug traffic.

Another characteristic of correctional institutions is that they have a split personality: the goals of prisons include confinement and control but also reform and rehabilitation. There is little doubt, however, that the primary purpose of prisons is to punish. As the following discussion will reveal, this may compromise efforts to treat and rehabilitate inmates.

Inmates in correctional institutions have access to television and the print media but few opportunities for contact with law-abiding people in the community. Similarly, except for church groups and workers with not-for-profit organizations, the public rarely visits correctional facilities. For the majority of the public, incarcerated offenders are "out of sight and out of mind." The lack of non-media sources of information about institutional corrections plays a significant role in the perceptions (and misperceptions) that citizens hold about prisons, their operations, and their residents. This may make the public more susceptible to media images of prisons and inmates than would otherwise be the case.

In an attempt to increase public involvement in corrections, the Correctional Service of Canada sponsors **Citizen Advisory Committees (CACs),** which operate in all federal correctional institutions and are composed of residents from the local community. CACs provide valuable community input into correctional policies and programs. They also serve as independent observers of the day-to-day activities of the CSC and act as community liaisons. A recent study (Trevethan, Rastin, and Bell, 2004) found that CACs were functioning well as impartial observers in correctional facilities. This study recommended (among other things) that attempts be made to increase the diversity of CAC members and to increase interactions between them and prison staff.

Managing Correctional Institutions: A Challenging Job

Perhaps one of the most challenging jobs in the entire criminal justice system is that of managing a correctional institution. The warden of a correctional institution faces the almost impossible task of responding to the often conflicting

total institutions settings in which the activities and regimens of inmates/residents are highly controlled

Citizen Advisory Committees (CACs) operate in all federal correctional institutions and are composed of volunteers who provide input into federal correctional policies and programs and liaise with the community

demands of the public, politicians, and various public interest and victim groups. A warden must also provide leadership for the correctional staff and meet the needs of inmates.

Many of the challenges that confront wardens are beyond their control. These include the sentencing patterns of the criminal courts (which may contribute to overcrowding), the placement decisions of case management teams, decisions by parole boards as to whether to release offenders from confinement, and decisions by politicians that may increase sentence lengths, reduce financial resources, and affect staffing levels and program offerings. Even so, the management style of an individual warden does have an impact on the climate within the institution and can affect the overall relationship that develops between the keepers and the kept. In addition, the mismanagement of an institution can create the conditions for a riot or disturbance.

Within the institution, the warden is responsible for a wide range of administrative tasks, as well as for mediating the often conflicting interests of the inmates, the correctional officers, and the treatment staff. The warden is also held to account for disruptions such as riots, suicides, and escapes. Additional challenges confronting prison managers are discussed below.

Meeting the Requirements of Legislation and Policy

The legal and policy framework of institutional corrections has become increasingly complex. Senior corrections personnel must ensure that the institutions they run are in compliance with those changes. Prison wardens and other correctional officials are increasingly the targets of legal actions initiated by inmates, the families of inmates who have been victimized or died in prison, and crime victims and their families.

Overcrowding

Judicial sentencing patterns are changing; the number of long-term inmates is increasing; parole boards are growing more reluctant to release offenders into the community; and fewer facilities are being built. As a result of all this, many institutions are operating at overcapacity. Currently, about 25 percent of federal inmates are required to share cell space (to "double-bunk"). At the provincial level, Ontario and British Columbia authorities have decided to create "big box" correctional facilities, and this has also contributed to overcrowding.

Overcrowding and double-bunking can strongly affect daily prison life by heightening tensions among inmates and between inmates and correctional officers. Overcrowding has been associated with a number of disturbances in federal and provincial facilities. It also taxes program resources and compromises security. The Office of the Correctional Investigator has called on the CSC to end the practice of double-bunking.

Federal and provincial/territorial systems of correction have attempted to reduce overcrowding by focusing on reintegration as a core component of case management. Ontario, however, has established "double-bunking" as a key component of its "no frills" custodial regime.

Ensuring Inmate Safety

Corrections officials are required to ensure the safety of inmates in their charge—an onerous task, especially in federal maximum security institutions.

focus

BOX 7.3

The Murder of Denise Fayant

Thirty hours after arriving at the Edmonton Institution for Women, twenty-one-year-old Denise Fayant was found strangled behind her bed with a bathrobe sash around her neck. She died two days later in hospital. An investigation into her death (which was first ruled a suicide) found that she had been slain by two inmates. One of them was her former lover, against whom Denise was scheduled to testify.

A subsequent inquiry conducted by an Alberta provincial court judge found that Fayant had repeatedly told corrections officials that they would be endangering her life by transferring her to the newly opened institution. Two inmates were later convicted and sentenced to additional federal time for her death.

Prison officials insisted that they had been assured by inmates in the prison that no harm would come to Fayant. Even so, the investigating judge concluded that Fayant died as a result of "callous and cavalier" actions on the part of Corrections Canada and that she was a "victim of a process intent upon implementing an untested concept to manage federally sentenced female inmates. She was the test. The process failed tragically and inhumanely. Her death was avoidable" (in Cowan and Sheremata, 2000). The report's findings were quickly seized on by critics of the CSC policy for federally sentenced women, who argued that the concept of punishment had been removed from the response to federal female offenders. They viewed this policy as "exorcising the concept of punishment from female corrections policy almost completely" (Bunner, 2000: 13).

In its first six months, the Edmonton Institution for Women experienced numerous inmate assaults on staff and suicide attempts by inmates, as well as seven escapes and the death of Fayant. Three years later, the CSC implemented the Intensive Intervention Strategy, which resulted in the expansion of regional women's facilities to accommodate women classified as maximum security.

Source: C.T. Griffiths, *Canadian Corrections,* 2nd ed. (Toronto: Thomson Nelson, 2004), 193. © 2004. Reprinted with permission of Nelson, a division of Thomson Learning: www.thomsonrights.com, FAX 800 730-2215.

In recent years, a number of high-profile incidents—including murders of inmates by fellow inmates—have placed increasing pressure on the CSC to ensure that inmates are safe. The issues surrounding inmate safety were highlighted by the murder of Denise Fayant (see Box 7.3).

Communicable Diseases and Prevention Strategies

One of the most serious challenges confronting correctional administrators is how to prevent the spread of communicable diseases—including HIV/AIDS, tuberculosis, and hepatitis B and C—in inmate populations. Infection rates in correctional institutions are alarmingly high, as the following data suggest:

- Nearly 2 percent of federal inmates are HIV-positive—a rate ten times that of the general Canadian population.

- The HIV infection rate among provincial inmates in Ontario is six times higher than that of the general population.

- Overall, nearly 24 percent of federal inmates are hepatitis C-positive.

- HIV and hepatitis C infection rates among women inmates are soaring; 14 percent of inmates at the Edmonton Institution for Women have HIV or AIDS.

- Aboriginal inmates in provincial/territorial and federal institutions have higher rates of HIV and AIDS than non-Aboriginal inmates.

- As many as 40 percent of federal inmates have hepatitis B or C and, in some institutions, the rate is as high as 70 percent (Brady, 1998; Calzavara et al., 1995; Correctional Service of Canada, 2003).

Although many offenders enter institutions with these medical conditions, others contract the diseases while incarcerated.

Although HIV can be transmitted via anal intercourse between inmates, the primary cause of the increasing HIV infection rate is intravenous drug use and the sharing of HIV-contaminated syringes and needles. Hepatitis B and C, which are blood-borne diseases, are transmitted by the pens, pencils, and wire instruments used by inmates for body piercing and tattooing. Inmates may also bring these diseases with them into the institution.

An in-depth study (Lines, 2002) found considerable variability in the prevention and intervention initiatives taken by federal and provincial/territorial systems of corrections. This study "graded" each jurisdiction on a number of criteria, including the availability and accessibility of bleach; the provision of condoms, dental dams, and lubricants; and access to methadone treatment programs and needle exchanges. British Columbia was awarded the highest mark ("B") and identified as a leader in the field. The CSC scored "B-," and Newfoundland "D." All other jurisdictions were given an "F."

The CSC is following a number of proactive strategies in its efforts to reduce the prevalence of illegal drugs and other high-risk behaviours, such as tattooing. These strategies include frequent searches; a urinalysis program; drug dogs; video surveillance; and ion scanners that can detect drug residue on visitors and on inmates returning from absences in the community. The effectiveness of these strategies is uncertain. Surveys of correctional officers have found that most do not believe that drug strategies are successful (Robinson, Lefaive, and Muirhead, 1997).

To reduce the spread of HIV/AIDS, all federal correctional institutions and many provincial facilities are providing inmates with condoms and bleach kits. In addition, the CSC is providing methadone maintenance to inmates who are addicted to heroin. The Ontario Medical Association (2004), the federal Correctional Investigator, and the Canadian HIV-AIDS Legal Network, among others, have called on corrections authorities to implement clean needle programs similar to the ones offered on the streets of many Canadian cities. As well, the CSC has established regulated tattoo parlours in a number of federal institutions in an effort to reduce the spread of disease.

focus

BOX 7.4

Point/Counterpoint: Drug Prevention Strategies

Point

- Providing inmates with bleach kits reflects the reality that drugs are available to inmates.
- Providing inmates with condoms allows "safe" consensual sexual relations between inmates and reduces the risk of HIV infection.
- Methadone maintenance reduces the use of hard drugs in the inmate population and helps inmates overcome their addiction.

Counterpoint

- Providing bleach kits encourages drug use, which is not only a violation of institutional rules but is also against the law.
- Providing condoms encourages inmates to violate institutional rules and does not prevent non-consensual sexual relations.
- Methadone is itself addictive and does little to eliminate drug dependency.

Source: Adapted from C.T. Griffiths, *Canadian Corrections*, 2nd ed. (Toronto: Thomson Nelson, 2004), 196–97.

Box 7.4 outlines some benefits and drawbacks of these strategies for preventing or reducing the spread of communicable diseases.

On the Line: Correctional Officers at Work

Although the warden directs the overall operations of the prison, it is the correctional officers who have the most contact with the inmates on a daily basis. Correctional officers have a crucial role in prisons, yet their work has long been characterized by low prestige, poor pay, inadequate training, and high turnover. Prisons make extensive use of high technology, such as video surveillance and various warning devices (static security), but they still rely heavily on correctional officers to implement policies and regulations and to control the inmates (dynamic security).

Correctional officers are also key to rehabilitation efforts. In many provincial correctional facilities, they facilitate treatment groups. At B.C.'s Fraser Regional Correctional Centre, for example, specially trained correctional officers conduct core programs, including the Violence Prevention Program and the Substance Abuse Management (SAM) program. The extent to which the role and status of correctional officers may hinder the effectiveness of correctional programming has not been explored.

Recruiting and Training of Correctional Officers

Over the past twenty years, federal and provincial corrections systems have made strong efforts to improve staff training and development programs.

What do you think?

Should bleach kits, condoms, and methadone be provided to inmates? Would you support the creation of needle exchange programs in Canadian prisons? Explain.

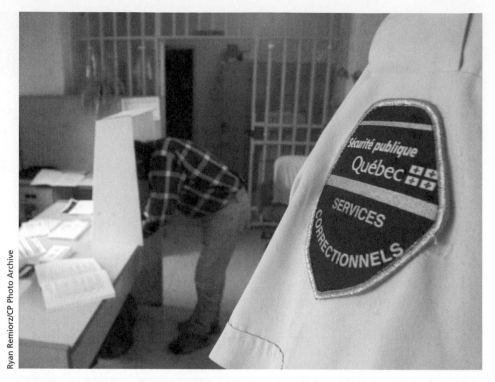

Ryan Remiorz/CP Photo Archive

An inmate in a Quebec correctional institution votes under the watchful eye of a correctional officer.

CSC Officers The CSC tries to recruit only individuals who are graduates of universities, community colleges, or accredited correctional training programs. New CSC officers are trained at one of five regional staff colleges across the country. Trainees must complete a twelve-week correctional training program that combines classroom instruction with self-directed learning and that covers such topics as the mission statement and core values of the CSC, interpersonal and communication skills, security procedures and strategies, and self-defence.

The CSC has special procedures for selecting and training staff to work in institutions for federally sentenced women. It also operates a career management program designed to upgrade the knowledge and skills of in-service correctional officers. Despite these efforts, only 25 percent of correctional officers surveyed rate the quality of the training they receive as "good," "very good," or "excellent," and 40 percent rated the training as "poor" (Samak, 2003: 29).

Provincial/Territorial Officers Each province/territory has its own procedures and standards for recruiting and training new correctional officers. In most provinces, new recruits receive classroom-based as well as on-the-job training. In Ontario, candidates for the position of correctional officer must complete, at their own expense, a program known as COSTART (Correctional Officer Staff Application, Recruitment and Training). Stages in the admission process include meeting mandatory qualifications, passing initial tests and evaluations, and undergoing a personal interview, followed by additional testing. Candidates who successfully complete the admission process are invited to enroll in a six-week pre-employment training program. Vacancies in provincial correctional facilities are filled from a list of those candidates who have completed the pre-employment training. These new recruits undergo a one- to three-week orientation in the facility to which they are assigned.

In B.C., prospective applicants for corrections work must meet a number of criteria. They must then attend the Justice Institute of British Columbia and complete various courses there, including Managing People in Conflict, Role and Duties of the Adult Correctional Officer, Professional Ethics and Standards of Conduct, and Multicultural and Diversity training.

The CSC and the provincial systems are making efforts to increase the cultural and ethnic diversity among correctional officers. For example, the CSC offers field placements to visible minority university students to promote public service careers. And in B.C., local Aboriginal band members shadow employees at the Kwikwexwelhp Institution, a federal medium security facility east of Vancouver, to help them define their own employment needs before they apply for positions.

Officer–Inmate Relationships

Much like the offenders themselves, correctional officers must adapt to prison life. New officers must learn to "read" individual inmates and develop responses to "being tested" by the inmates. They must also become familiar with the intricacies of the inmate social system and the various "scams" that inmates become involved in (such as gambling and the distribution of contraband). And they must learn how to interact with inmates without being "taken in." To ensure daily stability and order, correctional officers need to build accommodative relationships with inmates.

Research indicates that compared to administrators and treatment staff, correctional officers are more punitive, less empathetic, and less supportive of rehabilitation. As might be expected, officers in minimum security facilities tend to hold more favourable attitudes toward offenders than those in medium and maximum security institutions. More positive attitudes tend to be held by older officers with long experience on the job (Larivière and Robinson, 1996).

Should male correctional officers serve in women's institutions? This has long been debated. A 2001 report prepared for the CSC recommended that men no longer work as prison guards in women's institutions in light of continuing complaints from inmates about invasions of privacy and sexual harassment. The same report documented violations of a 1998 protocol for males working in correctional institutions that house female inmates. Examples: failure to announce the entry of frontline staff into prison units; not pairing male officers with a female officer at night or when entering the living units; and the use of male staff to extract prisoners from cells (Bronskill, 2001).

Occupational Stress

Correctional officers experience a considerable amount of stress in carrying out their tasks. The sources of this stress include safety concerns, shift work, and administrative policies that are vague and/or unrealistic. Problems with coworkers are often a greater source of stress than inmate-related problems

(Hughes and Zamble, 1993). A survey of nearly 2,500 federal correctional officers focusing on working conditions gathered the following data:

- 70 percent to 80 percent considered their work "stressful."
- Levels of stress increased with years of service.
- Levels of stress were higher among officers who worked in prisons with larger inmate populations.
- 60 percent to 65 percent of officers indicated that their work had a negative impact on family life "very often" or "often."
- Levels of stress experienced by correctional officers were comparable to those experienced by police officers (Samak, 2003).

In addition, large numbers of female (nearly 60 percent) and male (45 percent) correctional officers had experienced some type of harassment on the job. The most frequently identified type of harassment was psychological, although 30 percent of female correctional officers reported being the victims of sexual harassment. Interestingly, the most frequently cited sources of the harassment were correctional supervisors rather than coworkers or inmates, although female officers were more likely than male officers to identify coworkers as the source of harassment.

Doing Time: The World of the Inmate

status degradation ceremonies
the processing of offenders into correctional institutions

Offenders who are sent to correctional institutions encounter a world unlike any other. Through a series of **status degradation ceremonies,** the offender is psychologically and materially stripped of possessions that identify him or her as a member of the free world. These possessions are replaced by an identification number, prison-issue clothing, and a list of institutional rules and regulations designed to control every aspect of life inside the prison as well as the inmate's contact with the outside world. A typical daily routine for an inmate in a federal correctional facility is presented in Box 7.5. Note that this schedule may not apply to inmates who are in segregation or in specialized units, or who are involved in a temporary absence or day parole program. It does, however, illustrate the extent to which the lives and movements of inmates in the institution are regulated.

The process by which the status of convicted offenders is transformed from "citizen" to "inmate" is very important. This process may have a significant (and not always positive) impact on the offender. Note here as well that there are no "status restoration ceremonies" when an offender has finished serving his or her prison term. Rather, as the discussion of release from confinement in Chapter 8 will reveal, offenders can encounter considerable difficulties in moving from the world of the prison back into the community.

pains of imprisonment
the deprivations experienced by inmates confined in correctional institutions

All inmates must find ways to cope with the **pains of imprisonment:** the loss of security, autonomy, contact with family and friends, and independence to move freely. Although inmates are admonished by the prison officials to "do your own time," the conditions of confinement can make it difficult for an inmate to stay isolated from and unaffected by the prison environment.

focus

BOX 7.5

Inmate Daily Routine, Federal Institution

06:45	Inmates are counted—must be up and dressed
07:00	Breakfast
08:00	Go to program, work or back to the cell
11:45	Return to cell for inmate count and lunch
13:00	Go to the program, work or back to the cell
16:30	Return to the cell for inmate count and then supper
18:00	Go to recreation, cultural events, self-help groups
22:30	Night inmate count
23:00	Lock-up

Source: *A Profile of Visible Minority Offenders in the Federal Correctional System—Research Report 144*, p. 7, Correctional Service of Canada, 2004. Reproduced with the permission of the Minister of Public Works and Government Services Canada, 2006.

Following is a description of what it means to "do your own time," developed from one scholar's extensive correspondence with inmates in Canada and the United States:

> It means keep yourself separate from everything and everybody. Don't comment, interfere, or accept favours. Understand that you are "fresh meat" and need to learn the way of the joint. You have to deal with the "Vikings" (slobs, applied to both guards and cons), "booty bandits" (someone looking for ass to fuck), and the "boss," "hook," "grey suit" or "cookie" (all terms for prisons officials of various ranks) without "jeffing" (sucking up) to the staff. You have to deal with other cons who want you as a "punk" or a "fuck boy." Anybody can be carrying a "shank" (home-made knife) made out of a toothbrush and a razor blade or a piece of sharpened steel. Probably the more innocent someone looks, the more you have to worry. (in Thompson, 2002: 15–16)

A common feature of all correctional institutions is the existence of an inmate social system, which is centred on the **inmate code.** Among the rules of the inmate code: do your own time; don't rat on other inmates; don't trust anyone; don't weaken and don't whine; be tough, be a man; and don't be a sucker. Originally, the inmate code was viewed as providing inmates with a common front against correctional officers and the administration. In recent years, however, there has been a significant erosion of inmate solidarity. Daily life in many correctional institutions is characterized by the exploitation and manipulation of weaker inmates by the strong and by an underground economy of drugs, sexual

inmate code
a set of behavioural rules that governs interaction among inmates and between inmates and institutional staff

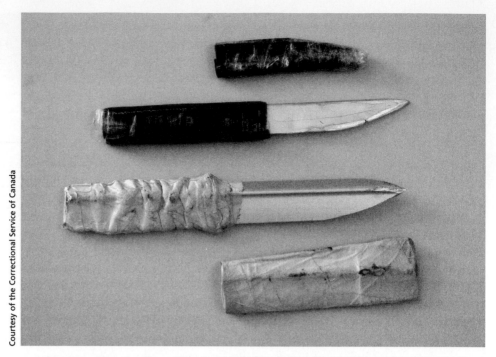

Courtesy of the Correctional Service of Canada

Home-made weapons ("shanks") seized in a correctional institution

services, and gambling. For the individual inmate, it is not the institution's staff but rather other inmates who pose the greatest threat to personal safety.

Violence and Victimization

For many inmates, violence and fear are facts of life. Male inmates in federal institutions have a greater likelihood of being murdered than males in the out-side community, and this risk has increased—along with violence in institutions generally—in recent years. A survey of 4,000 male inmates in federal institutions found that 50 percent did not feel safe from assault by other inmates and that 20 percent had been beaten or assaulted (cited in Makin, 1996). Younger inmates and those without "friends" inside were most likely to feel vulnerable to victim-ization by other inmates; they were also more likely to actually experience victimization (McCorkle, 1993). As the following inmate account illustrates, inmates may resort to physical violence to protect themselves:

> He was always winking, blowing kisses and always trying to talk me into letting him give me a blow job. Until one day when he grabbed a hold of my penis and said I want you. Up until then I let it ride but after dinner that night I caught him by the tennis court where no guards could be and I smiled and said so you want me and when he said yes I plant my foot upside his jaw and left him laying on the ground and that put an end to it (Struckman-Johnson et al., 1996: 73).

Recall from Chapter 5 the case of *R. v. Kerr*, in which the Supreme Court of Canada overturned the conviction of a prison inmate who used a knife (fatally) to defend himself from an attack by another inmate. Offenders who are vulnerable to attack by other inmates may be placed in protective custody.

Segregation is also used to isolate inmates who are a threat to themselves or to other inmates and staff.

To survive prison life and to reduce the likelihood of victimization, most inmates find a group of friends, who provide support and protection. Friendship groups may be formed on the basis of the inmate's sentence (for example, the "lifers' group"), racial identity, participation in religious or self-help groups, or pre-prison affiliations, including organized gangs.

All inmates share the same fate of being locked up and deprived of their freedom. That said, individual inmates employ a wide range of strategies to cope with the daily pressures of prison life. Some offenders succeed in "doing their own time" and not becoming involved in the inmate social system; others involve themselves heavily in illegal activities. A much smaller number seek to escape through self-mutilation and suicide. Corrections officials maintain a constant watch in their efforts to prevent inmates from harming themselves.

The suicide rate among incarcerated offenders is more than twice that of the general population. Suicide is the most common cause of inmate death. Inmates who commit suicide tend to have a history of violence and to be single, male, Caucasian, between twenty and thirty-four, and housed in a medium security facility (Larivière, 1997). In contrast to their male counterparts, female inmates tend to engage in self-injurious behaviour, including slashing and self-mutilation. Over a recent two-year period, however, there were six suicides in a correctional institution housing provincial and federal female offenders. This behaviour has been linked to childhood abuse and other pre-prison experiences (Snow, 1997).

Prisonization is a term used to describe the process whereby new inmates are socialized into the norms, values, and culture of the prison. The degree to which an offender becomes prisonized may depend on individual personality, pre-prison experiences, and the nature of the relationships he or she establishes with other inmates. Offenders who spend too long in confinement run the risk of becoming institutionalized: they are unwilling or incapable of functioning in the outside world.

> **prisonization**
> the process by which new inmates are socialized into the norms, values, and culture of the prison

Note that not all inmates are similarly affected by incarceration and the pains of imprisonment. The specific impact of incarceration on a given offender and the extent to which that offender successfully adapts to life "inside" may depend on a variety of factors. In fact, one of the drawbacks to using imprisonment as punishment is that it cannot be predicted how any one offender will respond to the experience. Incarceration may lead some offenders to start abandoning a criminal lifestyle; for others it may only solidify and strengthen their association with pro-criminal attitudes and values.

The spouse, children, parents, and extended family of a prison inmate may also experience the pains of imprisonment, even though they themselves are not incarcerated. One study found that the young and adolescent children of incarcerated mothers experienced disruptions in their lives, including placement in foster care and a lack of emotional and financial support from the mother. Also, these children were at high risk of becoming involved in the criminal justice system (Cunningham and Baker, 2003).

It is very difficult for the families of prison inmates to maintain contact during the period of incarceration. The correctional facility in which the inmate is incarcerated may be hundreds or even thousands of kilometres away from his or her home community and family. Long-term offenders—including those who receive twenty-five-year minimum sentences—are unlikely to be able to maintain their relationships with family and children. These offenders ultimately return to a much-changed community after a quarter-century absence without the benefit of family support. To encourage inmates to develop and maintain positive family relationships, the CSC operates the Private Family Visiting Program, which facilitates private and unsupervised visits between inmates and their families.

ASSESSING INMATE RISKS AND NEEDS

Classification

classification
the process by which inmates are categorized through the use of various assessment instruments

Classification is an ongoing process. It involves identifying the treatment needs of the inmate (along with the risks he or she presents) and assessing that inmate's progress in altering his or her attitudes and behaviours. The objective of classification is to place offenders in correctional settings most appropriate to their individual needs, while ensuring that the risks they pose are recognized and addressed.

The initial classification decision is made after an offender is sentenced in the criminal court. In each CSC region there are reception centres where offenders stay for a period after sentencing. At the provincial/territorial level, classification may occur either during the inmate's stay in the remand centre or on his or her arrival at the facility. Ideally, the assessment process should be continued throughout the inmate's confinement and following release.

The classification systems used by federal and provincial/territorial corrections generally include psychological, personality, and behavioural inventories that attempt to categorize offenders into certain types. The Offender Intake Assessment (OIA) and Custody Rating Scale (CRS) are used to assess and classify federal inmates. To classify provincial inmates, Ontario and several other provinces use the Level of Service Inventory—Ontario Revision (LSI–OR), a standardized interview that contains questions on offence history, substance abuse, employment history, and other offender-related issues.

The Risk/Needs Profile

A key focus of corrections systems is risk analysis, which involves taking steps to ensure that the general public is protected from offenders who might cause harm and thereby expose the corrections system to civil liability. More specifically, corrections officials use risk analysis to

- determine which facility the inmate should be confined in or moved to;
- determine the inmate's treatment needs;
- identify those inmates who require a higher level of support, intervention, and supervision upon their release;
- assist in release decisions; *and*
- defuse concerns of the general public and politicians about the release of high-risk offenders back into the community.

Research studies indicate that assessing both risks and needs improves the ability of corrections officials to predict which offenders will recidivate.

In assessing the degree of risk posed by an inmate, corrections personnel will generally consider **static risk factors** and **dynamic risk factors.** Static factors are inmate characteristics that cannot be altered through intervention. One example of a static factor is the inmate's criminal history, including prior convictions and compliance (or non-compliance) with the conditions of probation and/or parole. Dynamic factors are inmate characteristics that *can* be altered through intervention (such as level of education and cognitive skills). Many dynamic factors are criminogenic, which means that future criminal behaviour may occur if they are not addressed.

static risk factors
inmate characteristics that cannot be altered through intervention

dynamic risk factors
inmate characteristics that can be altered through intervention

Case Management

Case management is a continuous process that extends from initial entry into the system, through the period of confinement, to supervision upon release. The case management team is responsible for coordinating the administration and management of the inmate's sentence. The core of the case management process is the *correctional plan.* Based on the inmate's risk/needs profile, this plan determines the offender's initial institutional placement, training or work opportunities, and conditions of release.

A number of assessment tools have been developed to assist case managers in case planning, supervision, and decision making. These instruments are designed to assess the risk posed by the offender, to identify the treatment requirements of the offender, and to determine whether the offender is ready to

case management
the process by which identified offender risks and needs are matched with services and resources

"*I never should have tried to take my accounting to the next level.*"

be transferred to a lower level of security or released from confinement. In provincial corrections, the shorter sentences often preclude the use of these measures. To facilitate the development of inmate case plans, the CSC implemented Operation Bypass in 1999. Under this initiative's guidelines, the Offender Intake Assessment and the correctional plan must be completed within 70 days of the inmate's admission.

INSTITUTIONAL TREATMENT PROGRAMS

The CSC and provincial/territorial corrections systems operate a wide variety of core and specialized treatment programs in correctional facilities of all security levels. Treatment programs include education, vocational and trades training, institutional employment programs, and specific treatment interventions. Significantly, only about 10 percent of the $1.5 billion budget of the CSC is spent on treatment programs, including cognitive skills programs, trades training and education programs, programs for specific categories of offenders (such as sex offenders and mentally disordered offenders), and programs addressing substance abuse and family violence.

In recent years there has been an increased emphasis on specific groups of offenders, especially sex offenders, offenders convicted of violent crimes, and Aboriginal offenders. The following are illustrative of the range of treatment programs offered in federal institutions:

- *Education programs.* These are offered in all institutions operated by the CSC and include adult basic education (Grades 1 to 10), secondary education, vocational education, and college- and university-level programs. Inmates are required to cover the full cost of college and university courses.

- *Living skills programs.* These include six interrelated modules: (1) cognitive skills training, (2) living without violence, (3) parenting skills training, (4) anger and emotion management, (5) leisure education, and (6) community integration. These programs are designed to teach offenders to identify problems, rationally analyze them, consider alternatives for action, make good decisions, and anticipate the consequences of their decisions.

- *Cognitive skills programs.* These focus on developing decision-making, problem-solving, and goal-setting skills, as well as interpersonal skills.

- *Substance abuse intervention.* This includes a wide range of education and treatment programs; their specific approach and content may vary across different correctional facilities.

- *Sex offender programs.* These centre on identifying the specific behaviour patterns of the offender and the "triggers" for sex offending, and providing the offender with the skills to manage his or her behaviour so as to reduce the risk of reoffending upon release.

- *Family violence programs.* These are directed toward inmates who have a history of or an assessed propensity toward being abusive in family or close relationships. Offenders are provided with information and skills to alter their attitudes and behaviours.

In addition to these system-wide treatment programs, there are initiatives that have been developed in individual institutions. Following are two examples:

- *LINKS (Letting Inmates Network Their Knowledge of Substance Use).* This community service, temporary absence program involves inmates from Pittsburgh Institution (Ontario), who speak to community organizations, youth groups, and high schools about their history of involvement with drugs, their involvement in the criminal justice system, and how drugs have affected their adult lives.

- *Ma MaWi Wi Chi Itata/Stony Mountain Aboriginal Family Violence Program.* This is a pre-release family violence program for Aboriginal offenders at Stony Mountain Institution (Manitoba). It incorporates elements of traditional Aboriginal culture and healing practices in an attempt to help Aboriginal men stop being violent toward Aboriginal women.

Provincial corrections systems also offer a range of treatment programs for inmates. In Ontario, for example, there are education programs, vocational and occupational development programs, industrial programs, and various work programs, as well as specialized treatment programs such as the sex offender treatment programs at the Ontario Correctional Institute in Brampton and the Millbrook Correctional Centre.

The design and delivery of treatment programs in provincial correctional facilities is made difficult by the relatively short incarceration periods. Even offenders who receive a sentence of two years less one day are likely to spend only a few months in a controlled institutional setting prior to being classified to lower security work camps or facilities or released on a temporary absence. Yet these offenders very often have treatment needs similar to those of federal inmates. A large percentage of provincial inmates have deficiencies in education and job skills, as well as alcohol and substance addictions that need to be addressed.

One provincial facility that attempts to address the treatment needs of inmates is the Northern Treatment Centre in Sault Ste. Marie, Ontario. This is a medium security facility that offers offenders individualized treatment programs on an intensive, short-term basis. Its programs address substance abuse, anger management, and living skills. The centre also provides specialized programs for Aboriginal offenders.

Treating Special Offender Groups

Sex Offenders

Sex offenders are widely regarded as presenting the greatest treatment challenge to corrections personnel. Much is unknown about the causes of sex offending, and there is considerable ongoing debate over the effectiveness of sex offender treatment programs. Compounding this is the fact that sex offenders—to a much greater extent than other offender groups—often deny committing the offence or minimize the significance and impact of their behaviour by blaming the victim.

Most treatment experts do not speak in terms of "curing" sex offenders, but rather in terms of providing sex offenders with the necessary strategies to manage and control their deviant impulses. There is some evidence that sex offenders who receive specialized forms of treatment have a reduced chance of reoffending once they are released. Other groups of sex offenders, however—especially pedophiles—are considered virtually incurable. For these offenders, many of whom will be at high risk of reoffending for the rest of their lives, the focus on relapse prevention is especially important.

A wide variety of sex offender treatment programs are offered in federal correctional facilities in all regions of the country. In Quebec, Montée Saint-François, a minimum security facility, operates a program called VISA, which is designed to treat offenders convicted of incest. The CSC Regional Health Centre, near Vancouver, offers the Northstar program, which attempts to meet the treatment needs of mentally disabled sex offenders. Provincial/territorial corrections systems have also created facilities and programs designed specifically for sex offenders.

Some treatment groups are directed by the inmates themselves. At Mountain Institution, a federal prison near Vancouver, inmates convicted of sex offences formed the Phoenix Self-Help Group, a twelve-month program patterned on Alcoholics Anonymous. The program focuses on education, self-awareness, and encouraging offenders to confront and address behavioural problems.

Female Offenders

Corrections has been, and still is, a male-dominated arena in terms of both correctional staff and inmates. Until recently, correctional policies and programs were always designed for male offenders, with little thought given to the needs of women. Over the past two decades, however, increasing attention has been paid to the specific and unique needs of female offenders. This attention is reflected in the recent construction of small, regional facilities for federally sentenced women (see Box 7.2) and in the development of correctional programming that is community oriented, woman centred, and designed to improve the education, skills, and self-esteem of female offenders.

Specific treatment initiatives have been developed that are designed to address women's experiences with physical and sexual abuse, drug and alcohol dependency and substance abuse, and self-injurious behaviour, and to provide women with parenting skills and general life skills. A unique (and somewhat controversial) feature is the mother–child program, which allows infants and small children to live in the institution with their mothers for a specified period.

Aboriginal Offenders

In our discussion of "Who goes to prison?" earlier in the chapter, it was noted that Aboriginal people are overrepresented in federal and provincial/territorial prisons. In recent years, Aboriginal communities and organizations have become more involved in creating and staffing culturally appropriate programs to assist Aboriginal offenders. Aboriginal spiritual leaders are involved in teaching Aboriginal inmates, and in many federal institutions "sweat lodges" are used (Waldram, 1997). Native prison liaison officers and correctional workers provide assistance to Aboriginal offenders, and a number of wilderness camps for offenders are operated by Aboriginal organizations in collaboration with provincial corrections agencies. The Nishnawbe-Aski Wilderness Camp in Kenora, Ontario, for example, was established to treat Aboriginal offenders convicted of alcohol-related offences. In Nova Scotia, the Mik'maw Lodge is a facility for Mik'maq offenders on the Eskasoni Reserve in Cape Breton.

To better meet the needs of Aboriginal female offenders, the Okimaw Ohci Healing Lodge has been constructed on the Nekaneet First Nation in the Cypress Hills near Maple Creek, Saskatchewan. The lodge has a capacity of thirty women and up to ten children under the age of six, and its programs are centred on a healing model that incorporates Aboriginal culture and spirituality.

Aboriginal communities and organizations are also becoming more involved in the operation of correctional facilities. The federal Corrections and Conditional Release Act permits the CSC to enter directly into agreements with Aboriginal communities to provide correctional services. For example, Native Counselling Services of Alberta (NCSA), the oldest and perhaps best-known Aboriginal organization of its kind, operates a number of community and correctional programs for adult and young offenders. These programs include community correctional centres and wilderness camps.

The Effectiveness of Institutional Treatment Programs

It has long been debated whether institutional treatment programs work. Are those inmates who participate in treatment programs while they are confined more likely to be successful on release back into the community? Determining "what works" is very difficult.

The traditional measure for determining success is **recidivism rates**—the number of offenders who, once released from confinement, are returned to prison either for a technical violation of a condition of their parole or statutory release or for committing a new offence. The use of recidivism rates to measure program effectiveness has been criticized on a number of counts:

The grounds of an Aboriginal correctional facility

Courtesy of the Correctional Service of Canada

> **recidivism rates**
> the number of offenders who, once released from confinement, are returned to prison either for a technical violation of a condition of their parole or statutory release or for the commission of a new offence

- Legal criteria (which include subsequent contact with the criminal justice system) make no provision for the "relative" improvement of offenders. Offenders who previously committed serious crimes and are subsequently returned to confinement for relatively minor offences might be viewed as "relative successes" rather than as failures.

- It has yet to be determined how long after release from confinement the offender's behaviour is to be monitored. For example, many types of sex offenders remain a high risk to reoffend for a decade or longer following release.

- Recidivism rates are a result of detection of the released offenders by parole or police officers. In fact, the offender may have returned to criminal activity without being detected—a not unlikely scenario, given that clearance rates of the police for most non-violent offences are less than 30 percent and often as low as 5 percent.

- The success or failure of an offender upon release may be due in large measure to the level and types of supervision he or she receives. Among parole officers there are a variety of supervision styles; some have a more punitive orientation, while others focus on providing services and assistance.

- It is difficult to relate the offender's behaviour upon release to specific treatment interventions he or she received while incarcerated. There are many reasons why an individual may cease violating the law, including the efforts of a supportive family and/or spouse, success in securing stable employment, and maturation (Griffiths, 2004: 323–24).

Compounding the difficulties of assessing treatment effectiveness is the lack of controlled research studies to provide data on attitude and behavioural changes during confinement and after release. Notwithstanding the difficulties of measuring effectiveness, the research studies that have been done on correctional treatment programs in North America indicate that *some* treatment programs do work with *some* offenders (see Box 7.6).

focus

BOX 7.6

How Effective Are Correctional Treatment Programs?

Personal Skills Training

- Life-skills training programs appear to produce in some inmates positive changes that increase the likelihood of success upon release.

- Anger management programs are successful in providing some inmates with non-hostile ways in which to resolve conflict in interpersonal relationships as well as with strategies for managing anger-arousing thought patterns.

- Cognitive skills programs are a very successful intervention when appropriately matched with offender needs and have a positive impact on post-release behaviour, even among violent offenders.

Education Programs

- There is a significant relationship between completion of adult basic education courses and reduced recidivism upon release.

- Participation in university-level degree programs reduces the risk of reoffending for some offenders.

Vocational and Work Programs

- Uninterrupted participation in prison industry programs during confinement appears to help some offenders succeed upon release.

Sex Offender Treatment Programs

- Programs that are based on a cognitive-behavioural approach and that include a relapse prevention component reduce rates of reoffending among some types of sex offenders.

Generally, findings must be qualified with the terms "some inmates" and "many inmates." That said, the results do suggest that even in the total institutional world of the prison, correctional treatment can be effective.

A critical problem is the absence of evaluation studies of the effectiveness of correctional treatment programs. This problem is especially acute in provincial corrections, even though the majority of offenders sentenced to custody serve time in provincial/territorial facilities. This makes it difficult if not impossible to determine "what works" and "why" in correctional treatment (Latessa, Cullen, and Gendreau, 2002).

Complicating the issue is the fact that findings from research studies are not always consistent. While treatment programs for sex offenders have the potential to reduce reoffending among some offenders, there is debate over the ability of programs to "treat" sex offenders. For example, in the most comprehensive study of sex offenders conducted to date, offenders who participated in a highly regarded treatment program were compared to a group of sex offenders who did not participate in the program. No significant differences were found in the rates of reoffending between the two groups over a twelve-year period following release from custody (Hanson, Broom, and Stephenson, 2004). Increasingly, observers are calling for the development of "best practices" in corrections (also referred to as evidence-based practices), similar to the increasing emphasis on best practices in policing (Reitzel, 2005; and see Chapter 4).

Research studies have identified a number of principles that, if followed, increase the likelihood that treatment programs will succeed:

- *The risk principle.* Treatment interventions have a greater chance of success if they are matched to the risk level of the offender. This means that higher levels of treatment intervention should be reserved for higher risk inmates. Interestingly, lower risk inmates may be negatively affected by intensive treatment intervention.

- *The need principle.* Treatment interventions must address the factors (substance abuse, for example) that are related to the offending behaviour.

- *The responsivity principle.* Treatment interventions must be matched to the learning styles and abilities of individual inmates; "one size fits all" treatment approach must be avoided.

- *Professional discretion.* Correctional personnel must consider the unique attributes of individual inmates and apply the above three principles in an appropriate manner.

- *Program integrity.* Treatment programs must be delivered to inmates as originally designed, and the specific treatment model must be followed (Bourgon and Armstrong, 2005; Griffiths, 2004).

Doing Treatment in Correctional Institutions

Over the years, there has been general agreement that treatment programs designed to alter the attitudes, values, and behaviours of offenders have not been particularly effective. Many people have taken this lack of success to mean that people convicted of crimes are not interested in changing and would rather be lifelong criminals.

However, a significant number of people incarcerated in correctional institutions do want to change; but these people require a considerable amount of assistance both while confined and upon release. A close examination of treatment programs in correctional institutions indicates that there are a number of obstacles to the success of these programs. These obstacles include, but are certainly not limited to, the following:

- *The prison is a total institution.* Life inside correctional institutions bears little resemblance to life in the free world. Offenders are supervised twenty-four hours a day and are required to follow a strict institutional regimen. The dynamics that develop in the total institutional world of the prison, including the violence and exploitation that often characterize relations between inmates, may undermine the objectives of treatment programs, which include developing self-confidence, accepting responsibility for one's actions, and behaving in a positive and productive manner.

- *Post-release behaviour is hard to predict.* There is general agreement that life inside correctional institutions is considerably different from life in the free world. Except for a handful of offenders with histories of violence, this makes it extremely difficult to predict post-release behaviour based on behaviour inside prison. Corrections systems have developed a number of assessment instruments to help correctional officials predict how the offender will behave upon release.

- *Many prison inmates are "marginal."* With few exceptions, offenders in correctional institutions are marginal in their life skills, employment skills, education, and background. This makes inmates a challenging group of people to work with, especially since their attitudes and behaviours may have become well entrenched.

- *Attitudes and behaviours are difficult to alter.* Altering the attitudes and behaviours of *any* group of people, be they convicted offenders or law-abiding citizens, is a very difficult task. Remember that interventions require offenders to alter lifelong attitudes, values, and behaviours, all in an environment in which the offender is surrounded by people who may have little interest in giving up a criminal lifestyle. For an offender who has spent time in youth corrections and in adult facilities, this change may be very difficult indeed, especially if he or she is only familiar with life inside correctional institutions and knows very little of life outside them. Many offenders have no family support, friends, or opportunities in the free world and may feel more comfortable and secure among their friends in prison. This may make the transition from a criminal lifestyle to a law-abiding one that much more difficult.

- *Inmate needs must be matched with specific treatment interventions.* A key concept in corrections is **differential treatment effectiveness,** which points to the importance of matching the treatment needs of the individual offender with the appropriate treatment program, both during incarceration and following release from the institution. This is especially important for offenders convicted of high-risk offences such as sexual assault against children.

differential treatment effectiveness
the requirement that specific treatment interventions be matched to individual inmates

There is no doubt that some offenders, for whatever reasons, are not interested in turning their lives around. This reluctance or inability to change may be due to a number of factors. Some offenders have spent most of their youth and adult life in institutions. These offenders are often terrified of returning to the outside and generally have no family or support systems to help them when they are released. Others are committed to a criminal lifestyle and to being "in the life." Still others are in a state of denial and have not accepted responsibility for their criminal behaviour and the harm they have done to victims and to the community. Some offenders have learning disabilities that prevent them from gaining maximum benefit from treatment programs.

Within the inmate population, there is a wide variety of treatment needs. Plainly, this is a function of personality differences, types of offences committed, and cultural or ethnic identities. It can be anticipated, for example, that inmates convicted of sex offences and inmates convicted of property or drug offences will have different treatment needs. Similarly, Aboriginal inmates may not respond to treatment initiatives designed by non-Aboriginal professionals for non-Aboriginal inmates.

Program Fidelity

A crucial component in effective correctional treatment programs is program implementation (Gendreau, Goggin, and Smith, 1999). Correctional authorities must ensure that there is *program fidelity*—that is, that a treatment program is delivered in the way it was originally designed. Program fidelity requires a clear program manual and appropriate training and supervision for the staff who run the program. Little attention has been given to date to the role played by individual treatment staff in the success of a particular program, or to the factors that may hinder the efforts of treatment personnel. Research in the United States has found that treatment staff in prisons experience many of the same sources of stress as correctional officers (see above) and that this may compromise the delivery of correctional programming (Armstrong, 2004).

Additional factors that may impede efforts to rehabilitate offenders include geographically isolated and poorly designed correctional facilities, frequent transfers of inmates between facilities, the work orientations of correctional officers, and a lack of program resources (Micucci, 2004).

The comments of one federal inmate illustrate the problem of program fidelity: "The Substance Abuse Program I attended involved nothing more than watching a series of videotapes two days a week for 10 weeks. Our group viewed the tapes, then were dismissed. There was no discussion; there was no interaction between facilitator and group; there was no attempt to talk about anything. After the program was over I had an idea what abuse of drugs and alcohol would do to my body, but I had no idea why I ever wanted to use them" (Kowbel, 1995: A20).

ACCOUNTABILITY IN CORRECTIONS

Federal and provincial corrections systems (excepting that of Prince Edward Island) have established offices (a) to investigate complaints by prison inmates that have not been resolved at the institutional level and (b) to inquire into various aspects of the operations of the correctional system. These inquiries often address complaints by prison inmates about living conditions, inadequate health care, and staff conduct. In recent years the federal correctional investigator has investigated complaints about overcrowding and double-bunking, institutional grievance procedures, and the involuntary transfer of inmates from one institution to another.

focus

BOX 7.7

Ontario Ombudsman Investigations

Case 1

Mr. Z called our office and explained that he had been transferred from one facility to another because of allegations that he had done something to a staff person or his family. He said his inmate card, which correctional officers have access to, contained these allegations and that correctional officers were causing him difficulties as a result. An Ombudsman Representative spoke to the Security Manager at the receiving facility who explained that the information was on a computer-generated printout that stated, "management risk—management problem previous—must not return to the [facility]—offences against staff." The Security Manager explained that the transferring facility had advised that Mr. Z's brother, during a random shooting, had killed a good friend of an Operational Manager at the transferring facility. Mr. Z was transferred because inmates at the transferring facility liked the Operational Manager and there was concern that they might assault Mr. Z. The Security Manager said that, although the information on the inmate card appeared to be wrong, it would have to be corrected by someone at the transferring facility. The Ombudsman Representative made a number of enquiries to the transferring facility. Eventually, the Operational Manager whose friend had been killed offered to have the information corrected. The Security Manager at the receiving facility later confirmed that the information had been changed.

Case 2

Mr. V, an African Canadian inmate, complained that when he was transferred to a smaller correctional facility, his "afro pic" hair comb had been confiscated and culturally appropriate hair grooming products were not available from the facility's canteen list. Mr. V was concerned that his inability to properly comb his hair would negatively compromise his opportunity to present a positive image at his trial. Mr. V also noted that the restriction on culturally specific products had an adverse impact on an identifiable ethnic and cultural group within the inmate population of the facility. When our staff contacted the facility, we were told that the Superintendent had decided to remove these culturally specific products from the canteen list. We then contacted the Ministry's Anti-Racism Coordinator to discuss the situation. The Anti-Racism Coordinator ensured that the culturally specific products would be immediately returned to the canteen list. Further, the Ministry advised that guidelines had been changed to prevent an individual Superintendent from making a random arbitrary decision. Now Superintendents must proceed through the Ministry's Canteen Committee to obtain approval for removing any items from their facility's canteen list.

Source: Ombudsman Ontario, *2004–2005 Annual Report* (Toronto, 2005). Retrieved from http://www.ombudsman.on.ca/annrep05/st_mcscs.html. Reprinted with permission.

The findings and recommendations of the provincial ombudsman and the federal correctional investigator are not binding; they do, however, provide a basis for addressing problems. Box 7.7 describes two investigations that Ontario's ombudsman undertook in response to inmate complaints.

As you might expect, relations between provincial ombudsman's offices and correctional ministries and between the federal correctional investigator and the CSC are often strained. The federal correctional investigator has complained in annual reports that the CSC is too defensive and dismissive in its responses to recommendations for change. The increasing involvement of external investigators in responding to the complaints and concerns of inmates reflects a growing concern with the rights of the incarcerated.

CRIME VICTIMS AND INSTITUTIONAL CORRECTIONS

Victim involvement with the criminal justice system can continue after the verdict and sentence and even extend to situations in which the offender has been sentenced to a correctional institution. Correctional agencies in some jurisdictions provide victims with information on certain decisions made about inmates. This interaction is a two-way street in that victims can provide useful information about the inmate and the offence that might not be available otherwise. For example, if the victim submits a victim impact statement, correctional authorities need not rely so heavily on the offender's version of events. A victim who knows an inmate well can also provide useful social history and background information.

Some correctional agencies inform victims about inmate passes from the institution (usually called "temporary absences"), assist in preventing harassment of victims via letter or telephone, and take measures to protect victims who attend private family visits (formerly known as "conjugal visits"). In most cases, the victim has to contact the correctional agency and indicate that he or she wants to be informed about decisions. In provinces such as New Brunswick, however, officials proactively contact the victims of serious personal injury crimes to permit their input into decisions and to gather information to help assess risk to the community.

TRENDS IN INSTITUTIONAL CORRECTIONS

In the early twenty-first century, a number of trends are emerging that may significantly alter the dynamics of institutional corrections.

The Escalating Costs of Institutional Corrections

Although the number of offenders in custody has remained relatively stable over the past decade, the cost of housing inmates has been rising. In 2003–04, an average of $240.18 per day ($87,665.70 per year) was spent on federal inmates, compared to $141.75 per day ($51,738.75 per year) for provincial/territorial inmates (Beattie, 2005). The rates are higher for female offenders, averaging over $170,000 per year. This excludes the costs associated with the inmate's involvement in a specialized treatment program (Public Safety and Emergency Preparedness Portfolio Corrections Statistics Committee, 2004: 29).

These daily costs of confinement are on par with a one-night stay in a five-star hotel, and the annual costs greatly exceed the yearly costs of attending a top-flight undergraduate university or graduate school. This raises the question as to whether institutional correctional services are providing value for money in terms of protecting the public and helping offenders alter their criminally oriented attitudes and behaviours.

Public Pressure and the "Get Tough" Approach

There has been increasing pressure on the criminal justice system to "get tough"—that is, to impose longer sentences on criminal offenders and increase the pains of imprisonment. Historically, corrections systems have not enjoyed the same levels of public support as the police. This has made corrections more susceptible to pressure from politicians and public interest groups to adopt a "get tough" approach. Reflecting this approach is a policy of the federal solicitor general that offenders convicted of first and second degree murder should spend the first two years of confinement in a maximum-security institution. Typically, the heinous crimes of a very small number of offenders overshadow the positive accomplishments of program and treatment staff and of the inmates themselves.

Overcrowding

Increasing public pressure to get tough with criminal offenders has affected both legislation and the sentencing patterns of criminal court judges. At the same time, in response to fiscal crises, several provincial governments have closed institutions and consolidated inmate populations in larger facilities. Some parole boards, including, most dramatically, Ontario's, have reduced the rates at which they grant parole. One consequence is overcrowding, a fact of correctional life in the United States that is now emerging in Canada. Overcrowding increases the stress levels of inmates, may result in higher levels of violence, and presents challenges to correctional staff and administrators.

Although several new federal and provincial institutions have been opened in recent years, the available bed space has not kept pace with the numbers of inmates. In many federal institutions, two inmates share a room or cell originally intended for one. About 25 percent of federal prison inmates share cell space. As you might imagine, double-bunking can result in increased pressures on prison inmates and may contribute to violence.

Running directly counter to the trend toward incarcerating more offenders are the efforts of the Province of Quebec to reduce its institutional populations and to expand the use of non-custodial sanctions for non-violent offenders, with a concomitant increase in the emphasis on rehabilitation. Over the past several years, the Quebec government has closed several correctional facilities. Although this move was prompted at least in part by the need to reduce government expenditures, provincial officials have stated that "Quebec has decided to turn its back on the repressive model" and to use imprisonment only as a last resort (Ministère de la Sécurité Publique, 1996: 6).

Privatization

In recent years, provincial/territorial governments have expressed interest in private-sector involvement in the construction and operation of correctional facilities. Proponents of privatization contend that private companies can provide a service that is more cost effective than government-operated facilities, as well as more accountable to monitoring and review. Opponents argue that because their profits depend on increases in inmate populations, private prisons—or "prisons for profit"—are not motivated to rehabilitate offenders (doing so would reduce the likelihood of an offender's return to confinement).

Private prisons in the United States have had mixed results. There is some evidence that private facilities are often better run and more efficient, that private companies can operate institutions 15 percent more cheaply than governments, and that they may result in better performance than publicly run institutions. On the downside, the American experience indicates that the lower operating costs of private prisons result in fewer staff and lower wages, and that the savings are often offset by the requirement that privately operated institutions be monitored on a continual basis.

Creating Correctional Communities

In the late 1990s, there was a fundamental shift in thinking about how federal correctional institutions should be designed and operated. This shift was reflected in the redesign of several existing facilities and the construction of several new ones. The intent of these initiatives was to replicate, within the institution, many of the spaces and types of activities that inmates would encounter upon their release back into the community. Described below are three correctional institutions whose designs reflect this philosophy of "normalization":

- *William Head Institution (British Columbia).* This federal minimum security prison is built around three "units," each of which contains two "communities." Each community consists of four two-storey duplexes (eight five-man houses) and a community building with offices for the unit staff, common areas, and laundry facilities. The purpose of the small-group settings is to promote responsibility among the inmates and thereby alter the dynamics that exist in traditional institutions.

- *Frontenac Institution (Ontario).* Inmates in the Phoenix program of this minimum security federal facility reside in living units consisting of a living room, kitchen, dining room, bathrooms, and individual bedrooms. There are no bars on the windows, inmates have the keys to their own rooms, and the units are locked only between midnight and 4 a.m. The inmates in each unit are given a budget to buy food and must share household responsibilities, including cooking. The program is designed to teach inmates skills on how to live responsibly and make good decisions.

- *Grand Valley Institution for Women (Ontario).* Inmates of this federal facility live in cottages. Each woman has her own bedroom, which can be locked from the inside. (Some bedrooms have space for a baby crib.) There are no cells, staff do not wear uniforms, and the frontline staff are known as "primary workers" rather than as "guards" or "correctional officers" (Griffiths, 2004).

Competing Federal and Provincial Models of Correctional Practice

It has become evident that over the past decade, a split has emerged between the federal government and the provincial governments in terms of institutional correctional practice. This development can be traced in part to the election of neoconservative provincial governments in several provinces (most notably in Ontario in the 1990s).

At the federal level, correctional policy and practice has remained firmly entrenched in what can be termed a "liberal European model" of corrections, which places a high value on proactive intervention in the lives of offenders and on the involvement of social and justice agencies in responding to crime. This is reflected in the closing of the Kingston Prison for Women in 2000 and the construction of a number of smaller, regional facilities for federally sentenced women; in the development of Aboriginal healing lodges in partnership with First Nations; and in policies designed to increase the number of federal offenders under community supervision to 50 percent of the total federal offender population. The creation of correctional communities of the type discussed earlier is yet another reflection of the federal model of institutional corrections. There is some indication, however, that the policy and practice of the CSC may be moving more to right-of-centre with the appointment of a new Commissioner of Corrections in 2005 and a new emphasis by the CSC on community safety and the rights of crime victims. This bears watching.

At the provincial level, more conservative correctional policies have emerged, many of which mirror the more conservative, punishment-oriented American approach to corrections. Ontario, for example, has implemented significant reforms in corrections under the banner of a "no frills," "get tough" approach that has community safety at its core. Among other steps, the Ontario government has:

- closed thirty-one correctional facilities and created a small number of "super-jails," including Maplehurst, which is the largest correctional facility in Canada, with beds for 1,500 inmates;
- developed performance standards for jails that include tracking the number of escapes, disturbances, and suicides at each institution;
- established a policy of zero tolerance for inmate acts of violence toward correctional staff in prisons and in the community; *and*
- created provisions under the Corrections Accountability Act that have ended the automatic credit of remission and that require offenders to earn the privilege of early release by actively participating in programs and by demonstrating positive behaviour (see Chapter 8; see also Griffiths, 2004; Ontario Ministry of Community Safety and Correctional Services, online at http://www.mpss.jus.gov.on.ca).

The provincial government in British Columbia has also closed a number of correctional facilities and consolidated inmate populations in a smaller number of larger institutions.

The Ontario government has been exploring the potential for public–private partnerships in various facets of correctional operations. In 2001 it opened the

1,184-bed Central North Correctional Centre (CNCC), a "super-jail" near Penetanguishene. This was Canada's first private prison, although there are a number of such facilities for young offenders.

The province of Quebec has explicitly rejected a conservative, American-style approach to corrections and is continuing to pursue a correctional policy centred on the European model, which emphasizes prevention, promotes the use of alternative measures to incarceration, and uses custody only as a last resort.

Summary

This chapter focused on the operation of federal and provincial/territorial corrections systems. Although only a relatively small percentage of convicted offenders receive a sentence of confinement, correctional institutions are an important component of the criminal justice system and present numerous challenges for both the correctional personnel and the inmates. The federal government and the provinces/territories operate a range of facilities at all security levels. Large numbers of inmates have limited education and job skills; a disproportionate number of them are Aboriginals and persons of colour.

Life inside correctional institutions was discussed, as well as the inmate social system and the various occupational challenges of correctional officers. Specific attention was given to the various ways in which the risks and needs of prison inmates are assessed. The various treatment programs offered in correctional institutions were discussed. The chapter concluded with discussions of accountability in corrections and of several key trends in institutional corrections.

Key Points Review

1. The vast majority of inmates will ultimately be released back into the community.
2. In early Canada, imprisonment was not used for the specific purpose of punishing offenders.
3. The number of offenders admitted to federal and provincial/territorial correctional institutions is declining.
4. A large majority of those who are incarcerated reside in provincial/territorial correctional facilities.
5. The majority of inmates in provincial/territorial institutions (80 percent) and nearly half of the inmates in federal institutions (48 percent) have been convicted of non-violent offences.
6. Aboriginal inmates are a disproportionate percentage of inmates in Canadian correctional institutions.
7. A number of small regional facilities house female offenders across the country.
8. Most federal female inmates have histories of poverty, abuse, and long-term drug and alcohol dependency.
9. Most incarcerated offenders have deficiencies in education and job skills.
10. Wardens of correctional institutions have the difficult task of preventing or reducing the spread of communicable diseases such as HIV/AIDS and hepatitis B and C.

11. Correctional officers play a key role in managing inmates and often experience considerable stress in carrying out their duties.

12. Inmates must find ways to adapt to the total institutional world of the prison.

13. A key focus in institutional corrections is the assessment of offenders' needs and of the risks offenders pose.

14. Correctional facilities, especially federal institutions, offer a wide range of core treatment programs.

15. Some offender groups, including female offenders, Aboriginal offenders, and sex offenders, have unique and challenging treatment needs.

16. Studies on the effectiveness of correctional treatment programs indicate that *some* treatment programs do work with *some* offenders.

17. There are a number of obstacles to the delivery of effective treatment programs in correctional institutions.

18. Among the trends in institutional corrections are the following: public and political pressures to "get tough" with offenders; overcrowding in prisons; the privatization of correctional institutions; and competing models of institutional practice between the federal government and several of the provincial governments.

Key Term Questions

1. What is the **two-year rule**?
2. What is the status of an offender who is on **remand**?
3. What are the main characteristics of **total institutions,** and how does this concept help us understand the dynamics of prison life?
4. What are **Citizen Advisory Committees (CACs)** and what role do they play?
5. What is meant by the terms **pains of imprisonment, status degradation ceremonies,** and **inmate code**?
6. What process does the term **prisonization** describe?
7. What does **classification** mean in the context of inmate treatment and what steps does it involve?
8. What is the difference between **static risk factors** and **dynamic risk factors,** and how does each relate to risk assessment and correctional treatment intervention?
9. What process does the term **case management** describe in the context of inmate treatment?
10. What is **differential treatment effectiveness** and why is it a key concept in corrections?
11. Discuss what is meant by **recidivism rates** and then note the difficulties of using these rates to measure the effectiveness of correctional treatment.

References

Armstrong, G.S. (2004). "Does the Job Matter? Comparing Correlates of Stress among Treatment and Correctional Staff in Prison." *Journal of Criminal Justice*, 32(6):577–91.

Beattie, K. (2005). "Adult Correctional Services in Canada, 2003/04." Catalogue no. 85-002-XPE. *Juristat*, 25(8). Ottawa: Canadian Centre for Justice Statistics, Statistics Canada.

Bourgon, G., and B. Armstrong (2005). "Transferring Principles of Effective Treatment into a 'Real World' Prison Setting." *Criminal Justice and Behavior,* 32(1):3–25.

Brady, M. (1998, May 16). "Pens and Needles: Prison Population Growth Has Cranked Up the Risk of Infectious Disease Transmission." *Financial Post,* R2.

Bronskill, J. (2001, April 14). "Men Should Not Guard Women, Panel Says." *National Post,* A4.

Bunner, P. (2000, March 13). "A Philosophy on Death Row: Corrections Boss Ingstrup May Not Survive the Growing Clamour for Punitive Justice." *Report Newsmagazine,* 26(49):12–13.

Calzavara, L.M., C. Major, T. Myers, J. Schlossberg, M. Millson, E. Wallace, J. Rankin, and M. Fearon. (1995, September–October). "The Prevalence of HIV-1 Infection among Inmates in Ontario, Canada." *Canadian Journal of Public Health,* 86(5):335–39.

Canadian Centre on Substance Abuse. (2004). *Substance Abuse in Corrections.* Ottawa.

Canadian Human Rights Commission. (2003). *Protecting Their Rights: A Systemic Review of Human Rights in Correctional Services for Federally Sentenced Women.* Ottawa. Retrieved from http://www.chrc-ccdp.ca/legislation_policies/consultation_report-en.asp

Correctional Service of Canada. (2003). *Infectious Diseases Prevention and Control in Canadian Federal Penitentiaries 2000–01.* Ottawa.

———— (2001). *Community Strategy for Women on Conditional Release.* Discussion Paper. Ottawa.

———— (1997). *Mission of the Correctional Service of Canada.* Ottawa: Ministry of Supply and Services Canada.

Cowan, P., and D. Sheremata. (2000, February 9). "Death in Experimental Prison Unit—'She Was Helpless.'" *Edmonton Sun.* Retrieved online.

Cunningham, A., and L. Baker. (2003). *Waiting for Mommy: Giving a Voice to the Hidden Victims of Imprisonment.* London, ON: London Family Court Clinic.

Gendreau, P., C. Goggin, and P. Smith. (1999). "The Forgotten Issue in Effective Correctional Treatment: Program Implementation." *International Journal of Offender Therapy and Comparative Criminology,* 43(2):180–87.

Goffman, E. (1961). *Asylums: Essays on the Social Situation of Mental Patients and Other Inmates.* Garden City, NY: Doubleday.

Griffiths, C.T. (2004). *Canadian Corrections,* 2nd ed. Toronto: Thomson Nelson.

Hanson, R.K., I. Broom, and M. Stephenson. (2004). "Evaluating Sex Offender Treatment Programs: A 12-Year Follow-Up of 724 Offenders." *Canadian Journal of Behavioural Science,* 36(20):87–96.

Hughes, G.V., and E. Zamble. (1993). "A Profile of Canadian Correctional Workers." *International Journal of Offender Therapy and Comparative Criminology,* 37(2):99–113.

Johnson, S. (2004). "Adult Correctional Services in Canada, 2002/03." Catalogue no. 85-002-XPE. *Juristat,* 24(10). Ottawa: Canadian Centre for Justice Statistics, Statistics Canada.

Kowbel, R.D. (1995, January 13). "Self-Help Programs Are Part of Jailhouse Games." *The Globe and Mail,* A20.

Larivière, M.A.S. (1997). *The Correctional Service of Canada 1996–97 Retrospective Report on Inmate Suicides.* Ottawa: Correctional Service of Canada.

Larivière, M., and D. Robinson. (1996). *Attitudes of Federal Correctional Officers Towards Offenders.* Ottawa: Research Division, Correctional Service of Canada.

Latessa, E.T., F.T. Cullen, and P. Gendreau. (2002). "Beyond Correctional Quackery—Professionalism and the Possibility of Effective Treatment." *Federal Probation,* 66(2):43–49.

Lines, R. (2002). *Action on HIV/AIDS in Prisons: Too Little, Too Late. A Report Card.* Montreal: Canadian HIV/AIDS Legal Network. Retrieved from http://www.aidslaw.ca

Makin, K. (1996, June 7). "Inmates Fear Attacks; Drug Use Common." *The Globe and Mail,* A1, A5.

McCorkle, R.C. (1993). "Living on the Edge: Fear in a Maximum-Security Prison." *Journal of Offender Rehabilitation,* 20(1/2):73–91.

Micucci, A. (2004). "It's About Time to Hear Their Stories: Impediments to Rehabilitation at a Canadian Provincial Correctional Facility for Women." *Journal of Criminal Justice,* 32(6):515–30.

Ministère de la Sécurité Publique. (1996). *Vers un Recours Modere aux Mesures Penales et Correctionnelles.* Quebec: Author.

Moloughney, B. (2004). "A Health Care Needs Assessment of Federal Inmates in Canada." *Canadian Journal of Public Health,* 95 (Supp. 1), S1–S62.

Ontario Medical Association. (2004). *Improving Our Health: Why Is Canada Lagging Behind in Establishing Needle Exchange Programs in Prisons?* OMA Position Paper. Toronto.

Ouimet, R. (Chairman). (1969). *Toward Unity: Criminal Justice and Corrections. Report of the Canadian Committee on Corrections.* Ottawa: Queen's Printer.

Public Safety and Emergency Preparedness Portfolio Corrections Statistics Committee. (2004). *Corrections and Conditional Release Statistical Overview.* Ottawa: Public Safety and Emergency Preparedness Canada.

Reitzel, L.R. (2005). "Best Practices in Corrections: Using Literature to Guide Interventions." *Corrections Today,* 67(1):42–45, 70.

Robinson, D., P. Lefaive, and M. Muirhead. (1997). *Results of the 1996 CSC Staff Survey: A Synopsis.* Ottawa: Research Branch, Correctional Service of Canada.

Samak, Q. (2003). *Correctional Officers of CSC and Their Working Conditions: A Questionnaire-Based Study.* Montreal: Union of Canadian Correctional Officers.

Simonds, M. (1996). *The Convict Lover.* Toronto: Macfarlane Walter & Ross.

Snow, L. (1997). "A Pilot Study of Self-Injury amongst Women Prisoners." *Issues in Criminological and Legal Psychology,* 28:50–59.

Struckman-Johnson, C., D. Struckman-Johnson, L. Rucker, K. Bumby, and S. Donaldson. (1996). "Sexual Coercion Reported by Men and Women in Prison." *Journal of Sex Research,* 33(1):67–76.

Thompson, S. (2002). *Letters from Prison: Felons Write about the Struggle for Life and Sanity behind Bars.* Toronto: HarperCollins.

Trevethan, S. and C.J. Rastin. (2004). *A Profile of Visible Minority Offenders in the Federal Correctional System.* Ottawa: Research Branch, Correctional

Service of Canada. Retrieved from http://www.csc-scc.gc.ca/text/rsrch/ reports/r144/r144_e.shtml

Treventhan, S., C. Rastin, and A. Bell. (2004). *Report on the Evaluation of Citizens' Advisory Committees: CSC Perspective.* Ottawa: Research Branch, Correctional Service of Canada.

Waldram, J.B. (1997). *The Way of the Pipe: Aboriginal Spirituality and Symbolic Healing in Canadian Prisons.* Peterborough, ON: Broadview Press.

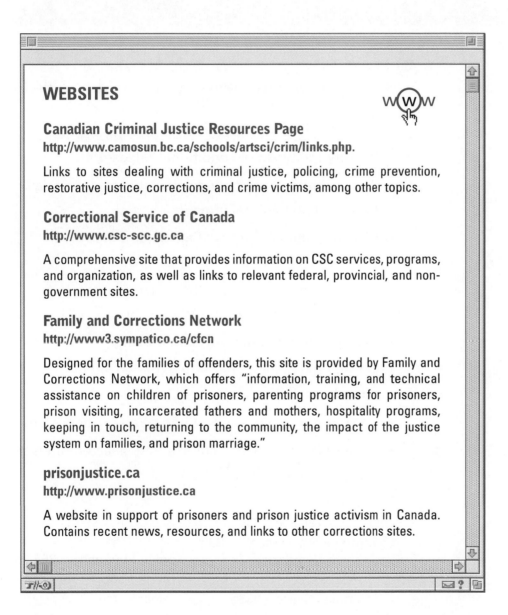

WEBSITES

Canadian Criminal Justice Resources Page
http://www.camosun.bc.ca/schools/artsci/crim/links.php.

Links to sites dealing with criminal justice, policing, crime prevention, restorative justice, corrections, and crime victims, among other topics.

Correctional Service of Canada
http://www.csc-scc.gc.ca

A comprehensive site that provides information on CSC services, programs, and organization, as well as links to relevant federal, provincial, and non-government sites.

Family and Corrections Network
http://www3.sympatico.ca/cfcn

Designed for the families of offenders, this site is provided by Family and Corrections Network, which offers "information, training, and technical assistance on children of prisoners, parenting programs for prisoners, prison visiting, incarcerated fathers and mothers, hospitality programs, keeping in touch, returning to the community, the impact of the justice system on families, and prison marriage."

prisonjustice.ca
http://www.prisonjustice.ca

A website in support of prisoners and prison justice activism in Canada. Contains recent news, resources, and links to other corrections sites.

chapter
8

Release and Re-entry

learning objectives

After reading this chapter, you should be able to

- Describe the parole process.
- Identify the types of conditional release.
- Discuss the release options available to provincial/territorial and federal inmates.
- Discuss victims' rights initiatives and the role of victims in the parole hearing process.
- Discuss the major issues that surround parole as a correctional practice.
- Discuss parole board decision making.
- Identify the pains of re-entry experienced by offenders.
- Discuss the issues surrounding parole supervision.
- Discuss the challenges confronting selected offender groups on conditional release.

key terms

cold turkey release   336
community assessment   333
community notification (CN)   366
conditional release   333
detention during a period of statutory release   349

one-chance statutory release   349
pains of re-entry   358
reintegration   357
release plan   333
statutory release   348

The vast majority of offenders confined in correctional institutions are ultimately released back into the community. Most offenders return to the community under some form of **conditional release.** Canadians often hear about parole in the media, typically after a parolee has committed a violent crime. In fact, most offenders who are released from custody under some type of conditional release successfully complete their period under supervision; of those who do not succeed, only a very small number commit a violent offence.

THE PAROLE PROCESS

Parole is best viewed not as an event, but as a process. The parole process actually begins in the criminal court when the judge sets the length of the prison term, thereby defining the maximum period the offender will serve. If the offender is sentenced to five years, the prison system has no authority to keep him or her confined beyond five years, *no matter how dangerous he or she is assessed to be.* The most restrictive option available to the courts is judicial determination (discussed in Chapter 6), which gives judges the authority to delay parole eligibility until one-half of the sentence (as opposed to the typical one-third) has been served.

The staff in correctional institutions participate in the parole process by helping inmates to develop a **release plan.** In provincial/territorial facilities, this function is performed by staff variously called inmate liaison officers (Ontario), parole coordinators, or conditional release coordinators (British Columbia). In federal institutions, case management officers and institutional parole officers prepare release plans and other materials that will be used by parole boards in their deliberations. The release plan contains information about where the prospective parolee will live, employment prospects, and any arrangements for community-based support (such as residence in a halfway house or residential drug treatment facility). Release plans must be vetted by probation or parole officers in the community into which the offender will be released.

A key component of the release plan is the **community assessment.** Prepared by the probation or parole officer, this report evaluates the feasibility of the offender's proposed community plan in terms of the level of supervision required and the availability of community resources. Among the areas examined in the assessment are the proposed residence, education/employment/treatment plans, family and other support networks in the community, the extent to which the offender accepts responsibility for and understands the offending behaviour, information supplied by the victim(s), and any recommended special conditions the parole board may attach to the parole. Parole board members use the information contained both in the release plan and the community assessment to determine whether an inmate should be granted a conditional release and, if so, the special conditions that should be attached to it.

Correctional officials also collect background information on prospective parolees, including reports about infractions of institutional rules and participation in correctional programs, and reports created to aid in classification and program selection. Third-party reports can include the following: police

conditional release
a generic term for the various means of leaving a correctional institution before warrant expiry whereby an offender is subject to conditions that, if breached, could trigger revocation of the release and return to prison; parole is one type of conditional release

release plan
a plan setting out the residential, educational, and treatment arrangements made for an inmate who is applying for conditional release

community assessment
an evaluation of the feasibility of the release plan, the level of supervision required, and the availability of community resources

reports about the offence; victim impact statements used at sentencing; a pre-sentence report compiled by a probation officer; a court transcript of the judge's reasons for the sentence; a printout of prior criminal history; reports gathered from agencies having previous contact with the inmate (such as mental health assessments and medical reports); a community assessment written by a parole officer who has interviewed family members and other community contacts; psychological assessments; and reports from any counselling sessions. Friends and family of the inmate may write letters of support as well. In recent years, efforts have been made to increase the reliability, clarity, and completeness of case file material so that the parole board will have all the pertinent information. The parole process for provincial offenders in British Columbia is illustrated in Figure 8.1.

CONDITIONAL RELEASE

Purpose and Principles of Conditional Release

As set out in Section 100 of the Corrections and Conditional Release Act, the purpose of conditional release is to contribute to the maintenance of a just, peaceful, and safe society through decisions on the timing and conditions of release that will facilitate the rehabilitation of offenders and their reintegration into the community as law-abiding citizens.

The act also sets out a number of principles to be followed by the National Parole Board and provincial parole boards as they pursue the objectives of conditional release. Following are the three most important of these principles:

1. The protection of society must be the primary consideration in every case.
2. Parole boards must consider all available information that is relevant to the case, including recommendations from the sentencing judge, the results of assessments completed during the case management process, and information from victims and the offender.
3. Parole boards should make the least restrictive determination required to ensure the protection of society.

The third principle in the preceding list is generally the one least understood by politicians and the general public. Another important section of the Corrections and Conditional Release Act (s. 102) states:

The [National Parole] Board or a provincial parole board may grant parole to an offender if, in its opinion,

(a) the offender will not, by reoffending, present an undue risk to society before the expiration according to law of the sentence the offender is serving; *and*

(b) the release of the offender will contribute to the protection of society by facilitating the reintegration of the offender into society as a law-abiding citizen.

FIGURE 8.1   Parole Flow Chart

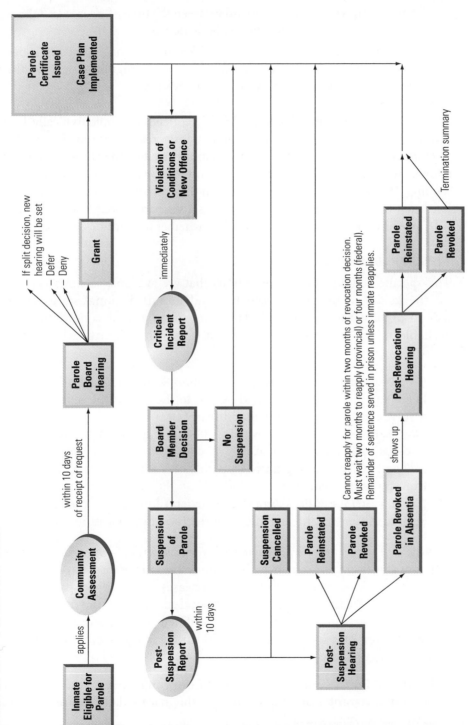

Source: Copyright © 1998 Corrections and Community Justice Division.

In practice, this means that conditional release can be granted under *either* of the following two conditions:

- The applicant is unlikely to reoffend between the time of release and the warrant expiry date (that is, the end of the sentence).
- There is a risk of reoffending but the risk can be managed through specific interventions—such as a residential treatment program—that would be unavailable except as part of a conditional release.

An offender who presents little risk of reoffending would typically have a favourable sociobiographical background and no previous criminal convictions. Offenders who present a high risk must demonstrate that they have taken steps to address those aspects of their lives that would increase the likelihood of reoffending. However, as the discussion later in the chapter will indicate, the general nature of the conditional release provisions gives parole board members a considerable amount of discretion in deciding whether to grant or deny parole.

Types of Conditional Release

The specific conditional release options that are available to inmates depend on the length of the prison sentence and its jurisdictional status (that is, federal or provincial/territorial). There are four types of conditional release available to inmates incarcerated in federal or provincial/territorial correctional institutions:

1. *Temporary absence.* This is generally granted by the institution and can begin soon after admission and extend to the end of the sentence. It may involve the use of electronic monitoring.
2. *Day parole.* This is granted by parole boards and is available as early as the one-sixth point in a sentence.
3. *Full parole.* This is also granted by parole boards and is available at the one-third point in a sentence.
4. *Statutory release.* This is granted to federal inmates by the Correctional Service of Canada and is available at the two-thirds point in a sentence.

Note that while inmates are *eligible* for various types of conditional release at different stages of their sentence, their application for release may not succeed. Figure 8.2 indicates that the numbers of releases on provincial parole, federal day parole, and federal full parole have been declining over the past decade.

Inmates may also be released from confinement without any conditions or supervision. This is referred to as **cold turkey release,** and it occurs in the following circumstances:

cold turkey release
discharge of an inmate from custody without conditions or supervision

- Provincial inmates may be discharged from confinement at the two-thirds point in their sentence (this is known as the "release date with remission" or "probable discharge date").
- Federal inmates are released at the warrant expiry date.

The various provincial/territorial and federal release options are discussed in depth later in the chapter.

FIGURE 8.2 Releases of Inmates from Correctional Facilities, by Type of Release, 1994–95 to 2003–04

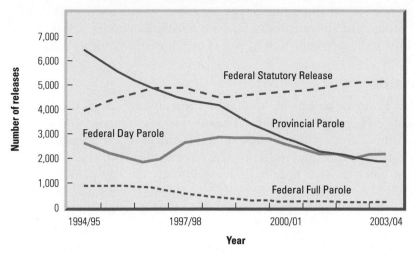

Source: Adapted from the Statistics Canada publication, *Juristat*, Adult Correctional Services in Canada, 2002/03, vol. 24, no. 10, Catalogue 85-002, October 27, 2004, p. 14.

Probation versus Parole: What Is the Difference?

Many people confuse probation and parole or are not clear on the difference. Parole involves two elements: early release and post-release supervision. Probation, as seen in Chapter 6, also involves supervision of offenders in the community, and both can involve conditions and restrictions on liberty. However, there are important differences:

1. Probation is ordered by the judge at time of sentencing, can be a term of up to three years in duration, and can follow a provincial prison term. Parole, in contrast, is granted only by a parole board and applies only to prison inmates. It allows offenders to serve a portion of their term of imprisonment in the community. Parole begins when a parolee is released and ends at the warrant expiry date. The length of time on parole is determined by the length of the prison sentence and by the timing of release by the parole board.
2. Violation of the conditions of the parole can result in a suspension of the parole, which means that the offender is returned to the correctional institution and may have to serve all or a portion of the remaining sentence. In contrast, violation of, or non-compliance with, a condition of probation cannot automatically lead to reincarceration. A new charge is laid, for breach of probation, and the probationer is brought back to court.
3. Parole conditions are easier to enforce than probation conditions.

To add to the confusion between probation and parole, some provincial/territorial inmates have a probation order attached to their sentence. For example, an offender sentenced to six months might be paroled after two months, be on parole for the remaining four months, and then start a proba-tion term. Recall from Chapter 6 that a probation order can be attached to a *federal* sentence *only* if the sentence is exactly two years' confinement. This probation order—the only instance in which a federal offender will be found on probation—follows completion of parole or release on statutory release.

Parole Boards

The decision to release an inmate into the community is made by the National Parole Board (NPB) or by the provincial parole boards in Quebec, Ontario, and British Columbia. These parole boards are independent of the corrections sys-tems. Board members are appointed by order in council and serve fixed terms.

Provincial parole boards hear inmate applications for day parole and full parole and hold post-suspension hearings for offenders who have had their parole suspended for violating the conditions of release or committing a new

Bob Krieger/Artizans.com

offence. Hearings are generally presided over by two parole board members. Provincial boards also hear the parole applications of federal inmates who are serving their sentences in provincial institutions on exchange-of-services agreements.

The National Parole Board is an independent division within the Ministry of the Solicitor General. NPB members are centralized in five regional offices and travel to the institutions for in-person hearings with federal offender parole applicants. The NPB also hears the conditional release applications of provincial inmates in those provinces which do not have provincial parole boards—all jurisdictions except British Columbia, Ontario, and Quebec—and in Yukon, the Northwest Territories, and Nunavut. The mechanics of parole hearings (both federal and provincial/territorial) are discussed later in the chapter.

RELEASE OPTIONS FOR PROVINCIAL/TERRITORIAL INMATES

Recall that the vast majority of inmates are confined in provincial/territorial correctional institutions. All provincial offenders are released before their warrant expiry date in one of three ways: temporary absence, parole, or discharge. Figure 8.3 illustrates the timeline for an eighteen-month sentence.

Temporary Absences

Temporary absences (TAs) are the most common type of conditional release for provincial inmates, although their use varies widely across jurisdictions. For example, British Columbia has eliminated TAs, while Ontario offers six types of temporary absence:

- humanitarian
- medical
- program/rehabilitative

FIGURE 8.3   Timeline for an Eighteen-Month Provincial Sentence

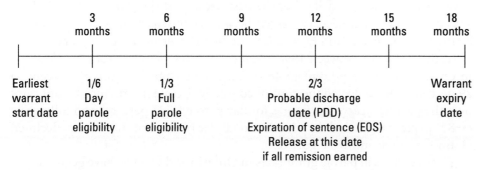

- stay in a community residential agency
- immediate temporary absence (ITA)
- extended temporary absence (ETA) (for longer term inmates)

Offenders released on a TA may be required to live in a community residential facility or live in their own residence and participate in programs offered by community agencies such as the John Howard Society. Under Alberta's Surveillance Supervision Program, for example, unemployed offenders must report daily to an attendance centre for programs or community service and be at home from 6 p.m. to 6 a.m. every day. As with any type of conditional release, violation of TA conditions can result in revocation and the offender's return to the institution. Ontario is among those provinces that use electronic monitoring (discussed in Chapter 6) as a TA condition.

Supporters of temporary absences cite the benefits of serving a sentence in a community setting; they also point out that their use reduces prison overcrowding.

Provincial Parole

Parole is a form of conditional release that is granted at the discretion of a parole board. Provincial inmates may apply for parole after serving one-third of their sentence in a correctional facility. The majority of provincial parolees are in the three provinces that have their own parole boards: Quebec, Ontario, and British Columbia. In the remaining provinces and in the three territories, inmate applications for day parole and full parole are made at hearings before one member of the National Parole Board (NPB). Provincial inmates who are granted day parole or full parole by the NPB are supervised by CSC parole officers; those who are granted parole by boards in British Columbia, Quebec, and Ontario are supervised by probation officers acting in the capacity of parole supervisors.

In accepting parole, provincial inmates forfeit all accumulated remission and so are subject to supervision until warrant expiry. Successful applicants must abide by specified conditions and report to a parole officer. In some provinces, parolees are required to observe a curfew and reside in a residence that is technically suitable for electronic monitoring.

Overall, the numbers of offenders being released on day parole or full parole have decreased substantially over the past decade. At the provincial level, there has been a 73 percent decrease, largely because of the sharp decline in grant rates by the Ontario Parole and Earned Release Board over the past ten years.

In the five years between 1998–99 and 2002–03, parole grant rates declined in Quebec (from 69 percent to 48 percent) and Ontario (from 33 percent to 29 percent); this reflected the continued steady decline in provincial grant rates over the past ten years. The figures for day parole (−35 percent) and full parole (−82 percent) granted by the National Parole Board have also declined (Johnson, 2004; and see Figure 8.2).

The decreases in parole grant rates in Ontario and Quebec have been accompanied by declines in applications for day and full parole in all jurisdictions over the past five years: for full parole, −33 percent in Quebec, −62 percent in

Ontario, and −38 percent for federal offenders (Johnson, 2004: 14). (Figures for the B.C. Board of Parole are not available.)

The precipitous drop in parole grant rates in Ontario has been largely the result of a more conservative approach to releasing offenders. This new approach has been criticized as placing the community at greater risk, in that provincial offenders who are not granted parole and who are released at their remission date—or at the end of their sentence—receive no supervision in the community. The Provincial Auditor of Ontario has noted an increase in the number of inmates who are waiving their right to a parole hearing; these inmates say they are not interested or that they consider parole "a waste of time" (in John Howard Society of Ontario, 2004: 3).

There is considerable disparity in the use of parole between Ontario and British Columbia (see Figure 8.4). This is reflected in the Grant/Deny rates of their parole boards and also in the trends in grant rates. A review of the release decisions of the two boards illustrates how these two jurisdictions utilize parole. Also of interest are the rates of parole revocation for the two jurisdictions.

Discharge

Generally, provincial inmates receive remission ("time off for good behaviour" or "good time") for up to one-third of the sentence. If there is no loss of remission stemming from institutional misconduct, the inmate is released from confinement after serving two-thirds of the sentence. Discharged inmates are not subject to supervision by a parole officer during the remaining one-third of the sentence. A release on discharge can occur by default (because the sentence is so short), by choice (because the inmate did not apply for conditional release), or because all applications for a TA or a day or full parole were denied.

FIGURE 8.4   Provincial Parole: A Tale of Two Provinces

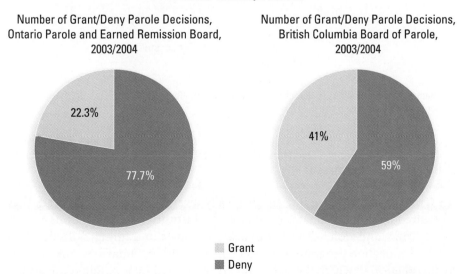

Parole Grants/Denials

Number of Grant/Deny Parole Decisions, Ontario Parole and Earned Remission Board, 2003/2004

22.3%
77.7%

Number of Grant/Deny Parole Decisions, British Columbia Board of Parole, 2003/2004

41%
59%

■ Grant
■ Deny

What do you think?

What factors might contribute to the differences in the parole grant rates between the Ontario Parole and Earned Remission Board and the British Columbia Board of Parole?

FIGURE 8.4   Provincial Parole: A Tale of Two Provinces (Continued)

Parole Grant Rates, Ontario Parole and Earned Remission Board, 1993–94 to 2003–04, and British Columbia Board of Parole, 1998–99 to 2003–04

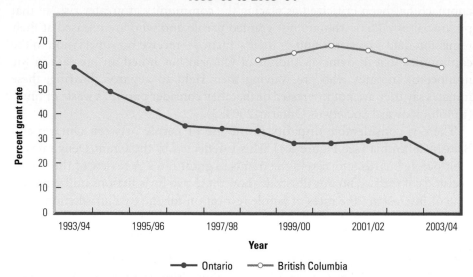

Parole Completion/Revocation

Parole Completion/Revocation, Ontario Parole and Earned Remission Board, 2003/2004

Parole Completion/Revocation, British Columbia Board of Parole, 2003/2004

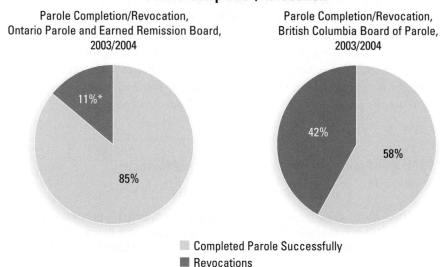

☐ Completed Parole Successfully
■ Revocations

*Remaining 4% committed further offences (3.0%), serious offences (1.1%) or involved termination of parole (.03%).

In Ontario, inmates are not automatically awarded remission; instead, they are required to earn it. Inmates must abide by institutional prohibitions against alcohol, drugs, and violence and must also participate in training, skills development, and education programs offered in the institution. Ontario is currently the only jurisdiction in which the parole board is involved in assessing remission.

RELEASE OPTIONS FOR FEDERAL INMATES

The planning process for releasing federal inmates begins during the first five days of the inmate's admission to the corrections system. To facilitate this process, each federal correctional institution has a reintegration manager. Early in the federal inmate's sentence, a reintegration potential rating is established, based on information gleaned from a number of assessments designed to identify the level of risk presented by the offender as well as his or her needs. This rating places the individual inmate in one of three categories: high, medium, or low reintegration potential. Most inmates with high reintegration potential will not require core programming. Box 8.1 sets out the various release options and eligibility criteria for federal offenders.

Figure 8.5 presents the conditional release eligibility dates for federal offenders (except those serving life sentences). The timeline for a six-year sentence is illustrated in Figure 8.6. In addition, there are conditional release eligibility dates for inmates who are serving life sentences. As noted earlier in the chapter, parole eligibility means exactly that: an inmate is *eligible* for parole. If parole is denied, the inmate can apply again after a specified period.

focus

BOX 8.1

Release Options for Federal Offenders

Temporary Absences

- TAs are usually the first type of release an offender may be granted.

- A TA may be escorted (ETA) or unescorted (UTA).

- TAs are granted so that offenders can receive medical treatment, renew contact with their family, undergo personal development and/or counselling, and participate in community service work projects.

- TA decisions are usually made by the prison administration.

Eligibility

- Offenders may apply for ETAs at any point in their sentence.

- UTAs vary depending on the length and type of sentence. Offenders classified as maximum security are not eligible for UTAs.

- For sentences of three years or more, offenders are eligible to be considered for UTAs after serving one-sixth of their sentence.

- For sentences of two to three years, UTA eligibility occurs six months into the sentence.

- For sentences under two years, eligibility for temporary absence is under provincial jurisdiction.

- Offenders serving life sentences can apply for UTAs three years before their full parole eligibility date.

- Federal offenders on UTAs may participate in work release programs, which usually involve working at a paid or volunteer job outside the institution and returning at night.

Day Parole

- Day parole prepares offenders for release on full parole or statutory release by allowing them to participate in community-based activities.

- Offenders on day parole must spend the evening hours in an institution or halfway house unless authorized to do otherwise by the National Parole Board.

Eligibility

- Offenders serving sentences of three years or more can apply for day parole six months prior to their full parole eligibility date.
- Offenders serving life sentences can apply for day parole three years before their full parole eligibility date.
- Offenders serving sentences of two to three years are eligible for day parole after serving six months of their sentence.
- Offenders serving sentences under two years are eligible for day parole after serving one-sixth of their sentence.

Full Parole

- The offender serves the remainder of the sentence under supervision in the community.
- The offender must report to a parole officer on a regular basis and advise him or her as to any changes in employment, involvement in treatment programs, or in personal circumstances.

Eligibility

- Most offenders (except those serving life sentences for murder) are eligible to apply for full parole after serving either one-third of their sentence or seven years.

- Offenders serving life sentences for first degree murder are eligible after serving twenty-five years.
- Eligibility dates for offenders serving life sentences for second degree murder range from ten to twenty-five years and are set by the court during sentencing.

Statutory Release

- Most federal inmates are automatically released after serving two-thirds of their sentence if they have not already been released on parole.
- Statutory release is not the same as parole because the release decision is *not* made by the National Parole Board.

Eligibility

- Offenders serving life or indeterminate sentences are not eligible for statutory release.
- The Correctional Service of Canada may recommend that an offender be denied statutory release if it is determined that the offender is likely to commit an offence causing death or serious harm to another person; commit a sexual offence involving a child; or commit a serious drug offence before the end of the sentence. In such cases, the National Parole Board may intercede and detain the offender until the end of the sentence or add special conditions to the statutory release plan.

Source: National Parole Board, *Facts: Types of Release* (Ottawa, 1997). Reproduced with the permission of the National Parole Board.

Temporary Absences

Because they serve longer periods of incarceration than their provincial counterparts, federal inmates tend to be released in gradual stages, beginning with escorted or unescorted temporary absences. Decisions about ETAs and UTAs are usually made at the institutional level. Inmates are also eligible for work releases and community development releases.

FIGURE 8.5   Overview of Conditional Release Eligibility Dates for Federal Offenders

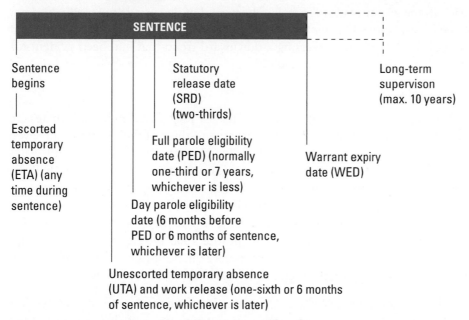

FIGURE 8.6   Timeline for a Six-Year Federal Sentence

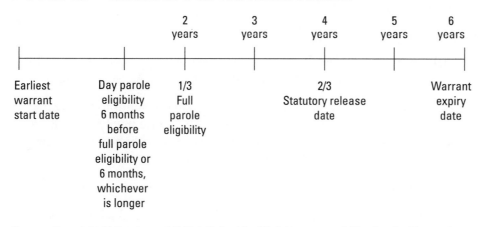

Day Parole

Non-violent offenders who are serving their first federal term are eligible for day parole at either six months into their sentence or the one-sixth point, whichever is longer. Most other offenders are eligible for day parole six months before their full parole eligibility date (with some variations for special categories of offenders such as lifers). Inmates on day parole typically live in a community residential facility (see photo, page 347), and most apply for full parole when they reach the one-third point in their sentence. Day parole is meant to aid transition and thus is seen as a short-term option to be used immediately prior to the granting of full parole. The period of day parole normally does not exceed six months. During the years 2002/03 to 2003/04, there was a 4 percent increase in day paroles, reversing a downward trend that had been evident for more than a decade (Beattie, 2005:13).

Full Parole

Full parole is the most common type of release for federal inmates. Inmates are eligible to apply for full parole after serving one-third of their sentence or seven years, whichever is less. The exceptions involve situations where a judge has made a "judicial determination," in which case the offender must serve half the sentence (but not more than ten years) before becoming eligible for parole.

Full parole is granted by the National Parole Board in 45 percent of considered cases. There was a slight upward trend in the federal parole grant rate in 2003/04, reversing several years of steady decline (Beattie, 2005, see Figure 8.2). On average, federal offenders serve about 40 percent of a sentence prior to being released on parole. Additional features of federal full parole include the following:

- A higher grant rate (72 percent) for female offenders than for male offenders (45 percent). Also, women serve a lower proportion of their sentence than men before being released on parole.
- An increase in the grant rate for Aboriginal offenders, although Aboriginal offenders serve a higher proportion of their sentence before being released on parole (Public Safety and Emergency Preparedness Portfolio Corrections Statistics Committee, 2004).

Accelerated Parole Review

Accelerated parole review (APR) is a relatively new practice in corrections. Section 126 of the Corrections and Conditional Release Act states that

> if the Board is satisfied that there are no reasonable grounds to believe that the offender, if released, is likely to commit an offence involving violence before the expiration of the offender's sentence according to law, it shall be directed that the offender be released on parole.

APR (on either day parole or full parole) applies to all inmates who

- are first-time federal offenders (the number of provincial admissions is not an issue);
- are not serving a life sentence;
- have not been convicted of murder or any CCRA Schedule I offence that was proceeded on by way of indictment;
- did not attempt or were not accessories to a CCRA Schedule I offence; *and*
- have neither been convicted of a CCRA Schedule II offence nor had their parole eligibility set by the judge at one-half of the sentence. (Note: Schedule I of the CCRA includes serious violent offences. Schedule I offences include attempted murder, impaired driving causing death, and manslaughter. Schedule II offences include serious drug offences.)

A community-based residential facility for offenders on day and full parole

Although inmates who meet these criteria must still serve one-third of their sentence before being eligible to apply, the parole approval process is streamlined. The NPB makes a "paper decision"—that is, no hearing is held and the decision is made by one parole board member.

Judicial Review

One of the most controversial provisions in the Criminal Code is Section 745. Often called the "faint hope clause," the judicial review provision allows offenders who are serving a life sentence with no eligibility for parole for at least fifteen years to apply to the court for a reduction of the parole eligibility period. The intent of this provision is to

- give inmates an incentive to participate in rehabilitation programs;
- assist corrections personnel in managing long-term offenders;
- allow offenders who no longer present a danger to the community to return to society and be productive citizens; *and*
- provide an option for addressing the unique circumstances of elderly offenders.

Inmates must serve fifteen years of their sentence before applying for judicial review. Offenders who have been convicted of multiple murders are not eligible to apply. All applications made under Section 745 are first screened by a superior court judge to determine whether there is a reasonable prospect of success.

The judicial review hearing must take place in the community where the crime was committed. A jury of community residents determines whether the parole eligibility date should be reduced (not whether the offender should be released). At the judicial review hearing, both the offender and the Crown counsel present information relevant to the application (including the offence that was committed, the offender's conduct during the fifteen years of confinement, and assessments of the offender's character). The onus is on the offender to persuade the jury to reduce the parole eligibility date. Both the offender and the Crown counsel may call and question witnesses and experts, and the offender must testify at the hearing. In its determination, the jury may

- make no change in the parole eligibility date, but set a date when another application can be made;
- make no change in the parole eligibility date, and deny the possibility of future applications;
- reduce the number of years of incarceration without eligibility for parole; *or*
- terminate the ineligibility for parole and make the offender immediately eligible to apply for parole.

If there is a decision to reduce the parole eligibility date, all members of the jury must be in agreement.

Only about one-quarter of eligible inmates have applied for a hearing, although as of 2004, 81 percent of those who did submit to a judicial review hearing had their period of parole ineligibility reduced (Public Safety and Emergency Preparedness Portfolio Corrections Statistics Committee, 2004: 93). To date, the recidivism rate of those offenders who were subsequently granted parole has been very low: of the ninety-seven offenders released, seventeen have been returned to custody. There is considerable variation across the country in the numbers of applications and the rates of applicant success. Section 745 continues to be the focus of considerable controversy (see Box 8.2).

Statutory Release

Some federal inmates choose not to apply for parole. Other inmates are denied parole each time they apply, while still others have had an earlier release revoked. When these offenders reach the two-thirds point in their sentence, they are eligible for **statutory release.** Under statutory release, federal offenders are subject to parole supervision until their warrant expiry. Statutory releases are an administrative decision of the CSC; as such, they do not involve the National Parole Board. In some cases, however, the CSC may ask the NPB to add a residency condition to the release.

On rare occasions, a statutory release may be denied. If the CSC believes there is a high likelihood that an inmate will reoffend if released, it may apply to the NPB for a detention hearing. The case must meet the following criteria:

- The inmate's offence was listed in Schedule I.
- The offence involved death or serious harm.
- There is good reason to believe that the inmate is likely to commit an offence causing death or serious bodily harm before the warrant expiry date.

What do you think?

1. What is your position on Section 745? Do you think it should be removed from the Criminal Code? Explain.
2. Review the points/counterpoints in Box 8.2. What are the strongest arguments made by each side in the debate?

statutory release
the release of federal inmates, subject to supervision, after two-thirds of the sentence has been served

focus

BOX 8.2

Point/Counterpoint: Section 745 (Faint Hope Clause)

Point

Supporters of Section 745 argue that it

- reflects enlightened corrections policy;
- provides for the rehabilitation of offenders, the majority of whom will not commit another crime after their release from confinement;
- does not guarantee that the offender will be released, but rather makes the successful applicant *eligible* for parole consideration; *and*
- requires that successful applicants who are granted parole remain under supervision for the rest of their lives.

Counterpoint

Detractors counter that Section 745

- undermines the potential impact of life sentences as a general and specific deterrent to crime;
- results in the revictimization of crime victims and their families during the judicial review hearing and subsequent parole hearings, as well as at the time of the offender's release on parole;
- makes a mockery of the twenty-five-year minimum sentence; *and*
- results in violent offenders being released back into the community.

Most inmates whose cases are referred to the NPB for detention hearings are serving longer sentences and have been convicted of a sexual offence and/or assault; one in four referred inmates is Aboriginal. Female offenders are rarely the subject of detention hearings because they tend not to have extensive criminal histories.

If the NPB is satisfied that the inmate meets these criteria, it imposes a measure called **detention during a period of statutory release.** In 2003–04, there were 303 detention reviews. Since 1991–92, 91 percent of initial detention reviews have resulted in a decision to detain. The majority of offenders who are detained are male. Over the past five years, only seventeen women have been referred for detention, and thirteen of these were subsequently detained. Aboriginal offenders, who account for 18 percent of the federal inmate population, compose 30 percent of all offenders detained over the past five years (Public Safety and Emergency Preparedness Portfolio Corrections Statistics Committee, 2004: 91).

An alternative to detention is **one-chance statutory release,** whereby an offender who violates the conditions of the statutory release is required to serve the remainder of the sentence in confinement. Yet another option is requiring that the released offender live under supervision in a community facility.

> **detention during a period of statutory release**
> confinement of high-risk federal offenders until the warrant expiry date

> **one-chance statutory release**
> a release option whereby offenders who violate the conditions of a statutory release are required to serve the remainder of their sentence in confinement

SPECIAL CATEGORIES OF OFFENDERS

"Lifers" and Parole

A life sentence does not necessarily mean the offender will spend the rest of his or her life in prison; rather, it means that the offender will be under some type of supervision—either in prison or on parole—until death (except in the rare cases where a pardon is granted). The exact amount of time "lifers" must serve before being eligible for parole depends on the offence for which they were convicted and the date on which they were sentenced. Every offender sentenced to a life term is subject to the laws that were in place at the time of the offence or at the time of the sentencing (whichever is more favourable to the offender). Today, the Criminal Code directs that such offenders are eligible to apply for parole:

- in the case of first degree murder, after twenty-five years;
- in the case of second degree murder, where the offender had previously been convicted of a culpable homicide, twenty-five years; *and*
- in the case of second degree murder, between ten and twenty-five years, as ordered by the judge at the time of sentencing.

Anyone sentenced to life for any other type of offence is eligible for parole after seven years.

Inmates serving life terms must apply to the NPB for ETAs unless the pass is for a medical reason. Once they are within three years of the parole eligibility date, lifers can apply—as would any other inmate—to the head of the institution. Lifers are not eligible for UTAs or day parole until they are within three years of their parole eligibility date. Of course, the NPB can turn down a lifer's application for parole. Such is likely to be the case for high-profile serial killers, who may remain in prison until death. Lifers are not eligible for statutory release.

Long-Term Offenders

As noted in Chapter 6, sentencing judges can designate certain federal offenders as "long-term offenders" and attach to the end of their sentence a long-term supervision order. This order, which can run as long as ten years, begins at warrant expiry and as such does not affect the inmate's parole eligibility.

The CSC is responsible for supervising long-term offenders, and the NPB can set standard and special conditions to which the offender must adhere during the supervision period. An offender who violates one or more conditions of the order may be readmitted to custody for up to ninety days or admitted to a halfway house or mental health facility. The CSC can also refer the matter to the NPB, which may cancel the suspension (and release the person), cancel the suspension and add new conditions, or recommend (with the approval of the provincial attorney general) that the offender be charged before a judge with "breaching long-term supervision" without reasonable excuse. This offence carries a maximum penalty of ten years.

Dangerous Offenders

An inmate who has been declared a "dangerous offender" (see Chapter 6) by a criminal court serves an indeterminate prison sentence and can be released only by the National Parole Board. Some 200 people—all but two of them men—have been declared dangerous offenders. The majority are sex offenders, and many have been diagnosed with an antisocial personality disorder (Bonta, Harris, and Carrière, 1998).

Statutory release is not available to dangerous offenders, and the NPB will not release a dangerous offender unless it is satisfied that the community would not be put at risk. In practice, the NPB rarely releases dangerous offenders.

VICTIMS AND CONDITIONAL RELEASE

The increasing involvement of crime victims in the criminal justice system extends to corrections systems as well. The Corrections and Conditional Release Act and provincial/territorial legislation have established a number of victims' rights, including the following:

- At their request, victims can be advised of eligibility dates, parole board decisions, and release status.
- Victim impact statements and concerns about post-release safety are considered by parole boards and the parole service.
- Victims can attend parole hearings.
- Anyone can apply to the Decision Registry for written copies of NPB decisions; some 2,000 requests for case-specific information are made each year, with victims representing the largest number of applicants.

In some jurisdictions, victims have the right not only to attend parole hearings but also to address the parole board. Under British Columbia's Victims of Crime Act (1995), for example, victims can present oral statements to the province's Board of Parole. Crime victims in Ontario also have the right to attend and participate in provincial parole board hearings. Victims must address their statements directly to the parole board, not the inmate. Victims can also attend and make verbal representations at NPB hearings.

Victims who do not wish to attend the parole hearing may submit video-taped or written statements. A victim can request that this submission not be divulged to the inmate. Following such a request, the parole board must tell the inmate it is in receipt of confidential information and advise the inmate of the general nature of that information.

Despite the recently established victims' rights initiatives, victims often lack information about the corrections process. They are often unaware, for example, that release decisions hinge on the determination as to whether or not the offender will reoffend prior to warrant expiry, and that the Corrections and Conditional Release Act directs parole boards to make the *least restrictive* determination required to ensure the protection of society. Crime victims are also often unaware that most sentences of incarceration imposed by Canadian criminal courts are for one month or less, and that for federal offenders, most federal sentences are for less than three years. Lacking this knowledge, victims have been surprised by encounters with their perpetrators only weeks after they witnessed the passing of sentence in court.

Crime victims can attempt to protect themselves from such surprises by asking the parole board and/or correctional officials to advise them of the timing of the release and other specifics. In Manitoba, correctional officials attempt to locate and contact victims. In Ontario, qualifying victims who register with the Victim Notification Service (VNS) receive a personal identification number (PIN) that allows them to access, via telephone, case-specific information about an offender in the provincial correctional system.

PAROLE BOARD DECISION MAKING

The determination of a parole board to grant or deny parole is one of the most important decisions made in the corrections process. In making this decision, the parole board must consider the following questions:

- Has the inmate served a long enough sentence relative to the severity of the crime?
- Has he or she been in prison long enough to have learned a lesson?
- Will he or she commit another crime if released?
- Would the community object to the release?
- Has the inmate taken full advantage of self-improvement programs?
- Has he or she complied with the rules while incarcerated?

Even inmates who present risks may be released if the parole board is satisfied that the conditions exist to "manage" that risk. This makes the parole hearing an exercise in behavioural prediction and short-term risk assessment.

The Parole Hearing

Inmates have a legislated right to a hearing before a parole board at specific times during their confinement. As noted earlier, inmates become eligible for day parole and full parole after serving one-sixth and one-third of their sentence, respectively. At the hearing, the inmate may be accompanied by an assistant. This person may be a lawyer, spouse, parent, employer, sibling, or friend. At federal hearings, the assistant is usually the inmate's case management officer. After the inmate has been interviewed, the assistant has an opportunity to speak directly to the parole board.

Prior to the hearing, the parole board members review the inmate's file and make notes on key points. Among the materials reviewed are the release plan (described earlier in the chapter), reports on the inmate's institutional performance and participation in treatment programs, classification documents, and a risk/needs assessment. During the hearing, the board members ask the inmate about the release plan. They also consider the community assessment, the recommendations of the probation officer (for provincial offenders) or institutional parole officer (for federal offenders), and any recommendations that special conditions be attached to the release. A great deal of attention is focused on the inmate's version of the offence. The board members look for some insight into why it was committed; the steps the inmate has taken to address factors associated with the criminal behaviour; evidence, in cases where the offence was committed to support a drug addiction, that the inmate has completed a substance abuse management program and a relapse prevention program; and inmate expressions of remorse and empathy for any victims of the offence.

After the question portion of the hearing, the inmate is asked to wait outside the hearing room while the board members deliberate. In some cases, the board defers a decision pending the gathering of further information. If the board decides to deny parole, it must provide the inmate with its reasons for doing so. These reasons (which must be presented in the form of a written statement) might include the inmate's failure to participate in or complete a treatment program, confirm acceptance into a halfway house or residential drug treatment facility, or develop educational or employment plans. Inmates who receive this information can take appropriate remedial action, thus improving their prospects at a subsequent parole hearing.

If parole is granted, a certificate of parole is prepared. This certificate specifies the mandatory conditions of the release as well as any special conditions the inmate must abide by. Mandatory conditions include reporting regularly to a parole officer, obeying the law, and securing permission from the supervising parole officer prior to leaving a specified geographic area. Special conditions are designed to reduce the risk of reoffending based on an understanding of the factors that precipitated past offences; they include such provisions as abstaining from intoxicants, completing a treatment program, and avoiding contact with children. A copy of the certificate of parole used by the British Columbia Board of Parole is reproduced in Figure 8.7. The observations of a journalist who attended an NPB hearing in the Ontario region are presented in Box 8.3.

The parole hearing can be a very daunting experience for inmate-applicants, who may have limited communication skills and be intimidated by the stature and questions of the board members. The often vast differences between the backgrounds and life experiences of the inmate applicant and the parole board members may make it difficult for the two parties to engage in a conversation about the factors related to the offence, the offender's experience and activities while in custody, and the proposed parole plan.

Criticisms of Parole Board Decision Making

The decision making of parole boards has been a subject of considerable controversy, especially when an inmate on conditional release commits a heinous crime (Roberts, 1998). The three most commonly voiced criticisms of parole board decision making are discussed below.

1. *The appointment of parole board members.* Parole board members are not required to have any special training or expertise in law, criminology, psychology, or corrections. In the past, board members were often appointed

FIGURE 8.7 Sample Certificate of Parole

AMENDMENT: YES / NO

Office of the Chair
#1810 - 4720 Kingsway
Burnaby, BC V5H 4N2
Telephone: (604) 660-8846
Fax: (604) 660-2256

CERTIFICATE OF PAROLE - BRITISH COLUMBIA BOARD OF PAROLE

Prisoners and Reformatories Act, Canada
Corrections and Conditional Release Act, Canada

F.P.S. No. _____
D.O.B. _____
C.S. No. _____

This is to certify that _____ who is serving a term of imprisonment in _____ _____ , for the following offences: _____ was granted parole effective on _____ which will expire on _____ provided the parole is not sooner suspended or revoked.

CONDITIONS OF PAROLE

1. Report immediately upon release to the parole supervisor on _____ at _____ _____ Phone No. _____ .

2. Report to the parole supervisor a minimum of _____ per _____ in person or more frequently as directed by the parole supervisor.

3. Remain within the area of British Columbia, namely: _____ **or as directed by the Parole Supervisor.**

4. Obey the law and keep the peace.

5. Immediately inform parole supervisor on being questioned or arrested by the police.

6. Carry the release certificate at all times and produce it on request by the peace officer or parole supervisor for identification.

7. Report to the police at the time and place as directed by the parole supervisor, if so directed.

8. Advise the parole supervisor on release of the address of residence, namely: _____ , and thereafter obtain permission from the parole supervisor for:

    (i) a change in address of residence

    (ii) a change in occupation including a change in employment, vocational or educational training or in volunteer work.

    (iii) a change in living arrangements or financial situation and a change in family situation, or

    (iv) a change that may reasonably be expected to affect compliance with conditions of the release.

9. Not own, possess or have control of a weapon as defined in section 2 of the Criminal Code except as authorized by the parole supervisor.

10. Special Conditions:

    a) ❑ Abstain from all intoxicants.

    b) ❑ Submit to a breathalyzer and/or urinalysis test on demand of the Parole Supervisor or Peace Officer where reasonable and probable grounds exist to believe the offender is violating the condition to abstain from intoxicants.

    c) ❑ Do not enter any premises where the primary commodity for sale is alcoholic beverages.

    d) ❑ Seek and maintain employment.

    e) ❑ Actively participate in, and complete such substance abuse treatment and/or counselling as you may be directed to attend by the Parole Supervisor.

    f) ❑ Actively participate in, and complete such sex offender treatment and/or counselling as you may be directed to attend by the Parole Supervisor.

    g) ❑ Do not have contact directly or indirectly with the victims:

    h) ❑ Do not associate with the following person(s), _____ except as approved by the Parole Supervisor.

    i) ❑ Do not associate with known criminals, except as approved by the Parole Supervisor.

    j) ❑ Do not operate a motor vehicle on a public road.

    k) ❑ Attend and complete _____

    l) ❑ Reside at _____

    m) ❑ Not to be in the company of children under the age of _____ unless accompanied by a responsible adult.

    n) ❑ Abide by the attached electronic monitoring conditions.

    o) ❑ Abide by a twenty-four hour curfew except for the following purposes (employment, education, counselling etc.) and as approved through the written permission of the Parole Supervisor. Such permission is to be carried on your person at all times when not in your place of residence outside of normal curfew hours.

    p) ❑ Agree to permit staff members of the Corrections Branch to enter your residence at any time to verify equipment operation and otherwise ensure compliance with the terms of the Parole Certificate.

FIGURE 8.7   (Continued)

q) ☐ Other _____

_____

This Certificate of Parole is granted under the authority of the British Columbia Board of Parole subject to the Conditions set out on the Certificate. Failure to comply with any of these Conditions may result in the parole being suspended and/or revoked.

Issued at _____, BC   this _____ day of _____, _____.

_____   _____
Member of the BC Board of Parole   **Member of BC Board of Parole**

I have read and understand the contents of this document and accept all the above conditions and will abide by them. I acknowledge that I am the person to whom the statements contained herein refer and agree that the said statements are correct. I further acknowledge that I have been made aware of the forfeiture of earned remission upon my accepting this parole release.

Signed _____

Witnessed _____

Released from _____   on _____

Officer in Charge _____

cc: BC Board of Parole (*original*)
    Parolee
    Inmate Master File
    Operational Communications Centre, RCMP
    Community Supervisor

because of their political connections. Although in recent years there has been a decline in patronage appointments, concerns were expressed in 2005 when the Public Security Minister in Quebec appointed twenty new board members who were not on the list of nominees vetted by the parole board.

2. *The absence of clearly defined release criteria.* A source of difficulty for parole board members is that the Corrections and Conditional Release Act provides only very general criteria for conditional release decisions. Although board members have access to extensive information on the inmate (police reports, recommendations from the sentencing judge, risk/needs and community assessments, and so on), they receive no guidance on how to prioritize or weigh the importance of the information. In this context, a lack of empirical findings as to the causes of criminal behaviour may hinder effective decision making. In addition, the discretion exercised by board members may result in disparities in parole decisions across jurisdictions or even within the same jurisdiction.

3. *The lack of feedback about the results of decisions.* What happens to offenders while they are under supervision in the community and at the end of supervision? There are few, if any, mechanisms in place for parole board members to receive feedback on the ultimate outcomes of their decisions. Generally, parole board members become aware of an inmate's behaviour on conditional release only if the inmate commits a high-profile crime. This can give board members a skewed perspective and even lead to an increase in the denial of inmate applications for conditional release.

focus

BOX 8.3

Inside Hearings of the National Parole Board (Ontario Region)

Like Santa, a parole board is supposed to know who's been good or bad, and so by the time the hearing arrives, parole panellists (called directors) already know more about the prisoner than they perhaps care to—and a lot of it is not very nice.

On this particular day, three board directors—former prison warden Kenneth Payne, career correctional-service employee Sheila Henriksen and social worker John Brothers—have the final say.

Armed with documents describing the parole-seeker's criminal history, psychological assessments, education, family situation, other relationships, behaviour while in prison and the recommendation from Correctional Services Canada, the members try to evaluate what risks these individuals pose to society and determine if that risk is manageable in the community . . .

The first up to bat on this day is a 36-year-old Kingston man who was sentenced to life in prison on a charge of second-degree murder for killing a friend in a dispute over a woman.

At 8:30 a.m., the slight, frail-looking man is waiting outside the hearing with his case management officer and a university law student as the morning announcements play over the intercom. The atmosphere is weirdly like high school.

When the door to the hearing room opens, the brief window of opportunity has arrived that the convict has been waiting for, make-or-break time. The parole panel will soon begin its grueling interview. No holds are barred, and no part of a convict's life is off limits.

Sitting a couple of metres from the convicts, looking them in the face, panel members have to sift through what they're hearing and judge

what is sincere and what is contrived, remembering that people seldom get to this point in their lives by being totally honest.

The members take in the convict's appearance and mannerisms, dissect his answers, ask questions in different ways to get a better read and compare the answers to facts provided by the professionals.

They often caution the convicts against lying, because their replies must be consistent with what's in their files.

This morning, the murderer from Kingston slouches, his hair slicked back tightly like people wore in the 1950s. He is wearing dark clothes, a tweed sports jacket and unmatching light-coloured socks.

The case management officer sits at a table to his left . . . Right as the hearing starts, the convict withdraws his application for full parole. He says day parole will suffice.

The man has spent time in a number of jails and prisons since the murder. He stares straight ahead as the case management officer outlines his criminal record, all of it minor and non-violent up to the killing. He also relates how the convict was granted full parole twice, in 1992 and again in 1995, and violated it both times.

A doctor's report rates the probability he will reoffend within a year of release at 40 per cent, saying he suffers from an anti-social personality disorder.

A panel member asks why he withdrew his application for full parole. In a low, frail voice, the convict states the obvious: "In a realistic view, I don't think you guys would send me to full parole."

A member jokes, "You have already done some of our work for us." Mr. Payne asks the convict about the bad choices he has made through his life, and there are many. The convict says his worst was getting involved in a relationship with the woman he killed for and, as he puts it, his "negative thinking."

The focus shifts to what he might have learned from his failures. "I needed to change the way I view things," the convict says. "I used to go through distorted thinking patterns. I have a problem over-complicating things.

"I used to take on other people's problems and make them my own." In discussing an anger-management course he has just repeated, he is asked: "When was the last time you felt really angry?" "When I got the letter from the parole board that media would be at my hearing," he answers. He adds, "Nothing personal," as he turns toward the observers behind him.

Asked about the killing, the convict says he doesn't recognize the man who did it, that there are "some pretty blank spots surrounding that time."

Ms. Henriksen questions his integrity. "I have got a sense you have an ability to fool people," she said. He replies: "Sitting in the position I am in, it doesn't seem right for me to say, 'Trust me.'"

But the board chooses to trust him anyway. Following brief deliberations it grants the man once-a-month [unescorted temporary absence]. If he does well on those, the next step will be day parole and then full parole without any further hearing.

"The board is satisfied you have benefited from our programs," says Mr. Payne.

The convict thanks the directors and, as he leaves, passes the bank robber waiting outside. And the process repeats itself . . .

The day ends with the case of the Stratford father, a 29-year-old first time offender who smashed up his truck after a night of partying and nearly killed his passenger. His sentence was two years for criminal negligence causing bodily harm. He has served about a year.

If there is a common thread among these convicts it is the way they handle stress: Drugs and alcohol are their mainstays.

Oddly enough, the convict doesn't do a very good job of selling his case. Lucky for him it sells itself.

The case management officer gives an exemplary report on his prison behaviour, noting he attends night school and wants to pursue a trade in college.

The convict shakes as he appears before the panel members, who at times try to relax him.

One thing that works against him is a compelling victim impact statement. The victim is suing the convict. "I know he's mad but I have gone and tried to talk with him and all he does is yell at me or make rude gestures when he drives by," the convict explains. "I just wish it was me who got injured that night."

"All I know is I have two young kids I haven't seen in a month and I want to get back to them," says the man. "I can't wait."

Another quick verdict: immediate full parole. The directors deliver their judgment. And then they just hope.

Source: D. Campbell, "A Journalist Goes to Prison to See for Himself How Parole Boards Decide Which Convicts Are Good Risks and Which Ones Are Not," *Ottawa Citizen* (1997, November 3), A3.

THE REINTEGRATION PROCESS

Reintegration has been defined as "all activity and programming conducted to prepare an offender to return safely to the community as a law-abiding citizen" (Thurber, 1998: 14). The goal of reintegration is to avoid recidivism until warrant expiry and, ideally, over the long term as well. The short term is an especially crucial period in that most inmates who reoffend do so in the two-year period following their release from a correctional institution.

reintegration
the process by which an inmate is prepared for and released into the community after serving time in prison

Like parole, reintegration is a process rather than an event. For federal offenders, it begins with the institutional treatment programs described in Chapter 7 and continues with the development of a release plan that sets out where the inmate will live, work or attend school, and (if required) participate in a post-release treatment program. Ideally, there should be continuity between the inmate's participation in institutional programming and the services he or she receives while on conditional release in the community.

The Pains of Re-entry

In Chapter 7, the various pains of imprisonment experienced by offenders who are confined in correctional institutions were discussed. Upon their release from confinement, some offenders may also experience **pains of re-entry** (Griffiths, 2004). The relatively small number of offenders who serve lengthy periods of time in confinement may experience pains of re-entry upon release. Correctional institutions, particularly those at the maximum security level, are closed, highly structured, and artificial environments where an antisocial value system predominates and where inmates have few responsibilities. Upon release, these same individuals are expected to resume life in a community that values independence of thought and an ability to cope with the complexities of daily existence. Life on the outside is everything prison is not: unpredictable, fast-paced, and filled with choices.

In *The Felon*, John Irwin offers what has become a classic account of the re-entry process from the inmate's perspective:

> The ex-convict moves from a state of incarceration where the pace is slow and routinized, the events are monotonous but familiar, into a chaotic and foreign outside world. The cars, buses, people, buildings, roads, stores, lights, noises, and animals, are things he hasn't experienced at first-hand for quite some time. The most ordinary transactions of the civilian have dropped from his repertoire of automatic maneuvers. Getting on a streetcar, ordering something at a hot dog stand, entering a theater are strange. Talking to people whose accent, style of speech, gestures, and vocabulary are slightly different is difficult. The entire stimulus world— the sights, sounds, and smells—is strange. Because of this strangeness, the initial confrontation with the "streets" is apt to be painful and certainly is accompanied by some disappointment, anxiety, and depression. (1970: 113–14)

Imagine the difficulty you would have adjusting to a law-abiding lifestyle in the community if you were a parolee with a grade nine education, a poor record of employment, tenuous or nonexistent family support, a substance abuse problem, and/or few or no non-criminal friends (not to mention a criminal record). Unfortunately, earning a record of positive conduct inside the correctional institution (including completion of various treatment programs) may not adequately prepare you for the challenges of adapting to life in the community.

One parolee, who had been incarcerated almost continuously from age seven to thirty-two, related an incident to your author that illustrates the anxiety and panic that ex-offenders may experience in completing tasks that people in the

pains of re-entry
the difficulties that inmates released from prison encounter as they try to adjust to life in the community

outside, free community take for granted. The situation occurred during his first trip to the grocery store, soon after his release on day parole to live in a halfway house:

> I wanted to buy some groceries, so I went to Safeway. I must have been in the store for hours. There were so many choices, I had no idea of what to put in my cart. Finally, my cart was full and I pushed it up to the checkout counter. The store was really crowded, and I was so focused on deciding what to buy that I hadn't given any thought to the price of the things I was putting in the cart. I think I had about $50 in my pocket. When the cashier rang up the total, it came to over $150. When she told me the total, I just froze. Everyone was looking at me. I stood there for what seemed like an eternity and then, without saying a word, ran out of the store. At the bus stop, my heart was racing and I was sweating. I never went back to that store. And it was a long time before I went grocery shopping for more than one or two items. (Griffiths, 2004: 393–94)

According to a survey of federal and provincial correctional personnel, the most frequently mentioned problems facing offenders upon re-entry are a lack of education and job skills, the absence of family support, poverty, drug and alcohol problems, and low self-esteem (LaPrairie, 1996).

Inmates may spend a considerable amount of time thinking about, planning for, and speaking with fellow inmates about life on the outside. Often, thoughts of release focus on the quick acquisition of long-denied commodities such as heterosexual sex, fast food, and a cold beer. As the release date approaches, however, this positive outlook may be replaced by feelings of anxiety and even fear. These emotions may be especially intense for inmates who have failed on previous releases and/or who have spent a lengthy time in confinement. Of particular concern are "state-raised offenders"—those individuals who have spent most of their youth and adult lives in correctional institutions.

Planning a day without the rigid timetable of prison routine can be a daunting task. A newly released offender can feel like a stranger—inadequate, acutely self-conscious, and convinced that every person on the street can tell from appearance alone that he or she has been in prison. For a released inmate who has no friends on the outside who can be relied on for assistance, protection, and security, the institution may exert a stronger pull than freedom itself. As one ex-offender who had served over twenty years in prison stated to your author: "I have never had the intensity of friendships, the trust, the companionship, in the outside community that I had when I was incarcerated" (personal communication).

Among the challenges facing parolees is finding employment. Criminal records preclude entry into some professions, including those requiring the employee to be bonded (insured). Some employers may have less stringent requirements but still be reluctant to hire someone with a criminal record. The job-seeking parolee may not have suitable clothes for interviews or may not possess job-specific gear such as steel-toed boots and special tools.

Family reunification can be another source of stress. The longer the term of confinement, the less prepared the parolee is to resume family relations.

Financial necessity may force the parolee to live in the parental home (if one exists) longer than desired. Those parolees who reconcile with partners must find a new role in a family unit, which has been functioning without them. Learning to co-parent is a particularly difficult aspect of reintegration for many men.

In an attempt to cope with the pains of re-entry, a parolee may revert to such high-risk behaviour as heavy drinking, drug use, fighting, and spending time with old friends from prison. All the plans to "go straight" can crumble like a New Year's resolution in February. A significant minority of parolees commit a criminal offence within three years of their release.

It is likely that the pains of re-entry are experienced more acutely by offenders who have been incarcerated for lengthy periods of time. The majority of offenders in Canada are sentenced to relatively short periods of time in custody; for them, the key challenge is to avoid relapsing into criminal activities, rather than to adapt to a changed world. As well, offenders are confronted with the need to alter their routine activities, neutralize the potentially negative influence of their peer group, and begin to make different life choices. A tall order, to say the least, especially when the relatively short period of time in custody has provided little opportunity to participate in meaningful change experiences and when the offender has been surrounded by many fellow inmates who are actively involved in the criminal lifestyle.

For long-term offenders, the challenges are very real, and without assistance and support from a pro-social network, the chances for survival in the free community are appreciably reduced. This, even with the modern amenities (such as television) that are available to offenders in correctional institutions, which do little to raise self-esteem, develop interpersonal skills, and prepare the offender for the fast-paced world of the early twenty-first century.

Although the trauma of re-entering the community is likely greatest among those inmates who have served the longest periods of time in custody, offenders released from provincial custody may experience their own pains of re-entry in attempting to alter their behaviour patterns and friendship group and to attain a measure of stability in their lives. These efforts can be equally as challenging, particularly if the newly released offender does not have access to programs, resources, and an alternative, law-abiding family and friendship group.

Reintegration of Special Offender Populations

In Chapter 1 it was noted that special offender populations pose a challenge to the Canadian criminal justice system. This extends to the reintegration of offenders into the community following a period of custody.

Aboriginal Offenders

Aboriginals are overrepresented in provincial/territorial and federal institutions, but they also tend to receive shorter sentences than non-Aboriginals and to serve less time in confinement. However, Aboriginal inmates are less likely to succeed in their applications for conditional release, and as a consequence, they represent only about 10 percent of federal offenders on parole. In addition, Aboriginal parolees are less likely than non-Aboriginal parolees to complete parole without being returned to confinement. There are no differences in the rates of revocation for new offences among Aboriginal and non-Aboriginal offenders.

A number of factors contribute to the difficulties Aboriginal offenders encounter with parole. These include a lack of understanding of the parole process, a lack of assistance for Aboriginal inmates wishing to apply for parole, feelings of alienation, and a lack of self-confidence in the ability to successfully complete conditional release (LaPrairie, 1996). As well, in many remote and northern communities there are few full-time parole officers; as a result, much of the supervision must be carried out by volunteer parole officers who have little training and few resources.

Aboriginal inmates may also experience difficulties at parole hearings. The small numbers of Aboriginal parole board members across all jurisdictions can lead to the cultural insensitivity observed by Lisa Hobbs Birnie, a non-Aboriginal former NPB member in the Prairie Region:

> When I found myself sitting opposite Samuel Grey Hawk, or Amos Morning Cloud, or Josephy Brave Bear, or when I caught the shy, uncertain eyes of a Cree-speaking teenager from the far North attempting to follow, through an interpreter, our ritualistic procedures and answer our thoroughly middle-class questions, I felt a little like a fraud. It seemed incalculably unfair that these men had the misfortune to have to depend upon the decisions of people who might as well have come from another planet, as far as the similarities in culture and lifestyle were concerned. (Birnie, 1990: 197)

Initiatives designed to address the problems experienced by Aboriginal parolees include the use of elder-assisted parole hearings in some areas, active recruiting of Aboriginal parole board members, and cultural awareness training for board members (Correctional Service of Canada, 1998). Other remedies are contained in the Corrections and Conditional Release Act. Section 81 of the act provides for Aboriginal alternatives to incarceration and authorizes the federal government to enter into agreements with Aboriginal communities whereby the community will take over the "care and custody" of some Aboriginal inmates. Section 84 includes provisions whereby Aboriginal communities can be involved in the preparation of the inmate's release plan and in his or her re-entry into the community. In many cases, however, Aboriginal parolees take up residence in urban centres where they are not able to benefit from the support of their home community or band. Some urban centres offer specialized residential services for Aboriginal offenders.

Female Offenders

As a group, female offenders are better release risks than men but require different types of assistance while on conditional release. More specifically, women have greater health needs, are more likely to have experienced sexual or physical victimization prior to incarceration, and are more likely to be a caretaker parent for children (Conly, 1998). Finding employment may be a particular challenge for female parolees because they are less likely than male parolees to have completed their education, often have little job experience, and may have to find and pay for daycare. The Canadian criminologist Margaret Shaw (1991) has identified seven problems that are typically cited by provincially sentenced women offenders as they leave prison:

- problems with family members (including custody and access to children)
- lack of employment
- financial difficulties
- lack of affordable housing
- substance abuse
- physical and mental-health problems, *and*
- problems with criminal justice personnel

Although Shaw's work is more than a decade old, there is no indication that these challenges are any less for female offenders today.

Female parolees who are mothers may face a variety of challenges, including reestablishing contact with their children, finding suitable accommodation with sufficient space, and/or attempting to regain custody of children placed in care during the confinement. The likelihood of reestablishing a parental relationship with children will depend on the length of the absence, the ages of the children, and the quality of family relations before and during the incarceration. If the children have been in the care of a child protection agency (such as the Children's Aid Society in Ontario), the parolee will have to satisfy the agency that the children will not be placed at risk if they are returned to her. In some jurisdictions, parental rights can be terminated once children have been in care for two continuous years (now one year in Ontario for some children).

A review of the issues surrounding the reintegration of female offenders conducted by the Office of the Auditor General of Canada (2003a) found significant gaps in the delivery of programs and services for federal female offenders released into the community, particularly with respect to substance abuse programs and mental health services. The report recommended the expanded use of temporary absences and work releases, more emphasis on using Sections 81 and 84 of the Corrections and Conditional Release Act for Aboriginal offenders, attention to the issues surrounding employment for female offenders, and additional housing facilities for women with substance abuse issues.

Sex Offenders

No group of offenders has attracted more public interest than sex offenders. Their release from prison is often front-page news in the local press or even announced over the Internet. The successful reintegration of sex offenders may be assisted by relapse prevention programs, in which the offender identifies both the patterns of his offences and the distorted rationalizations used to justify them. The offender is then taught to avoid the situations in which he is most likely to reoffend, and in particular the sequence of events that, once started, he may not be able to stop. For example, an offender with a history of befriending prepubescent boys in parks must learn to resist the convoluted thinking that drove him to demonstrate that behaviour in the past.

While opinion is mixed over whether treatment can *reduce* the risk of sexual offending, there is an emerging consensus that techniques are required to *manage* the risks presented by this group. These techniques include treatment, drugs such as anti-androgens to reduce sex drive, community notification, registration, and supervision.

PAROLE SUPERVISION

Persons on conditional release are subject to differing levels of supervision by correctional officials. In the case of some provincial TAs, this supervision can take the form of periodic telephone calls to verify the offender's presence at home. At the other end of the spectrum, supervision can include a requirement that the parolee reside in a community-based residential facility with twenty-four-hour monitoring and frequent face-to-face meetings with a parole officer. This condition is often imposed on high-risk federal sex offenders who are on one-chance statutory release (discussed earlier in the chapter). The reporting requirements for parolees are set out on their certificate of parole (see, for example, Figure 8.7).

The Correctional Service of Canada operates a number of community correctional centres (CCCs) across the country and also contracts private operators (including not-for-profit organizations such as the 7th Step Society) to provide beds in community residential facilities (CRFs). Inmates must apply for residency and be accepted by centre staff. In each province/territory, there are parallel systems of residences that the government operates either directly or under contract with private operators. These CRFs are often called halfway houses. Most released offenders do not reside in halfway houses, but rather live on their own or with their families.

The Dual Role of Parole Officers

Following release from the institution, the parolee travels to his or her city of destination and reports to the local parole office. Parole officers supervise and monitor parolees for the period between release and warrant expiry. Most parole officers are employees of either the CSC or a provincial government. Non-profit agencies, such as the Elizabeth Fry Society, the John Howard Society, and the St. Leonard's Society of Canada, provide post-release supervision services to federal offenders on a contract basis. Many parole officers begin their careers as correctional officers and often have university degrees in such areas as social work or criminal justice.

Parole officers (and probation officers in Quebec, Ontario, and British Columbia) have a dual role in their relationship with the parolees under their supervision. The first role is as a resource person and confidant to counter the pains of re-entry that the parolee may experience. Supportive activities of parole officers can include job search advice, referral for counselling, and advocacy with welfare authorities. The second role involves surveillance and the enforcement of parole conditions—for example, contacting an employer to verify employment, checking from time to time to ensure employment stability, visiting the home to verify residence or compliance with curfews, arranging urinalysis tests, and checking with the police and/or the staff at treatment programs. This dual role is much the same as that for probation officers (see Chapter 6).

Ideally, a balance between the two roles is achieved, with more control and surveillance during the early phases of the release period and a greater emphasis on assistance later on. It is likely, however, that most parolees view their parole officer as more of a "watcher" than a helper, especially in the current era of corrections, during which risk management and protection of the community are paramount. The intensity of supervision will depend on the risk/needs of the parolee.

A review of the issues surrounding the reintegration of male federal offenders found that overall, the level of supervision of offenders in the community was improving (Auditor General of Canada, 2003b). However, the audit also found considerable variability in the provisions for managing those offenders who require high levels of supervision and that there was a need for improved training and in-service training for parole officers.

Innovations in Parole Supervision

Several innovative approaches to parole supervision are now being used across Canada. For example, Manitoba's Restorative Parole Project uses a restorative justice approach that involves crime victims and other community members in the development of an inmate's release plan. An Ontario initiative, the London Community Parole Project, uses trained volunteers to carry out a variety of tasks, including co-supervision of high-risk parolees with regular-duty parole officers.

Circles of Support

To serve high-risk sex offenders who have reached warrant expiry, the London Community Parole Project uses an innovative strategy called circles of support. Offenders participate in a circle of support voluntarily; there is no legal mechanism requiring them to be monitored. A circle of support is a team of five or

A circle of support

six volunteers who assist the offender with all facets of reintegration, including housing, employment, budgeting and financial management, spiritual development, and moral support. An individualized agreement called a covenant is negotiated in each case. Depending on the agreement, the offender may call only in times of stress or may have daily contact with the circle members.

Life Line Concept

While correctional planning for most inmates focuses squarely on the eventuality of release, half the battle for long-term offenders is coming to terms with the fact that they will be incarcerated for a significant period. Helping offenders achieve this goal is the Life Line Concept, which identifies four stages: adaptation (coming to grips with the reality of confinement), integration to the prison environment (living within the context of that reality), preparation for release, and reintegration into the community.

The Life Line Concept has three components: in-reach, community resources, and public education. In-reach workers (often "lifers" on parole) conduct long-term pre-release visits, assist in release planning and guided social reintegration, and (where applicable) provide liaison with the inmate's family. Most in-reach workers are employed by private agencies such as the St. Leonard's Society and visit long-term offenders in the federal facilities in their areas. For example, LINC (Long-term Inmates Now in the Community), which is based in Abbotsford, British Columbia, has two workers who serve more than 300 offenders in eight institutions. In-reach workers also assist with correctional plan development, attend parole board hearings to assist the inmate, and supervise escorted temporary absences as part of a guided social reintegration process.

Judicial Recognizance for Sex Offenders

When federal sex offenders are detained during the period of statutory release, they have a cold turkey release and are not obliged to inform law enforcement or correctional agencies of their location. The police may alert communities through the use of community notification (discussed below). Another option is to use Section 810.1 of the Criminal Code to compel the individual to enter into a judicial recognizance (often referred to, in this context, as a peace bond).

Judicial recognizance is most commonly used with pedophiles who have reached warrant expiry but still present a very high risk of offending against children under fourteen. The applicant—usually a police officer—need only have "reasonable grounds" to believe that the subject of the recognizance may commit one of the designated offences in the near future. In other words, it is applied proactively for offences that *may be* committed rather than in reaction to offences that *have been* committed.

The matter is heard in a provincial court. If the applicant is successful, the judge orders the subject of the recognizance to comply with set conditions. A sex offender would typically be prohibited from engaging in any activity that involves contact with persons under the age of fourteen, including visiting a daycare centre, school ground, playground, or any public park or swimming area where children are present or can reasonably be expected to be present. An offender who refuses to enter into the recognizance can be sent to prison for up to twelve months, while an offender who violates a condition of the order is liable for up to two years in prison. The order can be in effect for up to twelve months.

Intensive Supervision Program

Operated by the Correctional Service of Canada, the Intensive Supervision Program is designed for offenders on conditional release who have a history of violence and confinement in correctional institutions, who exhibit little motivation to change, who may have psychological disorders, and who present a high risk of reoffending. A typical candidate for the program is the high-risk offender who was denied parole and is subsequently granted statutory release. The objectives of the program are to ensure the safety of the community through a regimen of intensive supervision, to help offenders access resources and services in the community, and to manage the risk presented by the offender. Box 8.4 provides an inside look at the activities of a case management officer in the Intensive Supervision Program in Montreal.

Community Notification

community notification (CN)
the practice of announcing to the public that a high-risk offender has taken up residence in a specified area

The use of **community notification (CN),** while uneven in Canada, is widespread in the United States and has been widely employed in cases involving sex offenders released on parole. There are several models of community notification, but all involve a public announcement—usually made by the police—that a high-risk offender has taken up residence in the area after release from prison. CN can involve proactive measures, such as distribution of leaflets door to door, or it can be passive, involving the posting of the information on the Internet to be accessed by interested parties. Several American states, including Alaska and Florida, have posted the names of all sex offenders on the Web. In Canada, police departments periodically place public warnings on their websites.

CN is possible because the local police department is always notified by the CSC about the impending arrival of federal parolees to their jurisdictions. (In practice, most CNs involve federal offenders.) When first employed in Canada, CN was used on a sporadic basis by some police departments with some cases. Prompted in part by the fear that decision makers were exposed to civil liability, most provinces eventually developed policies to help guide police decisions to release identifying information. For example, British Columbia's policy, introduced in 1995, lists five factors to be considered in the decision to notify the community: the circumstances of the convictions or charges; previous violations of relevant probation or parole conditions; participation (if any) in treatment programs; any relevant psychiatric history; and current activities, including access to potential victims.

Most critics are assuaged by having the decisions about CN made by a committee. For example, in Newfoundland and Labrador they are made by the Community Notification Advisory Committee. The first such committee was established in Manitoba in 1995. There, correctional authorities are required to notify local police well in advance of the release of a sex offender classified as high-risk. After reviewing the matter, the police may choose to take the case forward to the Community Notification Advisory Committee, which is made up of representatives of the RCMP, the police services of Winnipeg and Brandon, Manitoba Justice, Manitoba Corrections, the CSC, and a specialist in

focus

BOX 8.4

The Intensive Supervision Program in Montreal: A Day in the Life of a Case Management Officer

The day starts early, before 8:00 a.m., often with a surprise visit to an offender to check whether he has respected his curfew. In spite of the rude awakening, he is in a relatively good mood—no doubt he's relieved to be at home for the check! Because of our clientele's shortcomings, a meeting like this one usually takes longer than a regular visit. I try to calm him down and get him to accept the limits placed on him. Then I can end the interview with some degree of reassurance, at least for the next 24 hours.

I will have six meetings today, although sometimes it can be as many as ten. They take up the whole day because of the distances involved. The Greater Montreal District is big, as is the Montreal Urban Community. I drive from Saint-Leonard to Hochelaga-Maisonneuve. I make a stop at the Salvation Army and then go over to the South Shore. I cover over 100 km of city streets and highways in one day.

During this mad rush my cell phone never stops ringing. An offender calls in. Then I confirm a meeting with another one. A resource person calls me to voice her concerns, and a colleague gives me information on one of my cases. A few minutes later, the head of the clinic gives me the latest. An offender's urinalysis results are positive, another has missed his curfew, and so on. I have to act quickly in every instance, assessing the risk of allowing the offender to remain in the community and deciding whether or not he should return to prison immediately. Usually I make that determination after a talk with the head of the clinic. I'll spare you the reasoning behind my final decision in one particular case, but today suspension is in order. So off I go to the local police station with a colleague from the Intensive Supervision Program. We work out with the police officers the procedures and the actions needed to ensure the suspension warrant is executed and the individual arrested as peacefully and safely as possible. The offender is under the impression that his parole officer is just paying him a visit and is surprised to find not only his officer but also a policeman at his door. I tell him the reasons for his suspension, he expresses his displeasure, and then off he goes back behind bars. The whole episode has taken up two hours of my time and I have fallen behind schedule. I'll have to continue my round a little later today.

It is clear from all of this, I am sure, that the caseload and management of the risk posed by the people under my supervision does not brook any delay. There is no way I can simply put things off until tomorrow. So I'll go back to the office at the end of the day to write up my log (more than 30 items some days). Then I'll get on with special reports on other suspensions and provide the community information needed by the institutional case management teams—our contribution to the risk management strategy established for offenders who may one day be in the Intensive Supervision Program.

Why live at this hectic pace, maintain a state of high alert, and accept the daily challenges posed by our clients in the Intensive Supervision Program? Because I like the professional challenge, of course, but also because I'm part of a super team and I know that I'm carrying out a vital mission.

Source: *Interactive Corrections*, Correctional Service of Canada, 1997. Reproduced with the permission of the Minister of Public Works and Government Services Canada, 2006.

medical/therapeutic interventions. This committee decides on the type and scope of the information to be released to the community. The options include the following:

- *Full public notification.* A province-wide warning to all citizens, as well as a news release to major media outlets that includes a photo and physical description of the offender and a list of past offences.
- *Limited public notification.* Similar to the above, but directed to a specific community or group and to media outlets in a specific area.
- *Targeted notification.* A warning to a specific community or group, but with no release of information to media outlets.
- *No notification.* The offender is deemed not to pose a significant risk to the community.
- *Other measures.* Police surveillance and a court order for no contact (www.gov.mb.ca/justice/safe/pubnote.html).

The protocol also involves listing the designation with the Canadian Police Information Centre (CPIC) and notifying past victims of the release. This is the model that will probably proliferate in Canada. However, the Federal/Provincial/Territorial Working Group on High-Risk Offenders (1998) suggested that the (CPIC) database, which already serves as a central registry for information on sex offenders, may carry out many of the same functions.

CN continues to be an intensely controversial issue: How is society to balance the rights of the community with the privacy rights of the offender (see Box 8.5)?

The difficulties surrounding CN were highlighted in a recent case in British Columbia. A sex offender with forty-two prior offences, including sexual and violent attacks against victims as young as nine, was forced out of a community after being released from custody. He had refused to participate in treatment programs during his time in prison. The RCMP had issued a CN, releasing his photo and warning that he was a high risk to reoffend. In the words of the town's mayor: "Basically, our community did what the judicial system wouldn't . . . We said, 'You have no rights. Get out of town.'" As the threats against him mounted, the offender decided to move to another community 70 kilometres away. "I'm in a small town right now," he commented, "and I've had the support of [the town] for long time here and I'm going to lose it over the media [coverage] . . . Then what do I do from here, go to someone else's community?" (CBC News, 2005).

ADDITIONAL PROVISIONS FOR SUPERVISION

The Criminal Code contains a number of provisions that can be used to impose conditions and supervision on offenders once they have completed their custodial sentence and/or parole. One is the long-term offender designation (Section 753 of the Criminal Code; and see Chapter 6). If certain criteria are met indicating that the offender will present a substantial risk of committing a

What do you think?

1. Would you want to be notified of the presence of a sex offender in your neighbourhood? Explain. If so, what would the knowledge cause you to do differently?

2. Does your province have a community notification law? Check the statute books, because several do and more are planned. Check out the website of your local police. Many now have community notification pages.

focus

BOX 8.5

Community Notification of Sex Offenders: Community Safety versus Offender Privacy Rights

Point

Supporters of CN contend that:

- Community notification alerts the neighbourhood to a potential risk, thereby reducing the likelihood of another offence.

- Public safety overrides any expectation the offender has for privacy.

- Information on convictions is in the public domain.

- The protection of potential victims is an important goal for society.

- Offenders have forfeited the expectation of privacy by virtue of their offences and the risk they pose.

- The criminal justice system has an obligation to inform the community of any risk to public safety, and failure to do so may result in civil liability suits.

Counterpoint

Opponents of CN counter that

- There is no evidence that CN is effective.

- CN can increase public fear, resulting in vigilantism.

- Individuals intent on reoffending can simply visit or move to another area.

- The stress of CN could increase a sex offender's propensity to recidivate in much the same way that stress increases relapse among substance abusers.

- CN makes it difficult for some offenders to reintegrate into the community because they can find it difficult to secure employment or accommodation.

- If the offender serves a sentence to warrant expiry, thus paying his or her entire "debt to society," CN could be construed as a punishment inflicted after a punishment (Griffiths, 2004: 413–15).

serious personal offence following release from custody, the sentencing judge can impose the designation of long-term offender and require the offender to be under the supervision of a parole officer for up to ten years.

The other provision is found in Section 810 of the Criminal Code, which can be used for offenders who have not been granted parole and who have served their entire sentence in custody. As previously noted, Section 810.1 may be used in situations where there is fear that the offender poses a risk to persons under fourteen. Under Section 810.2, prosecutors may ask the court to impose restrictions on persons who are considered a high risk to commit a violent offence in the community. The case of Karla Homolka (see Box 8.6; see also Box 5.6) illustrates how Section 810 of the Criminal Code can be applied to a federal offender who has completed an entire sentence in custody.

focus

BOX 8.6

The Section 810 Conditions for Karla Homolka

During her incarceration in federal correctional facilities, Karla Homolka (see Box 5.6) was denied parole on several occasions by the National Parole Board on the grounds that she still presented a danger to the community. Homolka was also denied release at her statutory release date (two-thirds into her sentence), the board having assessed that there was a strong likelihood she would commit a violent offence prior to the end of her sentence. As a consequence, Homolka remained in custody for her entire twelve-year sentence.

As her sentence expiry date approached, there was increasing political and public concern about her impending release. Prosecutors from Ontario and Quebec requested the imposition of conditions after her release from custody under Section 810.2 of the Criminal Code, which is applied to offenders who are deemed to be at risk to commit violent crimes in the community.

In June 2005, a Quebec judge ruled that Homolka had returned to her previous behaviour patterns and posed a threat to reoffend. By then it had come to light that Homolka had a relationship with a convicted killer, Jean-Paul Gerbet, whom she had met in the library of a federal prison in Quebec. The couple had exchanged weekly letters, pairs of clean underwear, and (at one point) a simple kiss. Gerbet is scheduled to be paroled in 2008 and will be deported to his native France upon his release from prison. This relationship appeared to convince the judge that Homolka was still a danger to reoffend: "Here we are in 2005 and she's come back to the same pattern," he said, referring to her relationship with Gerbet. "She's back with a partner who is acting in the same style as the old one."

The judge approved the imposition of restrictions under Section 810. This means that Homolka must maintain good behaviour and not own any weapons. She must also:

- report to the nearest police station on the day of her release and tell them where she is living and who her roommates are;
- notify the police of any change of name;
- report to the police station the first Friday of every month, or arrange for another time;
- give ninety-six hours' notice if she plans to move;
- give three days' notice if she plans to go away for more than a weekend;
- complete specific information about any travel plans;
- give police her travel plans if leaving Quebec;
- have no contact with people with a criminal record;
- have no contact with her former husband Paul Bernardo;
- have no contact with her former victims Jane Doe or Nicole T;
- have no contact with the families of the deceased victims;
- not possess drugs or illicit substances;
- not be in a job that gives her access to benzodiazepine, opiates, or barbiturates;
- have no job or volunteer position with people under sixteen;
- continue therapy and counselling; *and*
- provide police with a DNA sample.

In imposing the conditions, the judge informed Homolka that if any of the conditions were violated, he could renew the conditions after twelve months or she could face up to two years in jail. Critics argued that restrictive conditions were a final attempt to make up for the bad decision—made more than a decade earlier—to enter into a plea bargain with her (see Box 5.6). Homolka appealed the decision and, in late November 2005, a Quebec Superior Court Justice removed all of the 810 conditions that had been imposed on her, stating that sufficient evidence had not been presented that she posed a "real and imminent danger" to society as required by law.

Sources: I. Peritz, "Report Often, Judge Orders Homolka," *The Globe and Mail* (2005, June 4), A6; T. Blackwell, "Homolka Legal Leash Has Limits for Police," *National Post* (2005, June 2), A1, A4; A. Hanes, "Homolka is Free to Roam," *National Post* (2005, December 1), A1.

RECIDIVISM AMONG PAROLEES

Even the most institutionalized state-raised inmate does not leave a correctional institution with the intent of returning. Also, systems of correction have as a primary objective the reduction of recidivism among offenders released back into the community. Fortunately, most parolees successfully complete conditional release and do not reoffend prior to warrant expiry. Statistics on parole outcomes indicate the following:

- The majority of provincial/territorial and federal offenders released on parole successfully completed day parole and full parole (73 percent of federal offenders, 85 percent of offenders in Ontario, and 58 percent of offenders in British Columbia).

- Only a very small percentage of offenders who failed on day parole and full parole had committed a violent offence; most failures were due to breaches of parole conditions.

- Among those offenders who were released on statutory release—and who were generally considered to be a higher risk group—only 2 percent reoffended by committing a violent offence. Releases were most often revoked for breaches of conditions of release (Public Safety and Emergency Preparedness Canada Portfolio Corrections Statistics Committee, 2004).

However, as Figure 8.4 indicates, there are significant differences in provincial parole outcomes between Ontario and British Columbia, with B.C. having a significantly higher number of revocations (42 percent of cases) than Ontario (11 percent of cases). Possible reasons for this discrepancy include more lenient release decisions in B.C. and, perhaps, differences in parole supervision.

The media generally focus their attention on the relatively small number of released offenders who commit heinous crimes. The "silent majority" of offenders who successfully complete conditional release are invisible to the community.

Yet considerable criticism has been directed toward the way in which recidivism rates are calculated. The CSC lists the rate of repeat offenders returning to

Rick Eglinton/Toronto Star

A Peel Regional Outreach worker watches a client at work in the woodshop at St. Leonard's House, a counselling and training facility in Brampton, Ontario.

federal custody at 10 percent. Critics point out that this figure includes only those offenders who return to a federal institution within two years of being released. Furthermore, it does not include offenders who are admitted to a provincial correctional institution within two years of leaving a federal prison, nor does it include those offenders who return to *any* prison after being in the community for three years or longer (Baglole, 2004).

One of the difficulties in tracking offenders released from correctional institutions is that the federal and provincial systems of corrections often do not share information on repeat offenders; in other words, the CSC may not be aware that a federal offender has been readmitted to a provincial correctional facility. Also, there is limited information sharing between the provinces. This can result in a situation where an offender is released from a federal institution in Ontario and is subsequently incarcerated in a provincial institution in Manitoba. Then, if that offender is sentenced to custody a year or so later in a provincial institution in B.C., he or she will be recorded statistically in B.C. as a first-time offender.

And it is important to remember that recidivism rates are an imperfect indicator of the success of correctional treatment programs (see Chapter 7).

Factors Associated with Recidivism on Parole

Researchers have identified several factors that seem to be associated with the failure of offenders to successfully complete their sentences in the community on parole. These include:

- antisocial values, attitudes, and companions;
- interpersonal conflict;
- lack of social achievement (including education and employment);
- substance abuse;
- number of prior convictions and admissions to prison; *and*
- failure on a previous parole (Gendreau, Little, and Goggin, 1996; Griffiths, 2004: 382).

A review of the research on the effectiveness of programs in reducing recidivism among offenders re-entering the community found that vocational training and/or employment programs were effective in reducing rates of reoffending (Seiter and Kadela, 2003).

Research on specialized populations conducted in the United States has found, for example, that long-term, residential community treatment programs can be effective in reducing rates of reoffending among high-risk, drug-addicted offenders (Belenko et al., 2004; Incardi, Martin, and Butzin, 2004). And a follow-up study of a sample (N = 401) of high-risk violent offenders who were not reconvicted over a ten-year period following release from custody found that all of the offenders had experienced a specific event or situation that motivated them to make a conscious decision to stop offending. This decision was then reinforced by family life and, interestingly, by a decision to avoid social interaction. In other words, men in the sample stated that their success in not reoffending was due in part to avoidance of social settings generally, and that this helped them avoid the triggers for their criminal behaviour (Haggard, Gumpert, and Grann, 2001). These findings raise doubts about the strategy of emphasizing the importance of offenders establishing (or reestablishing) contacts with conventional society—at least for offenders with a history of violence.

However, research findings on the long-term outcomes for sex offenders released into the community are less positive. A twenty-five-year follow-up study found that three-fifths of offenders reoffended by committing a sex-related offence; the number was even higher (four-fifths of offenders) when all offences and undetected sex crimes were included. The highest rates of recidivism were among child sex abusers and exhibitionists; the lowest rates of reoffending were among offenders convicted of incest (Langevin et al., 2004). These findings have generated considerable controversy. Clearly, more research is needed regarding the factors that contribute to success on parole generally, and also regarding specific types of offenders.

Suspension of Parole

If a parolee is charged with a new offence or is in violation of a parole condition, the parole may be suspended and a warrant issued for the parolee's arrest. The inmate is admitted to the local jail. The parole board reassesses the situation and may hold a hearing at which it questions the parolee about the violation. The new information may indicate that the parolee is no longer an acceptable risk to be out of custody. (Perhaps he or she has been expelled from an alcohol treatment program that was a condition of parole.) In such a case, the board may revoke the parole entirely. Revocation of parole sends the inmate back to prison, and he or she must reapply for release after developing a new release plan.

A parole officer may suspend the conditional release for technical violations or if there is reason to believe that the offender has committed or will commit a criminal offence. If the parole officer does not cancel the suspension, the case goes forward to a revocation hearing before a parole board. Those who violate parole by absconding become the subjects of arrest warrants, and their status is recorded at the Canadian Police Information Centre (CPIC), which is a centralized, computer-based information system linked to municipal and provincial police forces and the RCMP. For these offenders, the sentence stops running until they are arrested. In Ontario, the Fugitive Apprehension Squad (FAS) pursues and arrests offenders who have breached their parole conditions and who are unlawfully at large.

Revocation and Termination of Parole

At a revocation hearing, the parole board has three options: cancel the suspension, revoke the release, or terminate the release (for federal and Ontario provincial parolees). The last option is selected when there is a need to end the conditional release for reasons beyond the control of the individual offender. Suppose, for example, that the community residential facility in which the offender is living burns down. If the board feels that the offender cannot be on conditional release without a residency requirement, it will terminate the release until new accommodation can be arranged.

In the federal system, two additional options are available: issue a reprimand to the offender and cancel the suspension; or cancel the suspension but order a delay in its taking effect for up to thirty days, which can be imposed only for a second or subsequent suspension. In neither case will the release be revoked. It is not known how many violations are not followed by suspension or how many suspensions are not followed by revocation.

Most offenders whose releases are revoked have committed *technical* violations—that is, they have violated the general and/or specific conditions attached to their conditional release. These include having a positive urinalysis test, being in an unauthorized area, or making contact with prohibited persons. About 10 percent of cases involve a revocation for a new offence. This does not mean that only 10 percent of offenders on conditional release commit new offences. It is not known how many offences are committed but not discovered by the police or corrections authorities; how many offences are discovered but classified as technical violations because such charges are more easily proved; or how many suspensions are cancelled for other reasons when in fact an offence has been committed.

In reality, it can be difficult to revoke a release for the commission of a new offence because a parole board hearing is not the appropriate venue for determining a person's guilt or innocence relating to a criminal charge. In most such cases, the offender has been charged with a crime but has not yet been convicted. A federal offender who has been convicted of an indictable offence automatically has his or her release revoked.

DOES PAROLE WORK?

Parole is widely used in Canada. In the United States there is an ongoing debate about its effectiveness. A number of American states have abolished parole and are sentencing offenders to fixed terms of custody. Most of the research on the effectiveness of parole has been conducted in the United States, Britain, and Australia. These studies suggest that parole supervision can extend the time that offenders remain offence-free in the community and reduce the number of times that offenders are convicted (Ellis and Marshall, 2000). A recent American study of the impact of parole supervision on rearrest rates found that parole supervision (at least in the jurisdictions surveyed) did not improve public safety or reduce the rates of reoffending among violent and property offenders (Solomon, Kachnowski, and Bhati, 2005).

There are several reasons why parole supervision may not be effective in reducing reoffending. Levels of supervision are often compromised by the heavy caseloads carried by parole officers. Those officers may supervise parolees in an inconsistent manner. In recent years, in response to political and public pressure, there has been a shift toward a more surveillance-oriented approach to supervision by parole officers.

Parole Outcomes and Consequences

The release decisions of the National Parole Board, in particular, have come under fire in recent years. In part, this has been driven by a number of high-profile incidents in which offenders on parole have committed heinous crimes. Many of these offences are committed by offenders who are on statutory release, which occurs when offenders have completed two-thirds of their sentence in custody. Recall that offenders on statutory release have a much higher rate of recidivism than offenders on parole, indicating that this is a much more difficult population to supervise in the community.

For example, a report of the CSC found that one-third of the offenders who were required by the National Parole Board to reside in a halfway house on statutory release escaped from these facilities. Often they committed new offences, many of which were violent (Beeby, 2004). This occurred in the case of Eric Fish (see Box 8.7).

REVISITING ACCOUNTABILITY

A recurring theme of this book has been the increasing accountability of criminal justice personnel and agencies. In the past two decades, there has been a trend toward civil liability. The preceding chapters have highlighted the various accountability issues that exist at each stage of the criminal justice process. Similarly, corrections systems have become the target of civil suits filed by the

focus

BOX 8.7

Murder in Vernon: The Case of Eric Fish

On August 4, Bill Abramenko, a seventy-five-year old resident of Vernon, British Columbia, was beaten to death in his home. The person subsequently charged for his murder was Eric Fish, who had been paroled to Howard House, a halfway house in the city operated by the John Howard Society. Fish had been convicted of a similar offence in Ontario twenty years earlier and had been released on a two-to-one decision of the National Parole Board. The dissenting parole board member contended that he represented a danger to the community, but was overruled by the other two panel members. On June 21, Fish walked away from Howard House. Six weeks later, he beat Abramenko to death during an apparent home invasion. Fish is also a suspect in the murder of another city resident, whose body was found in Okanagan Lake. The CSC convened a Board of Investigation to inquire into the circumstances surrounding the release and supervision of Fish. As of early 2006, the findings from this investigation had not been released.

Hundreds of residents of Vernon demonstrated outside the halfway house, demanding that it be closed. Howard House, linked to at least four murders, was closed by the CSC in early 2005 and subsequently bulldozed. Critics of this move argue that it was done to deflect attention away from the CSC and the NPB and that the decision to destroy the building deprived the community of a facility that could have been used for other purposes (Berner, 2004). The case focused the attention of politicians and communities on the release practices of the NPB and raised a number of issues surrounding the use of halfway houses for violent offenders.

What do you think?

As a community resident of Vernon, B.C., would you have supported the closing, and destroying, of Howard House as described in Box 8.7? Explain.

victims of offenders released on conditional supervision. Most of these cases have been settled out of court, with no fixing of responsibility or documentation that might provide insights into how such incidents might be avoided in the future.

It is instructive to consider accountability issues in the context of cases that had less than positive outcomes. As you read Box 8.8, keep in mind that the vast majority of offenders on conditional release do not commit violent crimes. Nevertheless, it is high-profile cases such as that of Clinton Suzack that have the greatest impact on victims, the community, other offenders, and the justice system. For this reason alone, it is important to examine these cases and, where negligence is found, to hold accountable the responsible justice personnel and agencies. It is also important to consider the *limits* of justice system accountability. Is it reasonable to expect the criminal justice system to guarantee the safety of all community residents?

focus

BOX 8.8

Accountability in Criminal Justice: The Clinton Suzack Case

The Offender

In September, 1992, twenty-seven-year-old Clinton Suzack pleaded guilty to seventeen charges in Sault Ste. Marie, Ontario. His record to that point dated back to 1981 and included robbery with violence, unlawful confinement, and assault. There was an Alberta-wide arrest warrant (not valid outside of Alberta) on charges of breach of probation and assault. He had a serious, longstanding alcohol problem and an explosive temper, but was bright and articulate.

The seventeen charges stemmed from five incidents, involving nine victims, that had been committed between 1987 and 1992. In December 1987, Suzack assaulted a bar employee who refused to serve him. He failed to attend court and did not comply with the conditions of his bail. In June 1991, he punched a man three times in the face at a house party and viciously assaulted another partygoer with a broken bottle. The injuries to the second man's neck required surgery and a stay in the hospital's intensive care unit. Later in the evening, after the ambulance left, Suzack attacked a woman in the bathroom, punched her several times, and broke her nose. Two men who intervened were also assaulted, but they managed to subdue Suzack, who was subsequently arrested.

Another assault occurred in October 1991. Then in January 1992 Suzack assaulted and threatened a cocktail waitress and a female bar patron. In April, he assaulted a female acquaintance following an argument.

The Crown Attorney

In September 1992, Suzack's lawyer negotiated an arrangement whereby the Crown would recommend a prison sentence of two years less a day in exchange for a guilty plea.

The Judge

The plea bargain was not binding on the sentencing judge. The judge said he was inclined to hand down a sentence of four-and-a-half years, but conceded to the joint submission of the Crown and the defence attorney. He called Suzack a "vicious, violent person" but saw the guilty plea as a mitigating factor and handed down a 729-day sentence. He also ordered three years' probation—the maximum available—to follow the prison term. In addition, the judge recommended that Suzack attend the Northern Treatment Centre in Sault Ste. Marie, although with the acknowledgment that the recommendation was not binding on the provincial correctional system.

The Correctional System

Suzack began his sentence in September 1992. He was classified as a high-risk inmate and sent to Millbrook, the most secure provincial institution. He repeatedly requested a transfer to a lower security facility, but none were willing to take him. His institutional misconduct record was one obstacle. He played tackle football when only touch football was permitted, refused an assigned job as a cleaner, was rude to a guard, and was found in his cell with two other inmates after being warned this was against the rules. For these infractions, he lost seven days of recreational privileges and three days of earned remission and was reprimanded twice.

While at Millbrook, Suzack took computer and woodworking courses. He attended Alcoholics

Anonymous (AA) meetings, participated in an anger management group, and engaged in educational upgrading. A report from the institution's psychologist noted the pattern of alcohol-related violence, but concluded that Suzack was "a bright, articulate individual with considerable potential [who] has demonstrated to staff that he has insights into many of his problem areas and motivation to make some changes." The release plan Suzack devised for presentation to the parole board included an intention to live with three friends in Sudbury, one of whom would employ him in his computer equipment company.

The Probation/Parole System

A Sudbury probation/parole officer, in a pre-parole investigation, confirmed the release plan by contacting the people with whom Suzack intended to live. The opposition of one of his crime victims to the release was noted in the report, as was the alcohol problem and outstanding warrant. As is typically the case, the officer had never met Suzack. He recommended that parole be denied.

The Parole Board

With a 729-day sentence, Suzack would be eligible for parole on June 2, 1993, having served one-third of the sentence (or eight months). If denied parole, he would be automatically released at the two-thirds point in his sentence (February 1994). Three members of the Ontario Board of Parole interviewed him on May 5, 1993. Before the hearing, they reviewed the pre-parole report, his prior record, and the institution's file, including the psychiatrist's report, training reports, and the reports on institutional misconducts. The Sault Ste. Marie police sent descriptions of most but not all of the offences for which Suzack had been convicted. The police "strongly opposed" parole, calling Suzack a "menace to society and a threat to the safety of the public."

The board denied Suzack parole, primarily because the proposed living arrangement would not meet his needs for community-based treatment for anger management and alcohol abuse. Also weighing against Suzack were his prior record, the severity of the current offences, his minimization of the role of alcohol in his offences, and his previous failures under community supervision (for example, his failure to comply with conditions of bail).

The Correctional System

One month later, Suzack lost fourteen days' recreational privileges for fighting. He then reapplied for a transfer to the Sudbury jail, which offered a temporary absence program.

The psychiatrist updated the first report by noting that since being denied parole, Suzack had completed an alcohol awareness program, a woodworking course, and an anger management program (reenrolling in the latter). Moreover, he was chairman of both the Native Sons Program and the institutional AA group, was doing schoolwork, and had volunteered in the prison chapel. While noting the seriousness of Suzack's offences, the report concluded: "He will gain little else by remaining at this facility . . . He has exhausted the relevant treatment services here and the positive structured plan he currently has in place may not be there on discharge [in February]."

Despite the positive report, Suzack was denied admittance on the grounds of institutional misconduct.

Salvation Army Rehabilitation Centre

To have any chance of parole, Suzack had to devise a better release plan. He applied to the Salvation Army program in nearby Hamilton. When interviewed by a staff member at the centre, Suzack indicated that all his offences stemmed from the one house party. The centre agreed to admit him into a ninety-day program once he was released on parole.

The Probation/Parole System

Suzack's new release plan had to be confirmed. The Sudbury probation/parole officer conducted another investigation and again recommended that parole be denied. A Hamilton probation/parole officer confirmed the availability of the Salvation Army program. He had never met Suzack and had little information about him (because of a computer error), and therefore he was not able to offer an opinion on the suitability of parole.

The Parole Board

Suzack got a second hearing, at the discretion of the board, because of his new release plan and recent program participation. All the above-mentioned documents were available to the board, along with a social work report and a letter from a Hamilton-based Aboriginal centre that offered Suzack the opportunity to explore Native culture. At the August 17 hearing, he was seen by three other board members (one of whom was Aboriginal). He described the programs he had taken, expressed a willingness to learn about Native culture, and articulated some insights into his past behaviour.

Board members asked Suzack about his alcohol problem, past treatment, misconducts, and outstanding charges, but little about the offences. Again, he implied that all the charges stemmed from one incident. The board granted him parole, citing his participation in all available institutional programs and the offer of admittance into the ninety-day treatment program operated by the Salvation Army in Hamilton, far away from his victims.

Probation/Parole System

On August 26, Suzack was released from Millbrook and took up residence at the Salvation Army Rehabilitation Centre in Hamilton. He was on the caseload of the Hamilton probation/parole officer who had written the pre-parole report. This officer met Suzack for the first time five days after his release on parole. By then, the officer had gathered background information about Suzack and had begun to question whether the Salvation Army program was an appropriate placement for

him. It was not a secure facility and was designed for motivated parolees who did not require close supervision.

Salvation Army Rehabilitation Centre

The Salvation Army Rehabilitation Centre in Hamilton is one of many centres contracted by the Ontario government to provide services to released offenders. Upon Suzack's admission, the director assumed case management responsibility, instead of delegating the case to a staff member. The following day, the director and the parole officer discussed their mutual concerns. A program involving AA and attendance at the Aboriginal centre was established.

On September 9, Suzack asked to be excused from an AA meeting. The request was denied. Suzack left the Salvation Army centre later that day. The director learned of, and reported, the parole violation on the following day. An arrest warrant was issued, and his previous victims were notified.

The Offender

Suzack eluded arrest until October 7, 1993, when he and an accomplice, Peter Pennett, fatally shot a Sudbury police constable in the back of the head after an altercation during a routine traffic stop. Convicted of first degree murder, Suzack was sentenced to life with no eligibility for parole for twenty-five years. At sentencing, the judge indicated that Suzack should serve his entire sentence in maximum security. Suzack was initially housed in a maximum security facility in Ontario and was later moved to Joyceville, a medium security correctional facility. He was transferred back to maximum security after being suspected in an escape plot. He was then moved back to Joyceville.

In the fall of 2001, after serving six years of his life sentence, Suzack was moved to William Head Institution, a medium security correctional facility at the southern tip of Vancouver

Island. This transfer was criticized by several police groups in Ontario, including the Ontario Association of Police Chiefs, as well as by municipal councils throughout the province. All called on the federal solicitor general to return Suzack to maximum security to serve his life sentence. As of early 2006, Suzack was still incarcerated.

The Victim

The estate of the police constable filed a civil suit against the various justice agencies that had been involved in making decisions about Suzack, including the provincial parole board and probation/parole service. The case was settled out of court for an undisclosed sum.

What do you think?

1. If you were a member of a panel reviewing the case in Box 8.8, would you recommend that any of the justice personnel and agencies involved in making decisions about Suzack be held civilly liable for their decisions? Explain.
2. If you believe that the decision-making process in the Suzack case was flawed, which decision makers were most responsible?

Summary

This chapter dealt with the release of offenders from correctional institutions and their re-entry back into the community. These are critical stages of the criminal justice process because the vast majority of inmates are eventually released from confinement. Parole is a process that begins at the offender's sentencing and continues with the inmate's preparation of a release plan during confinement, appearance before the NPB or a provincial parole board, and eventual release into the community via day parole, full parole, statutory release (federal offenders), cold turkey release (provincial/territorial offenders), or warrant expiry.

Like parole, reintegration is a process. Among its more important elements are the pains of re-entry experienced by offenders. Released Aboriginal offenders, female offenders, and sex offenders present unique challenges for parole officers and community-based service providers. Innovative parole initiatives, such as circles of support and the Life Line Concept, are designed to facilitate the reintegration of special groups of offenders. Most provincial/territorial and federal offenders released on parole successfully complete day parole and full parole. The Clinton Suzack case—which illustrates various accountability issues—is among the high-profile exceptions to this general rule.

Key Points Review

1. The vast majority of offenders who are confined in correctional facilities are ultimately released back into the community.
2. Parole is a process that begins at the sentencing stage and continues through to supervision in the community.
3. There are different types of conditional release, including temporary absence, day parole, full parole, statutory release, and cold turkey release.
4. There are key differences between probation and parole.
5. Judicial review is a controversial provision in the Criminal Code that allows offenders who are serving a life sentence with no eligibility for parole for at least fifteen years to apply to the court for a reduction of this period.

6. Crime victims have the right to be advised of eligibility dates, parole board decisions, and release status, as well as to attend (and in some cases speak at) parole board hearings.

7. The determination of a parole board to grant or deny parole is one of the most important decisions that is made in the corrections process.

8. Commonly voiced criticisms of parole boards' decision making relate to methods of appointing board members, the absence of clearly defined release criteria, and the lack of feedback about the results of decisions.

9. Reintegration is best understood as a process rather than as an event.

10. Many inmates experience pains of re-entry on their return to the community.

11. Aboriginal offenders, female offenders, and sex offenders on conditional release present unique challenges for parole supervisors and community-based service providers.

12. In carrying out their supervisory tasks, parole officers play a dual role: they support the parolee and at the same time enforce the conditions of release.

13. Circles of support, the Life Line Concept, and the Intensive Supervision Program are among the innovative approaches to supervising offenders in the community.

14. Community notification is a strategy for providing communities with information about inmates (especially sex offenders) who have been released.

15. The majority of provincial/territorial offenders released on parole successfully complete day parole and full parole.

16. There is ongoing debate as to whether parole "works."

17. Accountability of systems of correction is an emerging trend.

Key Term Questions

1. What is **conditional release** and what role does it play in the corrections process?

2. What is the role of the **release plan** and the **community assessment** in the parole process?

3. What is **cold turkey release**?

4. What is **statutory release** and how does it differ from parole?

5. What is **detention during a period of statutory release** and in what circumstances is it imposed?

6. What is **one-chance statutory release**?

7. Why is **reintegration** best described as a process rather than an event?

8. What are the **pains of re-entry** and what are some characteristics of offenders who are most likely to experience them?

9. What is **community notification (CN)** and how is it implemented?

References

Baglole, J. (2004, November 6). "Repeat Offender Rate Four Times Higher Than Reported," *Vancouver Sun*. Retrieved from http://www.primetimecrime.com/Recent/Courts/Sun%20Repeat%20offender.htm

Beattie, K. (2005). "Adult Correctional Services in Canada, 2003/04." Catalogue no. 85-002-XPE. *Juristat*, 25(8). Ottawa: Canadian Centre for Judicial Statistics, Statistics Canada.

Beeby, D. (2004, October 4). "1 in 3 Parolees Escaping from Halfway Houses." *Edmonton Journal*, A6.

Belenko, S., C. Foltz, M.A. Lang, and H-E. Sung. (2004). "Recidivism among High-Risk Drug Felons: A Longitudinal Analysis Following Residential Treatment." *Journal of Offender Rehabilitation*, 40(12):105–32.

Berner, D. (2004, August 25). "Why Blame the Halfway House?" The Tyee.ca. Retrieved from www.thetyee.ca/Views/2004/08/25/HalfwayHousetoClose

Birnie, L.H. (1990). *A Rock and a Hard Place: Inside Canada's Parole Board.* Toronto: Macmillan.

Bonta, J., A. Harris, and D. Carrière. (1998, October). "The Dangerous Offender Provisions: Are They Targeting the Right Offenders?" *Canadian Journal of Criminology*, 40(4):377–400.

CBC News. (2005, October 19). "'You Have No Rights. Get Out,' B.C. Town Tells Sex Offender." Retrieved from http://www.cbc.ca/story/canada

Conly, C. (1998). *The Women's Prison Association: Supporting Women Offenders and Their Families.* Washington, DC: National Institute of Justice, Department of Justice.

Correctional Service of Canada. (1998). *CCRA Five-Year Review: Aboriginal Offenders.* Ottawa.

Ellis, T., and P. Marshall. (2000). "Does Parole Work? A Post-Release Comparison of Reconviction Rates for Paroled and Non-Paroled Prisoners." *Australian and New Zealand Journal of Criminology*, 33(3):300–17.

Federal/Provincial/Territorial Working Group on High-Risk Offenders. (1998). *Information System on Sex Offenders against Children and Other Vulnerable Groups.* Ottawa.

Gendreau, P., T. Little, and C. Goggin. (1996). "A Meta-analysis of the Predictors of Adult Offender Recidivism: What Works!" *Criminology*, 34(4):575–95.

Griffiths, C.T. (2004). *Canadian Corrections.* 2nd edition. Toronto: Thomson Nelson.

Haggard, U., C.H. Gumpert, and M. Grann. (2001). "Against All Odds: A Qualitative Follow-Up Study of High-Risk Violent Offenders Who Were Not Reconvicted." *Journal of Interpersonal Violence*, 16(10):1048–65.

Incardi, J.A., S.S. Martin, and C.A. Butzin. (2004). "Five-Year Outcomes of Therapeutic Community Treatment of Drug-Involved Offenders after Release from Prison." *Crime and Delinquency*, 50(1):88–107.

Irwin, J. (1970). *The Felon.* Englewood Cliffs, NJ: Prentice Hall.

John Howard Society of Ontario. (2004). "Provincial Parole in Ontario: The Case for Renewal." *Fact Sheet #20.* Toronto.

Johnson, S. (2004). "Adult Correctional Services in Canada, 2002/03." Catalogue no. 85-002-XPE. *Juristat* 24(10). Ottawa: Canadian Centre for Justice Statistics, Statistics Canada.

LaPrairie, C. (1996). *Examining Aboriginal Corrections in Canada.* Ottawa: Solicitor General Canada.

Langevin, R., S. Curnoe, P. Fedoroff, R. Bennett, M. Langevin, C. Peever, R. Pettica, and S. Sandhu (2004). "Lifetime Sex Offender Recidivism: A

25-Year Follow-Up Study." *Canadian Journal of Criminology and Criminal Justice,* 46(5):531–52.

Office of the Auditor General of Canada. (2003a). "Correctional Service of Canada—Reintegration of Women Offenders." *Report of the Auditor General of Canada.* Chapter 4. Ottawa.

_____ (2003b). "Correctional Service of Canada—Reintegration of Male Offenders." *Report of the Auditor General of Canada.* Chapter 4. Ottawa.

Public Safety and Emergency Preparedness Portfolio Corrections Statistics Committee (2004). *Corrections and Conditional Release Statistical Overview.* Ottawa: Public Safety and Emergency Preparedness Canada.

Roberts, J.V. (1998). "The Evolution of Penal Policy in Canada." *Social Policy and Administration,* 32(4):420–37.

Seiter, R.P., and K.R. Kadela. (2003). "Prisoner Reentry: What Works, What Does Not, and What Is Promising." *Crime and Delinquency,* 49(3):360–88.

Shaw, M. (1991). *The Release Study: Survey of Federally Sentenced Women in the Community.* Ottawa: Ministry of the Solicitor General.

Solomon, A.L., V. Kachnowski, and A. Bhati. (2005). *Does Parole Work? Analyzing the Impact of Post-Supervision on Rearrest Outcomes.* Washington, DC: Urban Institute.

Thurber, A. (1998). "Understanding Offender Reintegration." *Forum on Corrections Research,* 10(1):14–18.

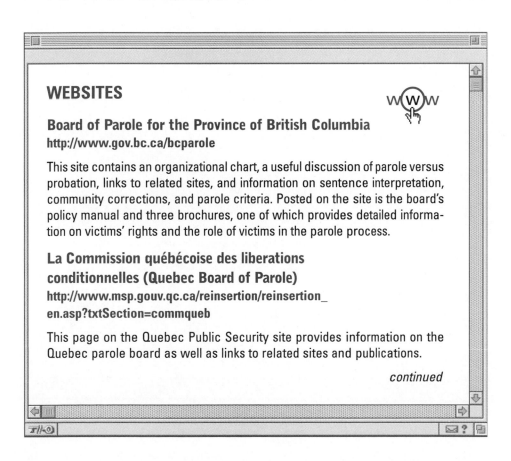

WEBSITES

Board of Parole for the Province of British Columbia
http://www.gov.bc.ca/bcparole

This site contains an organizational chart, a useful discussion of parole versus probation, links to related sites, and information on sentence interpretation, community corrections, and parole criteria. Posted on the site is the board's policy manual and three brochures, one of which provides detailed information on victims' rights and the role of victims in the parole process.

La Commission québécoise des liberations conditionnelles (Quebec Board of Parole)
http://www.msp.gouv.qc.ca/reinsertion/reinsertion_en.asp?txtSection=commqueb

This page on the Quebec Public Security site provides information on the Quebec parole board as well as links to related sites and publications.

continued

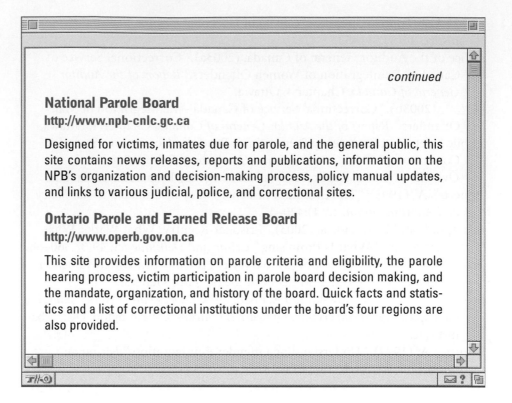

continued

National Parole Board
http://www.npb-cnlc.gc.ca

Designed for victims, inmates due for parole, and the general public, this site contains news releases, reports and publications, information on the NPB's organization and decision-making process, policy manual updates, and links to various judicial, police, and correctional sites.

Ontario Parole and Earned Release Board
http://www.operb.gov.on.ca

This site provides information on parole criteria and eligibility, the parole hearing process, victim participation in parole board decision making, and the mandate, organization, and history of the board. Quick facts and statistics and a list of correctional institutions under the board's four regions are also provided.

Key Terms

absolute discharge (6) a sentence wherein the accused is found guilty but does not gain a criminal record and is given no sentence

arrest warrant (3) a document that permits a police officer to arrest a specific person for a specified reason

avoidance behaviours (2) lifestyle changes designed to reduce one's risk of becoming a victim of crime

basic qualifications (for police candidates) (3) the minimum requirements for candidates applying for employment in policing

bias-free policing (4) the requirement that police officers make decisions on reasonable suspicion and probable grounds rather than on the basis of stereotypes about race, religion, ethnicity, gender, or other prohibited grounds

Canadian Charter of Rights and Freedoms (1) a component of the Constitution Act that guarantees basic rights and freedoms

CAPRA model (4) a problem-solving approach used by the RCMP

case management (7) the process by which identified offender risks and needs are matched with services and resources

circle sentencing (6) a restorative justice strategy that involves collaboration and consensual decision making by community residents, the victim, the offender, and justice system personnel to resolve conflicts and sanction offenders

Citizen Advisory Committees (CACs) (7) operate in all federal correctional institutions and are composed of volunteers who provide input into federal correctional policies and programs and liaise with the community

civil law (1) a general category of laws relating to contracts, torts, inheritances, divorce, custody of children, ownership of property and so on

classification (7) the process by which inmates are categorized through the use of various assessment instruments

clearance rate (4) the proportion of the actual incidents known to the police that result in the identification of a suspect

cold case squads (4) specialized police units that focus on unsolved crimes

cold turkey release (8) discharge of an inmate from custody without conditions or supervision

community assessment (8) an evaluation of the feasibility of the release plan, the level of supervision required, and the availability of community resources

community notification (CN) (8) the practice of announcing to the public that a high-risk offender has taken up residence in a specified area

community policing (4) a philosophy, management style, and organizational strategy centred on police–community partnerships and problem solving to address problems of crime and social disorder in communities

community service approaches (4) police strategies designed to increase police–public contact and cooperation in addressing problems of crime and disorder in communities

concurrent sentences (6) sentences that are amalgamated and served simultaneously

conditional release (8) a generic term for the various means of leaving a correctional institution before warrant expiry whereby an offender is subject to conditions that, if breached, could trigger revocation of the release and return to prison; parole is one type of conditional release

conditional sentence of imprisonment (6) a sentence for offenders who receive a sentence or sentences totalling less than two years whereby the offender serves his or her time in the community under the supervision of a probation officer

conflict model (1) the view that crime and punishment reflect the power some groups have to influence the formulation and application of criminal law

consecutive sentences (6) sentences that run separately and are completed one after the other

Constitution Act, 1867 (1) Constitutional authority for the division of responsibilities between the federal and provincial governments

contract policing (3) an arrangement whereby the RCMP and provincial police forces provide provincial and municipal policing services

core elements of community policing (4) the organizational and tactical strategies and external relationships of a police service that employs a community policing model

crime (1) an act or omission that is prohibited by criminal law

crime attack strategies (4) police patrol operations that are proactive and aimed at crime control

crime control (1) an orientation to criminal justice in which the protection of the community and the apprehension of offenders are paramount

crime displacement (4) the movement of criminal activity from one area to another that results from the implementation of an effective crime prevention program

crime prevention programs (4) initiatives designed to prevent or reduce crime and the fear of crime

crime rate (2) the number of incidents known to the police expressed in terms of the number of people in the population

Criminal Code (1) federal legislation that sets out criminal laws, procedures for prosecuting federal offences, and sentences and procedures for the administration of justice

criminal injury compensation (2) financial remuneration paid to crime victims

criminal law (1) that body of law which deals with conduct considered so harmful to society as a whole that it is prohibited by statute and prosecuted and punished by the government

dark figure of crime (2) the difference between how much crime occurs and how much crime is reported to or discovered by the police

defensive behaviours (2) specific anticrime measures designed to reduce one's risk of becoming a victim of crime

detention during a period of statutory release (8) confinement of high-risk federal offenders until the warrant expiry date

differential treatment effectiveness (7) the requirement that specific treatment interventions be matched to individual inmates

discretion (1, 4) the freedom to choose between different options when confronted with the need to make a decision (1); the ability of a police officer to choose among several possible courses of action in carrying out a mandated task (4)

diversion (5) programs designed to keep offenders from being prosecuted and convicted in the criminal justice system

Division Staff Relations Representative (DivRep) Program (3) program that provides RCMP officers with a way to express their concerns to management

DNA typing (4) the use of genetic information in case investigations

due process (1) an orientation to criminal justice in which the legal rights of individual citizens, including those suspected of committing a crime against the state, are paramount

dynamic risk factors (7) inmate characteristics that can be altered through intervention

electronic monitoring (EM) (6) a sentencing option involving the use of high technology to ensure that offenders remain in their residences except while working or for other authorized activities

fine option program (6) a program that provides offenders who cannot pay a fine with the opportunity to discharge, through community service work, all or part of the fine

hybrid (or elective) offences (5) offences that can be proceeded summarily or by indictment—a decision that is always made by the Crown

indictable offence (3) a serious criminal offence that often carries a maximum prison sentence of fourteen years or more

information (3) a written statement sworn by an informant alleging that a person has committed a specific criminal offence

inmate code (7) a set of behavioural rules that governs interaction among inmates and between inmates and institutional staff

intelligence-led policing (4) the application of criminal intelligence analysis to facilitate crime reduction and prevention

intermediate sanctions (6) dispositions designed as alternatives to incarceration that provide control and supervision over offenders

interoperability (1) the ability of hardware and software from multiple databases from multiple agencies to communicate with one another

judicial determination (6) order that a federal inmate serve half of the sentence before becoming eligible for parole

judicial interim release (or bail) (5) the release by a judge or JP of a person who has been charged with a criminal offence pending a court appearance

one-chance statutory release (8) a release option whereby offenders who violate the conditions of a statutory release are required to serve the remainder of their sentence in confinement

one-plus-one use of force standard (3) police officers have the authority to use one higher level of force than that with which they are confronted

pains of imprisonment (7) the deprivations experienced by inmates confined in correctional institutions

pains of re-entry (8) the difficulties that inmates released from prison encounter as they try to adjust to life in the community

plea bargaining (5) an agreement whereby an accused pleads guilty in exchange for the promise of a benefit

preferred qualifications (for police candidates) (3) requirements that increase the competitiveness of applicants seeking employment in policing

preliminary hearing (5) a hearing to determine whether there is sufficient evidence to warrant a criminal trial

pre-sentence report (6) a document, prepared by a probation officer for the sentencing judge, that contains sociobiographical and offence-related information about the convicted offender and may include a recommendation for a specific sentence

prisonization (7) the process by which new inmates are socialized into the norms, values, and culture of the prison

problem-oriented policing (POP) (4) a proactive strategy centred on developing strategies to address community problems

professional model of policing (4) a model of police work that is reactive, incident driven, and centred on random patrol

racial profiling (4) the targeting by police of individual members of a particular racial group on the basis of the supposed criminal propensity of the entire group

recidivism rates (7) the number of offenders who, once released from confinement, are returned to prison either for a technical violation of a condition of their parole or statutory release or for the commission of a new offence

recipes for action (4) the actions typically taken by patrol officers in various kinds of encounter situations

reintegration (8) the process by which an inmate is prepared for and released into the community after serving time in prison

release plan (8) a plan setting out the residential, educational, and treatment arrangements made for an inmate who is applying for conditional release

remand (7) the status of accused persons in custody awaiting trial or sentencing

restitution (2) a court-ordered payment that the offender makes to the victim to compensate for loss of or damage to property

restorative justice (1) an approach to justice based on the principle that criminal behaviour injures the victim, the offender, and the community

Royal Canadian Mounted Police Act (3) federal legislation that provides the framework for the operation of the RCMP

rule of law (1) the foundation of the Canadian legal system

search warrant (3) a document that permits the police to search a specific location and take items that might be evidence of a crime

security certificates (5) a process whereby non-Canadian citizens who are deemed to be a threat to the security of the country can be held without charge for an indefinite period of time

stare decisis **(5)** the principle by which the higher courts set precedents that the lower courts must follow

static risk factors (7) inmate characteristics that cannot be altered through intervention

status degradation ceremonies (7) the processing of offenders into correctional institutions

statutory release (8) the release of federal inmates, subject to supervision, after two-thirds of the sentence has been served

stay of proceedings (5) an act by the Crown to terminate or suspend court proceedings after they have commenced

summary conviction offence (3) a less serious criminal offence, which is generally heard before a justice of the peace or provincial court judge

suspended sentence (6) a sentencing option whereby the judge convicts the accused but technically gives no sentence and instead places the offender on probation, which, if successfully completed, results in no sentence being given

task environment (1) the cultural, geographic, and community setting in which the criminal justice system operates and justice personnel make decisions

total institutions (7) settings in which the activities and regimen of inmates/residents are highly controlled

two-year rule (7) the basis for the split in correctional jurisdiction

typifications (4) constructs based on a patrol officer's experience that denote what is typical about people and events routinely encountered

value consensus model (1) the view that crime and punishment reflect commonly held opinions and limits of tolerance

ViCLAS (4) a data system based on the analysis of victim information, suspect description, modi operandi, and forensic and behavioural data

victim impact statement (VIS) (6) submission to a sentencing court explaining the emotional, physical, and financial impact of the crime

victim–offender mediation (6) a restorative justice approach in which the victim and the offender, with the assistance of a mediator, work to resolve the conflict and consequences of the offence

working personality of the police (3) a set of attitudinal and behavioural attributes that develops as a consequence of the unique role and activities of police officers

zero tolerance policing (4) an order maintenance approach that utilizes high police visibility and presence and that focuses on disorder and minor infractions with the goal of reducing more serious criminal activity

Index